Peter Maa
1909

geophysical
signal analysis

Peter Maa

PRENTICE-HALL SIGNAL PROCESSING SERIES

Alan V. Oppenheim, Editor

Enders A. Robinson *and* Sven Treitel

geophysical
signal analysis

Prentice-Hall, Inc., Englewood Cliffs, NJ 07632

Library of Congress Cataloging in Publication Data

ROBINSON, ENDERS A
   Geophysical signal analysis.

   (Prentice-Hall signal processing series)
   "Revised and expanded version of material contained
in the 'Robinson-Treitel Reader' which appeared in three
editions during the years 1969–1973."
   Bibliography: p.
   Includes index.
   1. Seismic reflection method. 2. Digital filters
(Mathematics) I. Treitel, Sven, joint author.
II. Title.
TN269.R55   622'.159   79-20749
ISBN 0-13-352658-5

Editorial/production supervision and interior
   design by Steve Bobker
Cover design by 20/20 Services, Inc.: Mark Bergash
Manufacturing buyers: Joyce Levatino and Gordon Osbourne

Printed in the United States of America

10 9 8 7 6 5 4 3 2

Prentice-Hall International, Inc., London
Prentice-Hall of Australia Pty. Limited, Sydney
Prentice-Hall of Canada, Ltd., Toronto
Prentice-Hall of India Private Limited, New Delhi
Prentice-Hall of Japan, Inc., Tokyo
Prentice-Hall of Southeast Asia Pte. Ltd., Singapore
Whitehall Books Limited, Wellington, New Zealand

To our teacher, Norbert Wiener
(1894–1964)

contents

preface

This text, an introduction to geophysical signal analysis, is concerned with the construction, analysis, and interpretation of mathematical and statistical models. In general, it is intended to provide material of interest to upper undergraduate level students in mathematics, science, and engineering. Much of this book requires only a knowledge of elementary algebra. However, at some points a familiarity with elementary calculus and matrix algebra is needed.

The practical use of the concepts and techniques developed are illustrated by numerous applications. Care has been taken to choose examples that are of interest to a variety of readers. Therefore the book contains material of interest to those engaged in digital signal analysis in disciplines other than geophysics. We have tried to include sufficient detail to facilitate self-study by people not directly involved in geophysics.

We have made an effort to bring the methods into logical focus by showing their relationships. For the sake of clarity we have emphasized motivation instead of mathematical elegance. Throughout we attempt to state results directly and clearly, and to point out significant limitations. In so doing, the text does not have that conciseness which comes from a strictly mathematical approach, but instead has the interconnections and redundancies that make it more readable. In this respect we have been guided by the words of Prof. Paul A. Samuelson that "Short writing makes long reading."

Our approach is to base the mathematical equations on models of basic geophysical phenomena. The method of least-squares is the foundation of many of the methods developed in this book. Although the features of least-squares are well known by most people, considerable effort is required to obtain a fundamental understanding of this method in the context of applications. The major contribution of this book, we feel, is in the application of the least-squares method to filter design, as in the case of prediction-error filters, shaping filters, and spiking filters, and in the development of the minimum-delay (i.e., minimum-phase) concept and its relationship to physical phenomena such as the reverberations and multiple reflections in a layered system. Using these ideas a detailed discussion of the process of deconvolution is presented. We have also given an elementary treatment of the type of inverse wave propagation known in exploration geophysics as migration. Furthermore, we have categorized spectral estimation in terms of the three basic models (AR, MA, and ARMA), and have given a new estimation algorithm in the case of the ARMA model.

To a considerable extent, this book is a revised and expanded version of the *Robinson-Treitel Reader* (1969–1973). The *Reader*, compiled as a service to the petroleum industry by the Seismograph Service Corporation of Tulsa, Oklahoma, under the direction of Dr. Robert L. Geyer, is a collection of papers in the general area of geophysical signal processing published in journals by the present authors separately, jointly, and with others.

In writing this book we have drawn upon a background for which we are indebted to our colleagues, teachers, and students. Most importantly, we are indebted to our teacher, Professor Norbert Wiener (1894–1964) of MIT, who founded the sciences of statistical communication theory and cybernetics, and to Professor A. V. Oppenheim, also of MIT, who encouraged us to undertake this venture.

Much of the material in this volume reflects past collaboration with many of our colleagues, in particular, L. C. Wood (Chapter 1), J. F. Claerbout (Chapter 8), K. L. Peacock (Chapter 12), and P. R. Gutowski (Chapter 16).

We are indebted to the management and technical staff of the Amoco Production Company for their enthusiastic support of our research over almost two decades. We have benefited from a close working relationship with many of our associates, among whom we would like to mention J. B. Bednar, J. B. Cameron, R. E. Doan, S. N. Domenico, E. Douze, D. F. Findley, W. S. French, C. W. Frasier, R. L. Geyer, P. Hubral, J. H. Justice, M. A. Knock, S. Laster, D. Loewenthal, S. T. Martner, I. R. Mufti, E. R. Prince, R. L. Sengbush, J. L. Shanks, J. W. C. Sherwood, T. J. Ulrych, R. J. Wang, R. A. Wiggins, and C. I. Wunsch. Our special appreciation goes to Betty Danahey, Cindy Hambrick, and Gerrie Hillier for their patience in typing the manuscript. P. W. Becker and S. Bobker of Prentice-Hall, Inc. have rendered valuable service in the production of the book. To all these people we express our warmest thanks.

Enders A. Robinson
Lincoln, Massachusetts

Sven Treitel
Tulsa, Oklahoma

a summary of
seismic signal processing

Summary

Seismic prospecting for oil and gas has undergone a digital revolution during the past decade. Most stages of the exploration process have been affected: the acquisition of data, the reduction of these data in preparation for signal processing, the design of digital filters to detect primary echoes (reflections) from buried interfaces, and the development of technology to extract from these detected signals information on the geometry and physical properties of the subsurface. The seismic reflection is generally weak, and it must be strengthened by the use of signal summing (stacking) procedures. The determination of depths to a target horizon requires knowledge of the propagational velocities of seismic stress waves, and a wealth of technology has evolved for this purpose. Much of the exploration effort occurs in offshore areas, where reverberations in the water layer mask reflections from below. The method of predictive deconvolution has been most effective in its ability to attenuate these reverberations, making it possible to detect reflections from structures at depth. Seismic signal processing is neither pure science nor pure art, and offers a continuing challenge to the practitioners of both cultures.

Introduction

Massive amounts of seismic data are recorded and processed on a routine basis by the oil industry. Seismic surveys are carried out on a surface grid in order to build up a three-dimensional picture of the subsurface geology in a region, and each survey mile contains around 50 million bits of information, making modern data processing impossible without high-speed digital computers. Worldwide, the oil industry acquired and processed some 600,000 line-miles of seismic data during the year 1977, at a total cost of over $1 billion, according to figures released by the Society of Exploration Geophysicists in 1978.

Seismic signal processing can be divided into three categories: data acquisition, data processing, and data interpretation. Although this book will deal mainly with data processing, data acquisition and interpretation are covered where necessary. Modern reflection seismology methods are discussed as they are presently used to explore for hydrocarbon reserves. Similar processing techniques are used in earthquake seismology, nuclear blast detection, earth crustal studies, and architectural engineering. General exploration objectives include the mapping of subsurface geological structures, the detection of hydrocarbon accumulations, and the estimation of total energy reserves in an area. As shown in Figure 1-1, most reservoirs are associated with geological formations having convex upward structures (anticlines) and linear displacements (faults). In many parts of the globe, for example in the US Gulf coast areas, oil and gas are found in association with salt domes. Many deposits also relate to lateral changes in composition (stratigraphic traps), or to fossil reefs (reef traps). Time differences between reflected seismic signals map structural deformations, whereas amplitude changes of reflected signals may indicate the presence of hydrocarbons. The first part of this chapter describes some basic time and amplitude adjustments, while the latter part deals with design of digital filters.

Data Acquisition

Readers unfamiliar with the fundamentals of seismic exploration should consult the volume by Dobrin (1976). In this chapter, we confine ourselves to acquisition procedures used in the *common-depth-point* (CDP) seismic profiling method.

Figure 1-2 shows the essence of a seismic data gathering system. Disturbances created by seismic energy sources propagate through the earth, where interfaces between geological strata reflect spreading wavefronts. Arrival times of single-bounce echoes (primary reflections) at surface receivers permit the determination of depths and inclination angles of reflec-

2

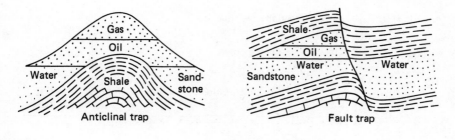

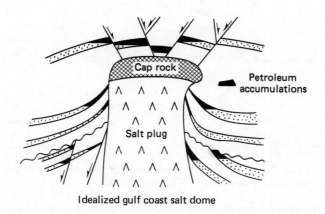

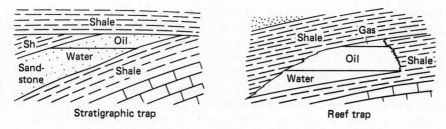

Figure 1-1. Typical structural configurations for trapping hydro-carbons. From *Geology of Petroleum*, second edition by A.I. Levorsen, San Francisco: W.H. Freeman & Co., 1967, reprinted with permission; and *Seismic Prospecting for Oil* by C. Hewlitt Dix, New York: Harper and Row, 1952, reprinted by the permission of the publisher.

tors when subsurface velocities are known. The receivers shown in Figure 1-2 actually represent a composite array (group) of transducers (seismom-eters), as illustrated in Figure 1-3. These groups may consist of up to 100 individual geophones laid out in various linear and spatial patterns, with

geophone

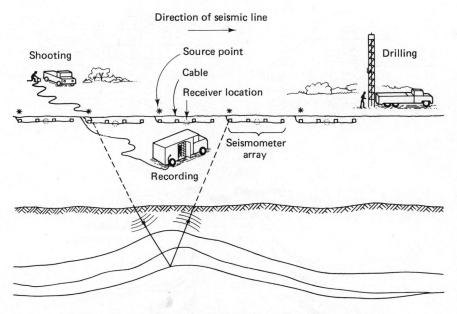

Figure 1-2. Configuration of source points and seismometer arrays for common-depth-point (CDP) surveying.

group intervals (distance between groups) ranging from 50 to 900 ft. Each time a source is activated it is common practice to record either 24, 48, or 96 group arrays (traces) on digital tape simultaneously as a single recording. The seismic "master" cable joining these groups typically ranges from 1 to 3 miles in length. Seismic surveys are conducted along parallel straight lines and individual lines may extend for distances of 1 to 100 miles, for total survey distances of 1 to 1000 miles or more.

Many different types of energy sources are used to generate seismic waves. Dynamite and other high-energy explosive sources provide the simplest and most efficient means of releasing energy, but environmental considerations have led to the development of many alternative sources: explosive air guns, electrical sparkers, vibrating chirp systems (Crawford et al., 1960), and so on.

A seismic source must provide good reflection signal-to-noise ratios at all times of interest. Weak sources are therefore laid out in arrays similar to receiver arrays, and signals generated by multiple source arrays are summed (stacked) together in a process called *vertical stacking*. Vertical stacking should not be confused with *horizontal stacking*, which sums traces lying in a common-depth-point plane. Source and receiver arrays cancel unwanted ambient noise, attenuate surface waves, smooth time

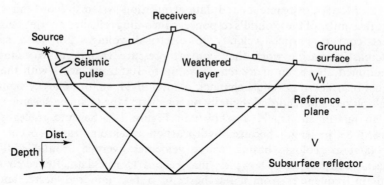

Depth Model with Low-Velocity Layer

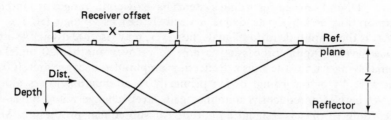

Depth Model After Static Corrections

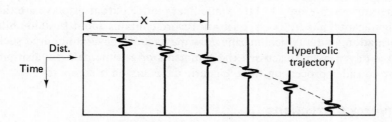

Static Corrected Reflection Record

Legend: Basic models of a reflection
seismograph system

Figure 1-3. Basic models of a reflection seismograph system.

variations caused by surface irregularities, and discriminate against scattered energy.

Modern instruments and data acquisition systems record fairly accurate facsimiles of the ground's response to incoming reflected energy. Seismometers convert particle velocity to electrical voltages for land surveys, whereas in marine work they convert pressure variations to voltage. As mentioned above, an array may have up to 100 transducers with the array signal recorded on a single channel. A total of 24, 48, or 96 array signals are multiplexed and recorded digitally on magnetic tape (a single seismic record) in an instrument truck, as shown in Figure 1-2. Seismic traces seldom exceed 6 s in length because hydrocarbon reservoirs rarely occur below 30,000 ft in geologic basins, where velocities average around 15,000 ft/s. Special chirp systems, however, may record 15 s of data. Reflected signals contain frequencies from a few hertz to a few hundred hertz, and field data are usually sampled at 1-, 2-, or 4-ms rates with alias frequencies (half the sampling frequencies) of 500, 250, or 125 Hz, respectively.

Digital seismic-recording systems have dynamic ranges around 80 dB. Exploration geophysicists define a decibel (dB) as $20 \log_{10} (A/A_0)$, where A/A_0 is the amplitude ratio. Signals, however, may rise 100 dB above ambient noise levels. Digital processing is able to recover another 20 or 30 dB of signal lying within the noise. Reflection amplitudes decay about 100 dB in the first 4 s of recording, owing primarily to attenuation losses along the travel path. Consequently, amplifier gain levels change many times during recording to preserve signal amplitude for subsequent processing. Modern gain systems include instantaneous floating point and binary-gain control. Binary-gain amplifiers record the times of gain changes to allow recovery of signal amplitude. Field instruments typically record 16 bits. Most processing programs require only 12 bits, while final plotter output displays use the most significant 8 bits. A 24-trace seismic record contains about 1 million bits of information, and a typical marine crew may acquire several hundred such records a day. Having discussed the rudiments of seismic data acquisition, we now consider processing these gigantic data sets on a computer.

Preliminary Corrections

Signal processing begins with the demultiplexing of field records. This results in a work tape with signal traces in sequential order. Trace data are preceded on tape by header information giving elevations, seismometer group intervals, sampling rate, word size, trace length, and similar information. The creation of a work tape in a format compatible with central computing center requirements is one of the largest and most frustrating processing tasks, despite industry attempts to standardize tape formats. Adjustment of times and amplitudes to correct for various physical phenomena follows demultiplexing and reformatting. We briefly discuss time adjustments before proceeding to the important topic of relative amplitude preservation.

Geophysicists divide time corrections into static and dynamic categories. A *static correction* consists of the application of a time shift, or translation, to an entire trace. In other words, a constant-time correction term is added or subtracted from all reflection times, regardless of record time or reflector depth. *Dynamic corrections*, on the other hand, vary with record time and therefore depend on reflector depth. Elevation changes and near-surface inhomogeneities severely degrade trace-to-trace continuity, and the purpose of automatic static computations is to remove time variations caused by anomalous conditions at the earth's surface.

A region of very low velocity extends from the earth's surface to a depth of several tens to hundreds of feet; at this point velocities change either gradationally or abruptly from values near 2000 to 5000 ft/s or more. Time delays associated with this "weathered layer" disrupt reflection continuity (i.e., trace-to-trace alignment) and pose a major obstacle to successful processing of data acquired on land. Seismic lines recorded at sea, however, do not usually require static adjustments because of the uniform water layer of constant elevation.

An assumption underlying all automatic correction programs is that a simple translation of a trace converts it into a model trace that would have been recorded had sources and receivers been vertically displaced downward to a reference plane with no weathering material present (see Figure 1-3). This time delay is assumed to be *surface-consistent*, that is, to be a sum of an "initiation" or source-related component and a contribution characteristic of a given surface or "receiver" position. The validity of these two assumptions is confirmed by the success of modern static correction programs (Taner et al., 1974), although near-surface layers behave as complicated filters whose impulse responses distort amplitude and phase characteristics of seismic wavefronts. Further discussion of static and dynamic corrections requires knowledge of seismic trace sorting procedures.

Traces are usually collected into one of four kinds of data sets or "gathers," depending on different objectives. For this purpose a diagram called a *stacking chart* is used (see Figure 1-4). Sources are activated sequentially in the field (Figure 1-2), with each initiation creating a *common-initiation* record. Surface positions of sources and receivers are displaced vertically on a stacking chart for clarity. Figure 1-4 illustrates common-initiation gathers of 24 traces. These diagrams define four principal trace gathers, called common source, common receiver, common offset, and common depth point. *Common-source gathers* consist of traces having the same source, *common-receiver gathers* consist of traces having identical receiver locations, *common-offset gathers* are traces with the same source-receiver distance ("offset"), and *common depth point gathers* are described below.

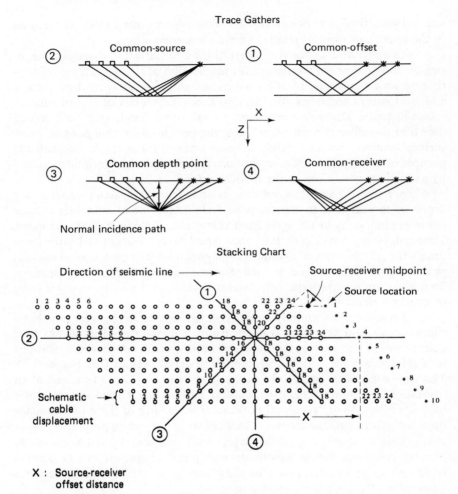

Figure 1-4. Source and receiver positions corresponding to four principle planes used for sorting seismic traces (Taner et al., 1974).

Horizontal (CDP) Stacking

Traces with a common depth point (CDP) have a common midpoint between source and receiver (Figure 1-4); in the case of horizontal interfaces, they also have common points on reflecting interfaces called *common reflection points* (CRP). Otherwise, common subsurface reflection points migrate laterally and spread apart as structures become more complicated; nevertheless, complex geometries frequently cause small dispersion of CRP locations

(Figure 1-5). The last two decades have shown that horizontal (CDP) stacking is a crucial step in seismic signal processing. Stacking consists of the simple sum of traces contained in a CDP plane to produce a single composited trace.

The surface position of the composited trace is then equated with the common source-receiver midpoint. The summing of CDP traces succeeds because primary CRP reflections are in phase and add constructively, whereas ambient noise and other seismic signals not in phase tend to cancel. Compositing increases reflection signal-to-noise ratios by factors approaching $\sqrt{N}$, where N is the number ("fold") of CDP traces summed. Horizontal stacks of 12-, 24-, and 48-fold are routinely produced.

CDP traces must be corrected for travel-time differences caused by varying ray path distances prior to stacking. The latter correction, called *normal moveout* (NMO), depends on depth (record time) to the reflecting horizon and is therefore classified as a dynamic correction. Normal moveout is defined as the increase in reflection time due to an increase in distance from source to receiver for a horizontal reflecting interface in a homogeneous medium of constant velocity. A simple expression for an NMO time increment, derived in Appendix 1-1, equation (1-19), is

$$\Delta T_{\text{NMO}} = T_x - T_0 = T_0\left(\sqrt{1 + \frac{x^2}{(vT_0)^2}} - 1\right) \qquad (1\text{-}1a)$$

with the approximation [Appendix 1-1, equation (1-20)]

$$\Delta T_{\text{NMO}} = \frac{x^2}{2T_0 v^2} \qquad (1\text{-}1b)$$

where T_0 is the two-way reflection time for the zero offset trace, T_x the two-way reflection time for a trace of offset distance x, and v the compressional wave velocity of the medium.

NMO correction involves the subtraction of a time increment ΔT_{NMO} from each record time T_x, with interpolation as necessary. This correction converts a trace of offset distance x into a zero-offset trace that would have been initiated and recorded at a common source-receiver midpoint (Figure 1-4). Equations (1-1) show the dynamic nature of the NMO correction, because even in this most elementary case it is a function of the two-way zero-offset reflection time.

Two facts contribute greatly to the success of CDP stacking. First, reflection time–distance curves (T_x vs. x) from complicated structures are approximated well by a simple hyperbolic relationship of the form

$$T_x^2 = T_0^2 + \frac{x^2}{v^2} \qquad (1\text{-}2)$$

[see Appendix 1-1, equation (1-17)].

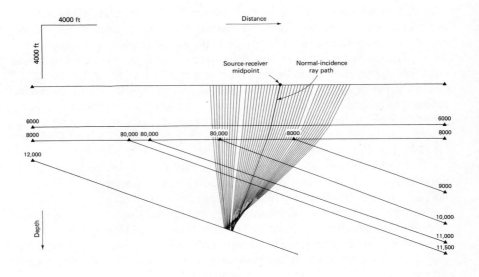

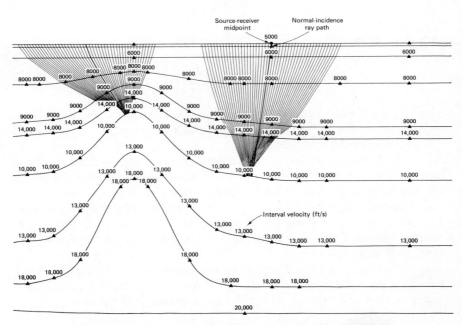

Figure 1-5. Ray-path diagrams for common-depth-point (CDP) traces. Normal incidence ray paths correspond to ideal CDP compositing.

Second, the purpose of NMO corrections is to align single-bounce ("primary") reflections prior to summing. Multiple-bounce ("multiple") reflections travel at lower average velocities than do primary reflections with the same arrival time, because velocity usually increases with depth. Therefore, multiple reflections having greater NMO are misaligned and partially attenuated in CDP stacking. Further multiple attenuation is achieved by *predictive deconvolution*, (see below).

NMO corrections and CDP compositing create new traces called *normal-incidence traces* (NIT). These correspond to identical source and receiver positions (i.e., zero offset). The zero-offset traces also have identical incident and reflected ray path segments, as shown in Figure 1-5. The NIT ray paths form right angles with reflecting horizons at points of reflection called *normal-incidence points* (NIP). Thus, CDP stacking produces a suite of NIT traces with reflection travel paths approximately normal to subsurface horizons (Figure 1-5).

Velocity Analysis

The most important variable in seismic prospecting is velocity, because distances to subsurface reflectors are calculated from observed travel times and known velocities. Seismic waves propagate with the velocity of sound in rock, and so the propagation velocity depends on chemical composition and local geology. Velocities increase with depth as a general rule, and vary from speeds of 1100 ft/s in air up to values approaching 21,000 ft/s in deep sedimentary basins. This information is obtained either through direct measurements in wells, or in the more usual case, is derived indirectly from seismic reflections with the aid of NMO relationships. We will not elaborate on well surveys where seismometers and sources are placed at varying depths in the well, because most velocity determinations make use of redundancy inherent in CDP surveys. Well survey , however, are always used when they are available. The objective of CDP compositing is to increase signal-to-noise ratios to a level sufficient to ensure reliable identification of primary events. Velocity as a function of time, however, must be known very accurately in order to apply proper NMO corrections prior to summing traces.

Hyperbolic characteristics of reflection time–distance curves provide a means for establishing the necessary velocity–time relationships by scanning CDP ensembles along hyperbolic trajectories for signal coherence. These scans establish a function to use in calculating NMO corrections. Reflection times as a function of distance do not satisfy hyperbolic relationships when more than one subsurface layer exists; nevertheless, this second-order approximation works very well, even in areas of complex structural geology, provided that parameters are determined correctly. Only reflections from a single plane interface in a homogeneous medium have the truly hyperbolic time–distance curve given by equations (1-1).

Reflection times for a horizontal reflector below a sequence of N horizontal layers with constant interval velocities can be described by an infinite power series of the form

$$T_{x,N}^2 = C_1 + C_2 x^2 + C_3 x^3 + \cdots \qquad (1\text{-}3)$$

as described by Taner et al. (1970).

A hyperbolic approximation analogous to the single-layer case results from retention of the first two terms,

$$T_x^2(N) = T_0^2(N) + \frac{x^2}{v_{\text{rms}}^2} \qquad (1\text{-}4a)$$

where

$$T_0(N) = \sum_{k=1}^{N} \frac{2Z_k}{v_k} \qquad (1\text{-}4b)$$

$$v_{\text{rms}} = \left[\frac{1}{T_0(N)} \sum_{k=1}^{N} v_k^2 t_k \right]^{1/2} \qquad (1\text{-}4c)$$

Here x is the offset distance, N the number of layers overlying the reflecting horizon, Z_k the thickness of the kth layer, v_k the interval velocity of the kth layer, t_k the two-way travel time in the kth layer, $T_0(N)$ the two-way travel time to the bottom of the Nth layer for the normal incidence trace (NIT), and v_{rms} the root-mean-square (rms) velocity. In the limit, equations (1-4) reduce correctly to describe a single layer, and then there is no difference between rms and interval velocity.

An expression can be derived from equation (1-4c) for calculating interval velocities v_N in the multilayered situation when rms velocities are known (Dix, 1955; Taner and Koehler, 1969). Velocity spectra described below are one way of measuring rms velocities. These average velocities $\bar{v}(N)$ are calculated in succession beginning from $\bar{v}(1) = v_1$, with the following relationship:

$$v_N = \left[\frac{\bar{v}^2(N)T_0(N) - \bar{v}^2(N-1)T_0(N-1)}{T_0(N) - T_0(N-1)} \right]^{1/2} \qquad (1\text{-}5)$$

where $\bar{v}(N)$ is the rms velocity to the bottom of the Nth layer and $\bar{v}(N-1)$ the rms velocity to the top of the Nth layer. The hyperbolic approximations (1-4) are accurate within 2 to 5% in geologic areas of simple structural deformations, that is, where inclination angles of interfaces do not exceed about 15°. The interval velocities (1-5) are often estimated with accuracies between 5 and 10% for use in stratigraphic studies and for detection of hydrocarbon accumulations.

A velocity vs. time display, called a *velocity spectrum* (Taner and Koehler, 1969), is generally used to determine the hyperbolic parameters, which are calculated from CDP traces assuming that travel times of reflections from a common reflection point lie along a hyperbola. The determination of velocity becomes a matter of scanning various hyperbolic trajectories for maximum reflection coherency. Spectra are generated by incrementing normal incidence travel times $T_0(N)$ and keeping them constant while incrementing v_{rms} at regular intervals between some minimum and maximum value. Each $[T_0(N), v_{rms}]$ pair defines a hyperbola, and coherency of data contained in a gate about this curve (Figure 1-6) is measured. Traces are scanned with various hyperbolas whose apexes are fixed at the origin [i.e., $x = 0$ and $T_0(N) =$ constant]. A velocity spectrum consists of a three-dimensional surface of coherency as a function of normal incidence time $T_0(N)$ and rms velocity v_{rms}.

This spectrum may be displayed as contour lines that represent the intersection of level planes of constant coherency with the coherency surface. Interpretation of velocity spectra requires skill and experience, because multiple reflections and other seismic events in addition to primary reflections tend to align themselves along hyperbolic trajectories. A spectral interpretation consists of the location of peaks on the coherency surface that correspond to primary reflections. These peaks are then suitably joined to obtain an average stacking velocity (v_{rms}) versus time $[T_0(N)]$ function display.

Coherence measurements are a crucial part of the determination of effective stacking velocities from multifold seismic data. The basic problem is to establish the similarity that exists between various time gates centered about hyperbolic trajectories (see Figure 1-6). The main task is to measure alignment. Crosscorrelation and semblance are two commonly used statistical measures.

Crosscorrelation functions may or may not be sensitive to amplitude changes between time gates, this sensitivity depending on normalization procedures. The following normalized coherency functions employing zero-lag values of autocorrelation and crosscorrelation functions is not sensitive to rms signal amplitude variations between channels:

$$ S = \frac{2}{M(M-1)} \sum_{i=1}^{M} \sum_{i>i'} \frac{R_{ii'}(0)}{\sqrt{R_{ii}(0)R_{i'i'}(0)}} \qquad (1\text{-}6) $$

where M is the number of CDP traces, $R_{ii}(0)$ the zero-lag value of the autocorrelation function of the ith trace, $R_{i'i'}$ the zero-lag value of the autocorrelation function of the i'th trace, and $R_{ii'}(0)$ the zero-lag value of the crosscorrelation function between the ith and i'th traces. This crosscorrelation measure varies between -1 and $+1$, where $+1$ corresponds to perfect sign coherency. Autocorrelation and crosscorrelation functions are described in Chapters 3 and 6.

Another useful quantity for measuring multichannel coherence is semblance, which was defined by Neidell and Taner (1971) as the normalized

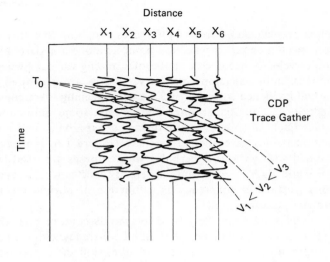

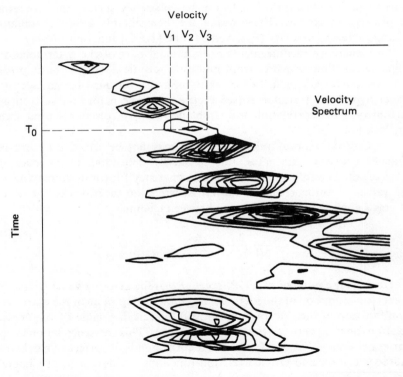

Figure 1-6. Velocity spectrum displaying coherency as a function of reflection time and rms velocity.

output/input energy ratio. Output energy is measured on a composited time gate obtained by summing input time gates. The semblance coefficient S_c is sensitive to channel amplitude differences and varies between 0 and 1, with 1 denoting identical signals. It, too, can be expressed in terms of zero-lag values of correlation functions,

$$S_c = \frac{\sum\limits_{i=1}^{M} \sum\limits_{i'=1}^{M} R_{ii'}(0)}{M \sum\limits_{i=1}^{M} R_{ii}(0)} \qquad (1\text{-}7)$$

These coherence measures are closely related, and the optimum statistic for extracting velocity information from CDP trace gathers may not yet have been found.

Relative Amplitude Preservation

Prior to 1970 seismic amplitudes were used almost exclusively as a qualitative tool for identifying seismic events. A subsequent development relating large-amplitude anomalies, called *bright spots,* with the possible presence of hydrocarbon accumulations has added a new and significant dimension to the search for oil and gas deposits. Exploration has been generally restricted to the location of structural features such as anticlines, faults, and salt domes (Figure 1-1) that are delineable with trace-to-trace differences in reflection time arrivals. Structural traps favoring the accumulation of hydrocarbons are drilled successfully about 20% of the time; however, amplitude information increases these percentages by helping to pinpoint changes in rock composition, layer thickness, and stratigraphic conditions.

Porous rocks at depth are usually filled with salt water, but may contain oil or gas. The bright-spot technique works best in locating gas reservoirs because they cause a greater variation of reflection amplitudes. The amount of energy reflected at an interface depends on the change in acoustic impedance (i.e., velocity–density product) across the interface. Gas-filled rocks have much lower velocities and therefore greater acoustic contrasts than do either oil- or brine-saturated rocks, and thus gas-filled rocks reflect a greater percentage of incident energy.

A simple expression (see Chapter 13) relates reflected and refracted amplitudes across an interface for the special case of plane waves incident on plane interfaces at normal incidence. In general, compressional (P)- and shear (S)-wave modes propagate in an elastic medium. Snell's law and Fermat's principle of minimum-time paths govern refraction, where incident energy splits into reflected and refracted P and S modes. No mode conversion occurs for normal incidence, and a simple normal-incidence reflection coefficient c relates incident and reflected trace amplitudes measured in units of pressure,

$$c = \frac{\rho_2 v_2 - \rho_1 v_1}{\rho_2 v_2 + \rho_1 v_1} \qquad (1\text{-}8)$$

where 1 is the medium containing the incident wave, 2 the medium containing the transmitted wave, ρ_i the density, and v_i $(i = 1, 2)$ the interval velocity.

A corresponding transmission coefficient t relates incident and transmitted amplitudes measured in units of pressure:

$$t = \frac{2\rho_2 v_2}{\rho_1 v_1 + \rho_2 v_2}$$

Appropriate equations for oblique incidence are much more complicated because mode conversion must be taken into account. Nevertheless, normal incidence coefficients are useful and quite accurate for stacked traces in areas having simple geological structures.

Gas has a much larger effect on reflected amplitudes than oil does, so that amplitude anomalies associated with gas/brine contacts are greater than those related to oil/brine interfaces. Normal incidence reflection coefficients for gas-filled sandstones encased in slow-velocity shales may approach 40%, as compared to 10% or less for brine-charged sandstones. Coefficients for oil-bearing sands exhibit intermediate values. Thus, hydrocarbons may produce amplitude anomalies around 12 dB. Polarity reversals also characterize hydrocarbon accumulations, because acoustic impedances across the upper interface of gas and, to a lesser extent, oil reservoir sands encased in shale decrease ($\rho_2 v_2 - \rho_1 v_1$ negative). In contrast, velocities and densities associated across the upper interface of a brine-filled reservoir sand encased in shale usually increase with depth ($\rho_2 v_2 - \rho_1 v_1$ positive). Thus, rapid lateral increases in amplitude and sudden changes in polarity, as shown in Figure 1-7, may indicate a hydrocarbon accumulation at depth.

Trace amplitudes may vary 100 dB during the first 4 s of recording. Hence, current digital instruments with dynamic ranges around 80 dB may be insufficient, but nevertheless they represent a significant improvement over the 40 dB analog systems in use several decades ago. The industry neglected amplitudes prior to the discovery of the bright-spot technique, and tended to destroy relative reflection amplitude relationships by improper use of automatic gain control (AGC) and trace average amplitude equalization procedures. Reflection amplitudes require some kind of time-dependent adjustment after corrections for gain recording functions have been made. This is because the human eye cannot assimilate dynamic ranges of 80 dB. The main factors contributing to reflection amplitude decay include attenua-

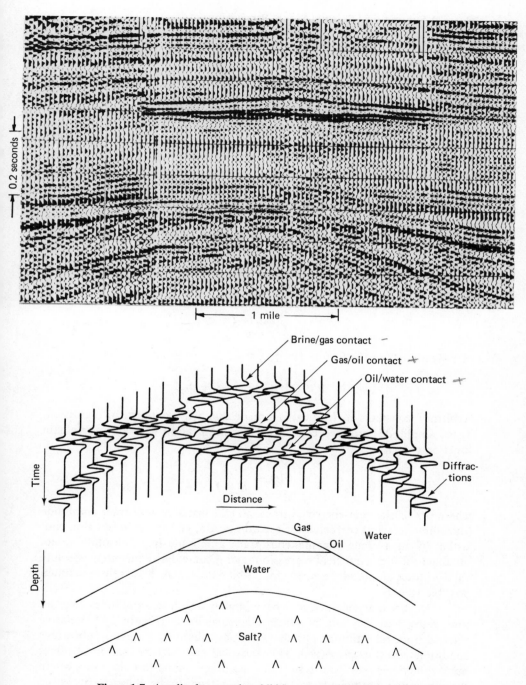

Figure 1-7. Amplitude anomaly exhibiting many seismic features of an idealized bright-spot associated with a hydrocarbon reservoir.

tion caused by reflection and transmission, diverging wavefronts, and frequency-selective absorption.

Techniques for the design of inverse gain functions to preserve relative reflection amplitude variations within and between traces fall into statistical and deterministic categories. Both approaches attempt to correct traces for average attenuation rates while preserving instantaneous variations caused by changes in subsurface acoustical impedances. *Deterministic approaches* define general models to describe many of the possible factors affecting amplitudes such as diverging wavefronts, frequency-selective absorption, reflection and transmission losses, source and receiver array effects, and so on. *Statistical approaches*, on the other hand, produce average gain functions based on collections of traces sorted by common range, source, receiver, and so on.

These average gain curves $g(t)$ may be exponential functions of the form

$$a_0 \exp{(a_1 t)}$$

and

$$\frac{a_0}{t} \exp{(a_1 t)}$$

or polynomials of the form

$$a_0 + a_1 t + a_2 t^2 + \cdots + a_N t^N$$

Arbitrary constants a_i $(i = 0, N)$ are determined by statistical regression. Trace amplitudes are then corrected by multiplication with an inverse gain function $g^{-1}(t)$,

$$G(t) = \frac{a(t)\bar{g}}{g(t)}$$

where $G(t)$ is the gain-corrected trace, $a(t)$ the instantaneous trace amplitude (including polarity) corrected for recording gain, $g(t)$ the gain function consisting of an average instanteneous trace amplitude (e.g., absolute value) obtained through statistical regression, and $\bar{g}$ some desired average absolute value of trace amplitude (e.g., 307 for 12-bit data, where 2047 is the maximum possible value).

In this manner all traces have similar absolute amplitudes over all time gates, and they can be displayed conveniently (Figure 1-8). Relative trace-to-trace reflection amplitude variations caused by changing subsurface conditions are thus preserved. The statistical approach is used most often, but it sometimes gives poor results in areas of low signal-to-noise ratio, where the regression coefficients $(a_0, a_1, \ldots, a_N)$ tend to be affected by noise. Deterministic models often yield better results in noisy areas.

Without
gain or NMO
corrections

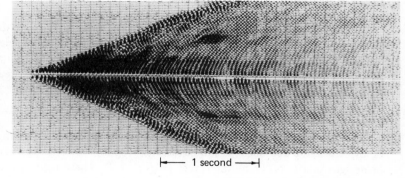

← 1 mile →

|← 1 second →|

Gain corrected
but without
NMO corrections

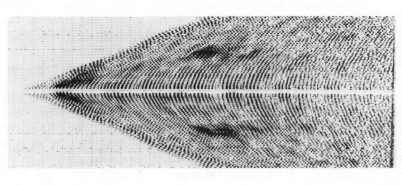

Gain corrected
and NMO
corrected

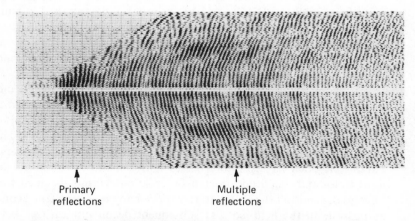

↑
Primary
reflections

↑
Multiple
reflections

Figure 1-8. Two common-depth-point (CDP) trace gathers, showing the effects of gain and normal moveout (NMO) corrections.

The bright-spot technique is used to locate hydrocarbons, to determine reservoir dimensions, and to establish fluid content, either oil or gas. These estimates ultimately provide reserve figures used in economic evaluations of prospects prior to lease sales. Bright spots possess diagnostic features in addition to their large reflection amplitudes and polarity reversals, all of which serve to pinpoint hydrocarbon indicators (HCI) as well as lithologic change indicators (LCI).

Large-amplitude events of limited lateral extent having no inclination or dip on a stacked section sometimes correspond to reflections from gas/brine, gas/oil, or oil/brine interfaces (Figure 1-7). These *contact events* constitute an important HCI. They are essentially horizontal on a stacked section because fluids tend to align themselves along gravitational equipotential surfaces, regardless of the complexity of geological structures. Contact events may indicate the presence of hydrocarbons and help to define reservoir dimensions. Slow velocities also characterize hydrocarbon accumulations. Their effect is to delay reflections from fluid contacts. Contact events from thick reservoirs often have a convex downward appearance ("velocity pulldown") on a stacked section. Consequently, many bright spots have a "fisheye" appearance, as shown in Figure 1-7. This effect occurs because reflections from the top of the reservoir are convex upward in accordance with the geological structure, whereas slower reservoir velocities cause contact events to be concave downward. Diffracted wavefronts from edges of reservoirs where hydrocarbons terminate add to this fisheye effect, and provide an additional HCI. Another criterion is the marked attenuation of amplitudes of reflections originating from horizons beneath reservoirs. Large transmission losses and strong reverberations associated with shallow accumulations attenuate or "mask" reflections from underlying strata and deeper reservoirs.

Modeling is still another important aspect of bright-spot interpretation. Here the objective is to assist geophysical interpretation by means of computer-simulated reflection amplitude anomaly patterns. This is done with synthetic traces computed from geological depth models (Figure 1-9). Sophisticated modeling procedures produce synthetic records, and model parameters such as layer thicknesses and velocities are varied iteratively until times and amplitudes match observations within specified tolerances. Success depends on the ability to record as many high-frequency components as possible in the field, so that subsequent deconvolution (see below) and source pulse-compression techniques can improve the resolution of thin layers.

Amplitude anomalies do not always indicate hydrocarbon accumulations. Reflected signal strengths depend on subsurface impedance contrasts, and many factors other than hydrocarbon accumulation can cause large impedance contrasts. Thin lenses of lava and tightly cemented layers of mate-

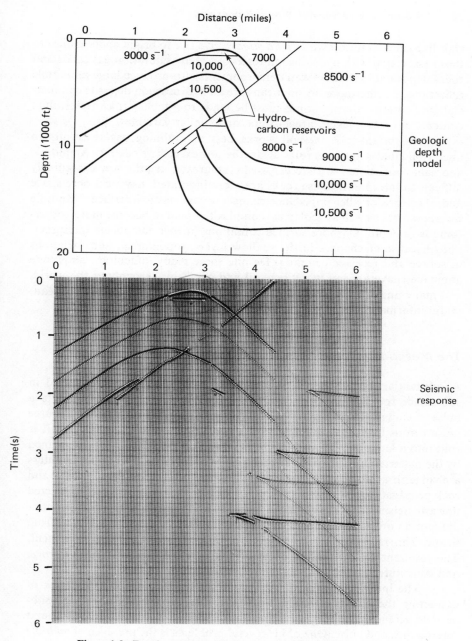

Figure 1-9. Depth model of a faulted anticline structure with hydrocarbon reservoirs. Idealized normal-incidence traces (NIT) show the corresponding seismic reflection times and amplitudes. (Courtesy of B.T. May, Amoco Production Co.).

rials like silt or lime known as *hard streaks* give rise to bright spots similar to those associated with hydrocarbon reservoirs. Low-saturation gas sands and rocks deposited in shallow-water environments also produce large-amplitude reflections. Furthermore, by no means are all hydrocarbon deposits commercial, and careful interpretations must be made to establish thicknesses, fluid content, saturation levels, areal extent, and similar variables.

The bright-spot method works best in outlining gas reservoirs in unconsolidated sand reservoirs at depths not exceeding 6000 ft. Amplitude anomalies associated with older rocks at greater depths are exceedingly difficult to interpret because rocks are more indurated, have less pore space and, therefore, smaller impedance changes across elastic interfaces. Multiple reverberations as well as geological complexities tend to become more bothersome with depth. Onshore surface conditions further complicate interpretations because of changes in the shallow layers, topography, and variations in source and receiver coupling. Despite these many difficulties, amplitude anomalies have defined many new oil and gas fields throughout the world, and many unexpected benefits have resulted from attempts to extract meaningful information from seismic amplitudes.

The Method of Predictive Deconvolution

A substantial fraction of the globe's deposits of oil and gas is buried in subsurface rocks covered by water. Typically, a seismic source imparts a pulse of energy into the water just a few feet below the surface. This source pulse travels from the water into the rock formations below it, where it is split into a large number of waves traveling along various paths determined by the material properties of the medium. Whenever such a wave encounters a change in acoustic impedance (which is the product of rock density and rock propagation velocity), a certain fraction of the incident wave is reflected upward. Seismic detectors situated near the water surface record the continual motion of the water under the impact of seismic waves impinging from below. This recording is performed digitally at a fixed sampling increment. The resultant set of discrete observations is called a *marine seismic trace*, and constitutes a sample of a time series.

The interpreter of such marine recordings is faced with the task of extracting the direct reflections that give him information about the subsurface geometry from a recording that contains a wealth of background interference and noise. One of his several problems is the presence of *multiple reflections* or *reverberations*. These slowly decaying wave trains usually arise in the water layer, which tends to act as a strong wave guide because it is bounded above and below by media of radically differing acoustic impedances. The water reverberation phenomenon came to light when it was observed that seismic traces recorded in water depths greater than 10 ft or so exhibit a marked sinusoidal, or "ringing" appearance.

During the past two decades very significant strides have been made in a continuing effort to remove reverberations from marine data. One of the more successful approaches is based on a rather simple theoretical model of a reverberating trace. The treatment given below is an abbreviated version of the discussion presented in Chapters 10 to 12.

Consider an ideal source located on the water surface emitting a unit spike (or unit pulse) at time $t = 0$, and assume that a pressure detector just below the water surface responds only to downward motion (see Figure 1-10).

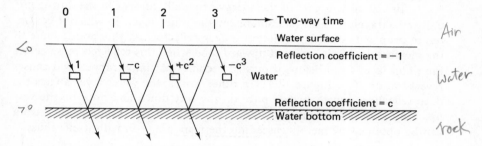

Figure 1-10. Reverberations in the water layer where ray paths have been drawn as slanted lines to illustrate time dependence.

Both the air/water and water/rock interfaces are strong reflectors. We restrict ourselves to plane wavefronts whose ray paths are perpendicular to the interfaces, although for the sake of clarity these paths have been drawn as slanting lines in Figure 1-10. Under such "normal incidence" conditions, we may associate a pressure reflection coefficient of -1 with the lower surface of the air/water interface, while we let the water-bottom reflection coefficient be c, whose magnitude must be less than unity from physical considerations. A source pulse generated in the water layer will reverberate between these two strong reflectors, although part of the energy will be propagated into the underlying rocks. Let the integer n represent one round trip, or two-way travel time in the water layer. Then the downgoing unit spike, which occurs at time $t = 0$, is followed at intervals of n time units by successive downgoing spikes whose values are $-c$, $+c^2$, $-c^3$, and so on. The z transform[1] of such a water-confined reverberation spike train is

$$C(z) = 1 - cz^n + c^2z^{2n} - c^3z^{3n} + \cdots \qquad (1\text{-}9)$$

[1]Geophysicists define the z transform as $C(z) = \sum_{n=-\infty}^{+\infty} c_n z^n$ rather than as $C(z) = \sum_{n=-\infty}^{+\infty} c_n z^{-n}$, as electrical engineers do.

Since $|c| < 1$, this convergent geometric series can be summed to yield

$$C(z) = \frac{1}{1 + cz^n}$$

An inverse filter to remove the water reverberations is therefore

$$A(z) = \frac{1}{C(z)} = 1 + cz^n$$

Because $|c| < 1$, it follows that both the water reverberation spike train $C(z)$ as well as the corresponding "dereverberation" filter $A(z)$ are minimum-delay (see Chapter 3).

In actuality, a part of the energy originally present in the downgoing unit spike travels into the medium below the water layer, in which it continues to propagate until it encounters a deep reflector. At this point, some of the incident energy is reflected upward, and when this reflected pulse enters the water layer from below, it in turn becomes partially trapped and causes reverberations (see Figure 1-11). In other words, the water layer affects the deep reflection returns twice—once on the way down and once on the way up. To a good approximation, the z transform of the resulting spike train can be obtained by merely cascading the response (1-9) with itself; thus,

$$C(z) = (1 - cz^n + c^2z^{2n} - c^3z^{3n} + \cdots$$

$$= \frac{1}{(1 + cz^n)^2} \qquad (1\text{-}10)$$

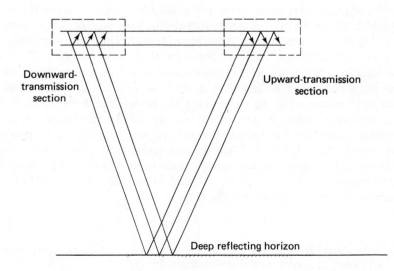

Figure 1-11. Cascading effect of the water layer on a reflection from great depth.

This reverberation spike train is also minimum-delay. An inverse filter to remove the cascaded set of reverberations is now given by

$$A(z) = \frac{1}{C(z)} = (1 + cz^n)^2$$
$$= 1 + 2cz^n + c^2z^{2n} \qquad (1\text{-}11)$$

or

$$A(z)C(z) = (1 + 2cz^n + c^2z^{2n})(1 - cz^n + c^2z^{2n} - c^3z^{3n} + \cdots)^2$$
$$= 1$$

We say that the filter $A(z)$ *deconvolves* the reverberation spike train $C(z)$ to the unit spike at zero delay.

The deconvolution method just described is of slight practical value because the reverberation patterns encountered in petroleum exploration are far more involved. Now it turns out that the minimum-delay property of the reverberation spike train $C(z)$ is quite general in the sense that the unit spike response of an arbitrary system of horizontally stratified layers also is minimum-delay (see Chapter 13). In practice, the source pulse is a broader time function, say b_t, rather than merely a unit spike. If this source pulse is reasonably sharp, as will be the case for an explosion of dynamite, we may expect b_t to have most of its energy concentrated near its front end (i.e., to be "front-loaded"). Front-loaded time functions tend to be approximately minimum-delay, and for the moment we assert that the source pulse b_t does in fact have this property.

We assume that the reverberation *pulse* train, say r_t, is the convolution of the reverberation spike train c_t of equation (1-10) with the pulse b_t,

$$r_t = c_t * b_t$$

source wavelet
b_t

where the symbol $*$ denotes convolution. We imagine that the marine seismic trace x_t arises from the linear superposition of a large number of deep reflections, (each of which has the characteristic shape of the pulse train r_t.) Let ϵ_t be a series of spikes whose amplitudes represent the value of the deep reflection coefficients, and whose times represent the two-way travel time to these reflectors. Our model of the marine seismic trace x_t is, therefore,

$$x_t = \overbrace{c_t * b_t}^{r_t} * \epsilon_t$$
$$= r_t * \epsilon_t$$

Next, we assume that the series ϵ_t is uncorrelated and random. In particular, this means that the series ϵ_t is totally unpredictable, in the sense that knowledge of the amplitudes and arrival times of the first k deep reflections does not permit us to make any deterministic statement about the amplitude and arrival time of the $(k + 1)$th reflection. Of course, we cannot prove that the actual earth has this property (and there are some demonstrable cases for which it does not), but the practical success of a deconvolution approach based on this model suggests that the random and uncorrelated representation of the series ϵ_t is generally reasonable.

On the other hand, the reverberation pulse train r_t is predictable if we assume, as we do here, that both the source pulse b_t as well as the reverberation spike train c_t are minimum-delay. Let ϕ_τ be the autocorrelation of the marine trace x_t. Then we have

$$\phi_\tau = E\{x_t x_{t+\tau}\} = E\{r_t r_{t+\tau}\} * E\{\epsilon_t \epsilon_{t+\tau}\}$$

where E is the expectation operator. But since ϵ_t is random and uncorrelated,

$$E\{\epsilon_t \epsilon_{t+\tau}\} = E\{\epsilon_t^2\} = P\delta_{\tau 0}$$

where P is the power in the series ϵ_t and

$$\delta_{\tau 0} = \begin{cases} 1 & \text{if } \tau = 0 \\ 0 & \text{if } \tau \neq 0 \end{cases}$$

is the Kronecker delta. Therefore,

$$\phi_\tau = PE\{r_t r_{t+\tau}\}$$

and P is a scale factor that does not affect the final result and will thus be neglected. We conclude that the trace autocorrelation ϕ_τ is equal to the autocorrelation of the reverberation pulse train r_t within an arbitrary scale factor. Furthermore, the minimum-delay property of r_t enables us to predict its reverberation component c_t if we compute a prediction operator for prediction distance n, where we recall that $n =$ two-way travel time in the water layer. If we delay the output of such a prediction operator by n time units and subtract it from r_t, we obtain the nonreverberatory component of r_t, namely the source pulse b_t. The linearity of the prediction operator allows us to apply it to the entire trace x_t, suppressing from the data the reverberatory components c_t.

Let a_t be such a prediction operator. For the simplest case, this operator is given by equation (1-11), but in practice a far more general approach results from the use of Wiener theory. Minimization of the mean square error between a desired output and an actual output yields a set of normal equations involving the trace autocorrelation coefficients ϕ_t. If we identify the desired output with an input advanced by n time units, the $(m + 1)$-length least-squares prediction operator a_t is the solution of the system

$$
\begin{bmatrix}
\phi_0 & \phi_1 & \cdots & \phi_m \\
\phi_1 & \phi_0 & & \\
& & \cdot & \\
& & \cdot & \\
& & \cdot & \\
\phi_m & \phi_{m-1} & & \phi_0
\end{bmatrix}
\begin{bmatrix}
a_0 \\
a_1 \\
\cdot \\
\cdot \\
\cdot \\
a_m
\end{bmatrix}
=
\begin{bmatrix}
\phi_n \\
\phi_{n+1} \\
\cdot \\
\cdot \\
\cdot \\
\phi_{n+m}
\end{bmatrix}
\qquad (1\text{-}12)
$$

The autocorrelation matrix of this system contains only the $(m + 1)$ independent elements $\phi_0, \phi_1, \ldots, \phi_m$, and these are arranged in such a manner that all elements on the main diagonal as well as any super- or subdiagonals are equal. This *Toeplitz structure* enabled Levinson to obtain an efficient recursion for the solution of the normal equations, which is described in Appendix 6-2 of Chapter 6. It is of interest to note that the case $n = 1$ leads to a set of normal equations arising in the linear prediction approach to speech compression (Makhoul, 1975).

The prediction operator coefficients $a_0, a_1, \ldots, a_m$ can be used to construct the prediction error operator for prediction distance n,

$$
1, \underbrace{0, 0, \ldots, 0,}_{n - 1 \text{ zeros}} -a_0, -a_1, \ldots, -a_m
$$

This prediction error operator is then convolved with the marine trace x_t to yield

$$
z_t = x_t - a_0 x_{t-n} - a_1 x_{t-n-1} - \cdots - a_m x_{t-n-m}
$$

The series z_t therefore represents the deconvolved marine trace, from which the reverberation spike train c_t has been removed. Alternatively, z_t is the prediction error series associated with the prediction error operator for prediction distance n, where $n = $ two-way travel time in the water layer.

The approach we have described is called the *method of predictive deconvolution*, and dates back to the work of the Geophysical Analysis Group (GAG) at MIT between 1952 and 1957. Figure 1-12 shows a selected portion of a marine seismic line that has been stacked. The vertical scale is two-way travel time, while the horizontal scale represents distance. In Figure 1-13 we may observe the output after every trace has been filtered with a predictive deconvolution operator. We note that a significant amount of reverberating energy has been removed from the input data. It is customary to follow the dereverberation procedure with a number of additional digital filter applications designed to compress the source pulse and to provide greater emphasis to the deeper reflections. This goal is accomplished with Wiener shaping filters (see below), which are designed for a selected number of gates on each trace. The variations in source pulse shape with travel time can be accounted

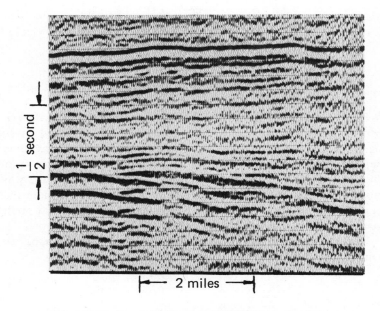

Figure 1-12. Example of composited (CDP) marine data.

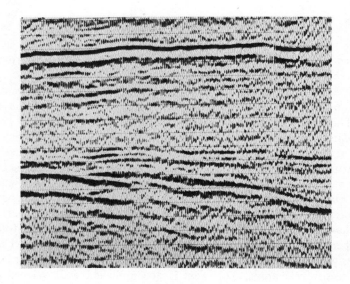

Figure 1-13. The seismic data shown in Figure 1-12 after application of predictive deconvolution.

28

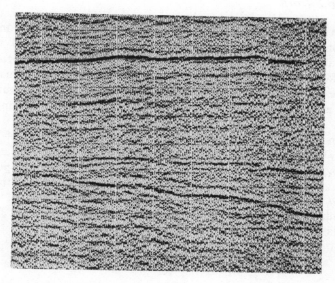

Figure 1-14. Seismic data shown in Figure 1-12 after time-varying Wiener filtering and pulse compression.

for and, in effect, the Wiener shaping filters are applied in a time-varying manner (see Figure 1-14).

> *Query:* Very nice indeed, but how often do real data respond to such treatment?

> *Answer:* To the extent that they obey our model's assumptions, namely, that:
>
> 1. The layered earth is a linear system.
> 2. The reverberation spike train and the source pulse are minimum delay.
> 3. The deep reflector reflection coefficient series is random and uncorrelated.

In actuality, of course, these assumptions may or may not be upheld. All we can say is that widespread application of the predictive deconvolution technique has demonstrated its ability to remove reverberations, and thereby to permit the identification of reflections from depth. In instances for which assumption (1-2) is inappropriate, viable alternatives exist, of which the homomorphic deconvolution approach (Oppenheim et al., 1968; Ulrych, 1971) is one.

Shaping Filters

We have already seen how the method of predictive deconvolution is based on Wiener's least-mean-square-error criterion. In geophysical work the need to alter the shape of a given signal pulse often arises. The problem can be attacked with zero-phase bandpass filters in the frequency domain. However, the amount of control one has on the shape of the output pulse is limited, and it is more expedient to design such shaping filters in the time domain.

In Chapter 6, we consider the problem of finding an $(m + 1)$-length filter $f_t = (f_0, f_1, \ldots, f_m)$, which shapes an $(n + 1)$-length input pulse $b_t = (b_0, b_1, \ldots, b_n)$ into an $(m + n + 1)$-length desired output pulse $d_t = (d_0, d_1, \ldots, d_{m+n})$ in such a way that the error energy between the desired output d_t and the actual $(m + n + 1)$-length output $c_t = (c_0, c_1, \ldots, c_{m+n})$ is minimized. Here we wish to summarize the results of Chapter 6. The actual output is the convolution of the filter with the input,

$$c_t = \sum_{s=0}^{m} f_s b_{t-s}$$

[handwritten: d_t - desired $(m+n+1)$; f - filter $(m+1)$; b_t - input pulse $(n+1)$]

The error energy, I, is

$$I = \sum_{t=0}^{m+n} (d_t - c_t)^2 = \sum_{t=0}^{m+n} \left(d_t - \sum_{s=0}^{m} f_s b_{t-s} \right)^2 \qquad (1\text{-}13)$$

The foregoing error energy is at its minimum value if its partial derivatives with respect to each of the filter weighting coefficients $f_0, f_1, \ldots, f_m$ equal zero. We have

$$\frac{\partial I}{\partial f_j} = \sum_{t=0}^{m+n} 2 \left(d_t - \sum_{s=0}^{m} f_s b_{t-s} \right) (-b_{t-j}) = 0$$

which gives

$$-\sum_{t=0}^{m+n} d_t b_{t-j} + \sum_{t=0}^{m+n} \left(\sum_{s=0}^{m} f_s b_{t-s} \right) b_{t-j} = 0$$

or

$$\sum_{s=0}^{m} f_s \sum_{t=0}^{m+n} b_{t-s} b_{t-j} = \sum_{t=0}^{m+n} d_t b_{t-j} \qquad (j = 0, 1, \ldots, m)$$

Now

$$\sum_{t=0}^{m+n} b_{t-s} b_{t-j} = \phi_{j-s}$$

and

$$\sum_{t=0}^{m+n} d_t b_{t-j} = g_j \qquad \textit{[handwritten: Crosscorrelation]}$$

where ϕ_j is the autocorrelation of the input pulse b_t and g_j the crosscorrelation between the input pulse b_t and the desired output pulse d_t. We thus obtain

$$\sum_{s=0}^{m} f_s \phi_{j-s} = g_j \qquad (j = 0, 1, \ldots, m) \tag{1-14}$$

This system of $(m + 1)$ linear simultaneous equations in the unknowns $f_0, f_1, \ldots, f_m$ can also be written in the matrix form

$$\begin{bmatrix} \phi_0 & \phi_1 & \cdots & \phi_m \\ \phi_1 & \phi_0 & & \\ & & \cdot & \\ & & \cdot & \\ & & \cdot & \\ \phi_m & \phi_{m-1} & \cdots & \phi_0 \end{bmatrix} \begin{bmatrix} f_0 \\ f_1 \\ \cdot \\ \cdot \\ \cdot \\ f_m \end{bmatrix} = \begin{bmatrix} g_0 \\ g_1 \\ \cdot \\ \cdot \\ \cdot \\ g_m \end{bmatrix} \tag{1-15}$$

where $\phi_{-j} = \phi_j$ because b_t is real-valued. We note that the normal equations (1-15) for the shaping filter reduce to the normal equations (1-12) for the predictive deconvolution filter if we identify the crosscorrelation vector $(g_0, g_1, \ldots, g_m)$ with the vector $(\phi_n, \phi_{n+1}, \ldots, \phi_{n+m})$. This is so because in the case of predictive deconvolution the desired output $d_t = r_{t+n}$, where r_t = reverberation pulse train. Hence,

$$g_j = \sum_t d_t r_{t-j} = \sum_t r_{t+n} r_{t-j}$$

$$= \sum_t r_{t+(n+j)} r_t = \phi_{n+j} \qquad (j = 0, 1, \ldots, m)$$

The method of predictive deconvolution is accordingly seen to constitute a particular realization of the shaping filter. Solutions of the more general system (1-14) are again readily obtained with the Toeplitz recursion described in Appendix 6-2 of Chapter 6.

An expression for the normalized minimum square error, E_N, results when the normal equations (1-14) are substituted into the error energy relation (1-13),

$$E_N = 1 - \sum_{t=0}^{m} f_t g_t'$$

where g_s' is the normalized crosscorrelation coefficient,

$$g_t' = \frac{g_t}{\sum_{t=0}^{m+n} d_t^2}$$

It follows that

$$0 \leq E_N \leq 1$$

and the extreme cases $E_N = 0$ and $E_N = 1$ correspond, respectively, to perfect agreement and to no agreement between the actual output c_t and the desired output d_t.

Concluding Remarks

In this chapter we have outlined how seismic data are acquired, interpreted, and processed. One of the major problems we have omitted is the processing sequence, that is, the order in which the corrections are made and the filters applied. For example: Should static corrections be estimated before or after NMO? Should predictive deconvolution be applied before or after CDP stacking? Should Wiener filters be applied before or after velocity analysis? Unique answers to such questions do not appear to be available.

Processing sequences depend on geological conditions and vary from area to area. Geophysical objectives tend to determine a particular sequence, whereas geological factors cannot always be considered. Land data require static corrections to adjust for surface irregularities, and therefore tend to undergo more complicated processing sequences than do marine data. As a general rule, however, both land and marine records are sorted into CDP trace gathers, subjected to velocity analysis, NMO-corrected, and stacked. Scaling, static corrections, and digital filtering alter this basic flow. A typical marine sequence might consist of the following steps:

Demultiplexing

Reformatting

Gain recovery

Sorting for relative amplitude scaling

Bandpass filtering

Predictive deconvolution

Wiener filtering

CDP sorting

Velocity analysis

NMO correction

CDP stacking

Wiener filtering†

Modeling, migration, and interpretation

†See Chapter 6.

   Static corrections complicate the processing of data acquired on land. The NMO correction procedure requires the use of an initial average velocity function in order to reduce trace-to-trace time variations of reflections. Automatic static correction programs cannot handle large time increments between traces, and this fact necessitates the use of an initial gross, average velocity function. Once static corrections have been determined, they can be applied before the NMO corrections to produce a final processed seismic line. If an average velocity function is used for preliminary NMO estimation, a typical land processing sequence might consist of the following steps:

   Demultiplexing
   Reformatting
   Gain recovery
   Sorting for relative amplitude scaling
   Bandpass filtering
   Predictive deconvolution
   Wiener filtering
   NMO correction
   Automatic static correction
   Residual automatic static correction

   A final seismic line on land might consist of the following steps after demultiplexing, reformatting, and amplitude scaling:

   Static corrections
   Bandpass filtering
   Predictive deconvolution
   Wiener filtering
   CDP sorting
   Velocity analysis
   NMO correction
   CDP stacking
   Wiener filtering†
   Modeling, migration, and interpretation

   New technology continually alters and modifies the flow of these sequences. Such innovations present a never-ending challenge to the ingenuity of geophysicists involved in the processing of exploration seismic data.

   †See Chapter 6.

the normal moveout correction

Normal moveout (NMO) is the term applied to the increase in reflection time due to an increase in the distance from source to geophone (under the assumption that the reflecting horizon is horizontal). The normal moveout can be calculated, for the case of constant velocity, very simply by reference to Figure 1-15. We wish to compare T_0, the travel time from S to C to S,

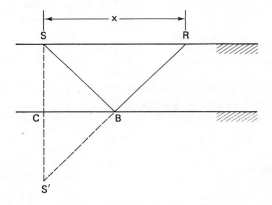

Figure 1-15. Vertical section showing normal moveout (NMO) determination.

and T_x, the travel time from S to B to R. We define the normal moveout ΔT_{NMO} by

$$\Delta T_{\text{NMO}} = T_x - T_0 \qquad (1\text{-}16)$$

We then apply the Pythagorean theorem to the triangle $SS'R$ (S' is the image point of source point S) to obtain

$$(S'R)^2 = (SS')^2 + (SR)^2 \quad \text{or} \quad (vT_x)^2 = (vT_0)^2 + x^2$$

which gives the *normal moveout* (NMO) *equation*:

$$T_x^2 = T_0^2 + \frac{x^2}{v^2} \qquad (1\text{-}17)$$

where v is the velocity, assumed to be constant throughout the medium, and x the offset distance (the horizontal distance from source S to receiver R). Extracting the square root of both sides of equation (1-17), we have

$$T_x = \sqrt{T_0^2 + \frac{x^2}{v^2}}$$

which is

$$T_x = T_0 \sqrt{1 + \frac{x^2}{(vT_0)^2}} \qquad (1\text{-}18)$$

Inserting (1-18) into (1-16), we obtain

$$\Delta T_{\text{NMO}} = T_0 \left[\sqrt{1 + \frac{x^2}{(vT_0)^2}} - 1 \right] \qquad (1\text{-}19)$$

$(a+b)^n = a^n + n a^{n-1} b + \binom{n}{2} a^{n-2} b^2$

To find an approximation for ΔT_{NMO} we expand the square root in (1-19) with the binomial theorem, keeping only the first two terms:

$$\Delta T_{\text{NMO}} \doteq T_0 \left[1 + \frac{1}{2} \frac{x^2}{(vT_0)^2} - 1 \right]$$

$1 + \frac{1}{2} \cdot (1)^{-\frac{1}{2}} \cdot \frac{x^2}{(T_0 v)^2} \cdots - 1 \Big]$

or

$$\boxed{\Delta T_{\text{NMO}} \doteq \frac{x^2}{2T_0 v^2}} \qquad (1\text{-}20)$$

This approximation is valid when the spread distance x is small compared with twice the depth to the reflector, vT_0. Formula (1-20) gives the normal moveout in terms of spread distance x, the center reflection time T_0, and the (constant) velocity v.

For v constant $\quad x \downarrow \quad \delta T_{NMO} \downarrow$

$T_x = T_0 + \delta T$

$\therefore \; T_x^2 = T_0^2 + \left(\frac{x}{v}\right)^2$

$T_x^2 - T_0^2 = \left(\frac{x}{v}\right)^2$

$(T_x - T_0)(T_x + T_0) = \delta T \, (T_0 + \delta x + T_0) = 2T_0 \delta T + (\delta T)^2$

$\qquad \simeq 2 T_0 \delta T$

$\therefore \delta T = \frac{1}{2T_0} \left(\frac{x}{v}\right)^2$

causal feedforward filters

Summary

The digital computer is a versatile tool that may be used to filter signals. Conventional filtering is performed by means of analog-type electronic networks, the behavior of which is ordinarily studied in the frequency domain. Digital filtering, on the other hand, is more fruitfully treated in the time domain. A digital filter is represented by a sequence of numbers called weighting coefficients. The output of a digital filter is obtained by convolving the digitized input signal with the filter's weighting coefficients. The mechanics of digital filtering in the time domain are described with the aid of discrete z-transform theory. These ideas are then related to the more familiar interpretation of filter behavior in the frequency domain. An important criterion for the classification of filters is the notion of minimum phase lag. The chapter ends with a simple presentation of this concept.

Introduction

The use of digital computers to process digitized signal recordings as a research tool in the development of filtering methods is now well established. The first part of this chapter presents an expository treatment of the theory of digital filtering in the time domain. We chose to do this because few elementary discussions of these topics exist. In the latter part of the chapter, we attempt to relate time-domain and frequency-domain filter theory, and are led to a consideration of the concept of minimum phase lag. Our main point here is that it is the phase lag, not the phase, that must be considered over the positive frequency range when classifying filters.

Digital Filtering of Signals

A *continuous signal*, that is, some continuous recording of data versus time, may be converted into a sequence of numbers. Each number represents the reading, or amplitude, of the signal at a specific time instant. The time points are chosen to be equally spaced, so that the time interval between two consecutive readings of the signal is always the same, for example, 1 ms.

The process of converting a continuous signal into a sequence of numbers at equally spaced time points is called *digitization*. Formerly, digitization was carried out by a scale and the eye. Today there are various automatic and semiautomatic devices available for this purpose. Figure 2-1 shows a portion of a digitized signal. The vertical arrows represent the amplitudes of the signal at the indicated time points. Instead of using the time scale appearing on the signal, it is more convenient to use a time index *t* so selected that the time increment is one unit. In this way the time index *t* associated with any reading x_t is a whole number, or integer. For example, a portion of the data of Figure 2-1 becomes

Time[a] (s)	*Time Index,* t	*Signal Reading,* x_t
1.000	0	$x_0 = 10$
1.001	1	$x_1 = 20$
1.002	2	$x_2 = 10$
1.003	3	$x_3 = 0$
1.004	4	$x_4 = -10$
1.005	5	$x_5 = 0$

[a]Time increment = 0.001 s.

A *digital filter* is represented by a sequence of numbers called *weighting coefficients*. A digital filter is said to be *causal* provided that its present output (at time *t*) depends on only present and past inputs (at times $t, t - 1, t - 2, \dots$). The simplest possible case is a digital filter with a constant weighting coefficient a_0; that is, a *constant filter*. Its action is shown schematically by the block diagram in Figure 2-2, where ovals indicate input and output and

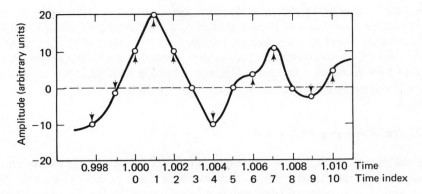

Figure 2-1. Portion of a digitized signal.

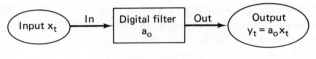

Figure 2-2

a rectangular box indicates the filter. The variable t is the time index, where $t = 0, 1, 2, 3, \ldots$.

Alternatively, we may illustrate the action of the filter a_0 by a table (where we have let $a_0 = \frac{1}{2}$):

Time Index, t	Input: x_t	Output: $y_t = \frac{1}{2}x_t$
0	10	5
1	20	10
2	10	5
3	0	0
4	−10	−5
5	0	0

We see that this filter is causal because its output at time t depends only on its input at time t.

Next, we introduce the concept of a digital filter that produces a unit delay. This is the *unit-delay filter*. Let us represent such a filter by the symbol z. Thus, we have the block diagram shown in Figure 2-3. In terms of the readings, we have $y_1 = x_0, y_2 = x_1, y_3 = x_2, \ldots$, or:

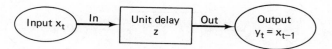

Figure 2-3

Time Index, t	Input: x_t	Output: $y_t = x_{t-1}$
0	10	Time lag is one unit
1	20	10
2	10	20
3	0	10
4	−10	0
5	0	−10
6	—	0

$x_{-1} = 0$

We see that this filter is causal because its output at time t depends only on its input at time $t - 1$.

The symbol z used here has a special mathematical meaning; z represents a mathematical operator that produces a *unit delay*. We thus call z the *unit-delay operator*. If we write z more explicitly as z^1 (i.e., z to the first power), the exponent 1 represents the delay.

What happens when we connect two unit-delay filters in series? That is, suppose that we have the situation shown in Figure 2-4. By a series connection we mean that the output from filter 1 is the input to filter 2. Now, filter 1 produces a unit delay, so its output is x_{t-1}, as shown in Figure 2-5.

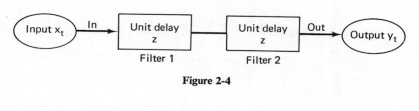

Figure 2-4

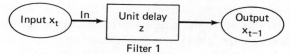

Figure 2-5

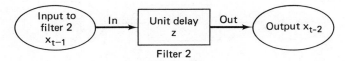

Figure 2-6

Next we use the fact that the input to filter 2 is the output from filter 1, so that the input to filter 2 is x_{t-1}. Because filter 2 produces a unit delay, its output is x_{t-2} (Figure 2-6). In terms of the readings, we have

Time Index, t	Input: x_t	Output from Filter 1	Input to Filter 2	Output x_{t-2}
0	10	—	—	Time lag is
1	20	10	10	two units
2	10	20	20	10
3	0	10	10	20
4	−10	0	0	10
5	0	−10	−10	0
6	—	0	0	−10
7	—	—	—	0

We see that this filter is causal because its output at time t depends only on the input at time $t - 2$.

In summary, we see that *two unit-delay filters in series are equivalent to a filter with a two-unit delay; that is, if the input is x_t, the output is x_{t-2}.* A delay of two units is thus represented by the mathematical operator z^2 (i.e., z to the second power), the exponent 2 representing the delay (Figure 2-7).

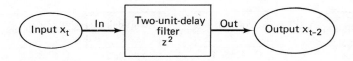

Figure 2-7

By the same reasoning, it is easy to see that *n unit-delay filters in series are equivalent to a filter with an n-unit delay;* that is, *if the input is x_t, the output is x_{t-n}.* A delay of n units is represented by the mathematical operator z^n (i.e., z to the nth power), the exponent n representing the delay (Figure 2-8).

One may ask what happens when $n = 0$? Because the exponent represents the delay, we see that in this case the delay is zero, and so input is equal

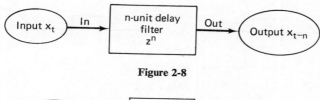

Figure 2-8

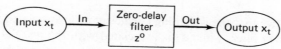

Figure 2-9

to output (Figure 2-9). Thus, the filter z^0 represents the *identity filter,* and in keeping with ordinary algebra we let $z^0 = 1$. We see that the *n*-unit delay filter is causal for any nonnegative value of n (i.e., for $n \geq 0$) because the output at time t depends only on its input at time $t - n$.

The constant filter a_0 (described previously) may be represented more explicitly by $a_0 z^0$. The series (or cascaded) combination of a constant filter a_1 followed by the unit-delay filter z gives the filter $a_1 z$; that is, the filter a_1 *connected in series with the filter z* is shown by Figure 2-10, or (with, say, $a_1 = \frac{1}{4}$)

Time Index, t	Input: x_t	$a_1 x_t$	Output: $y_t = a_1 x_{t-1}$
0	10	2.5	—
1	20	5	2.5
2	10	2.5	5
3	0	0	2.5
4	−10	−2.5	0
5	0	0	−2.5
6	—	—	0

We see that this filter is causal because its output at time t depends upon its input at time $t - 1$.

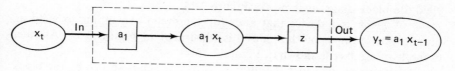

Figure 2-10

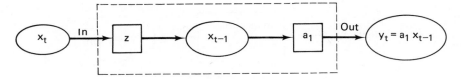

Figure 2-11

We notice that the weighting coefficient a_0 is associated with the constant, *zero*-delay filter, while a_1 is associated with the *unit*-delay filter. It is evident that the series combination of the z filter followed by the a_1 filter gives the filter za_1 (Figure 2-11), or (with $a_1 = \frac{1}{4}$)

Time Index, t	Input: x_t	x_{t-1}	Output: $y_t = x_{t-1}a_1$
0	10	—	—
1	20	10	2.5
2	10	20	5
3	0	10	2.5
4	−10	0	0
5	0	−10	−2.5
6	—	0	0

Hence we see that the filter a_1z is equivalent to the filter za_1.

Up to this point, we have connected two digital filters in series. Now we wish to introduce a *parallel connection.* Such a connection will be illustrated in our block diagrams by the connecting element, as shown in Figure 2-12. This figure illustrates that a parallel connecting element, when taken

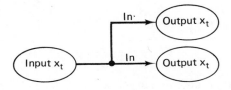

Figure 2-12

by itself, yields the same output on each line of a fork. The combination of the filter a_0 and the filter a_1z *connected in parallel* to the same input x_t would yield the block diagram shown in Figure 2-13.

A *mixer* is a device that adds (or subtracts) two inputs to yield an output. An example of this device is illustrated in Figure 2-14. Another example of a mixer is shown in Figure 2-15.

Let us now consider the filter $a_0 + a_1z$, which is shown by the block diagram given in Figure 2-16. The output of the subcomponent a_0 is a_0x_t.

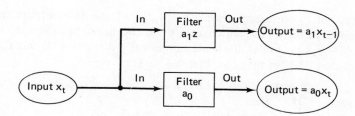

Figure 2-13

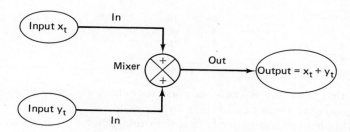

Figure 2-14

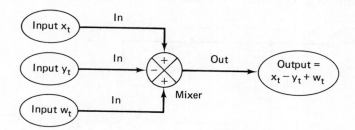

Figure 2-15

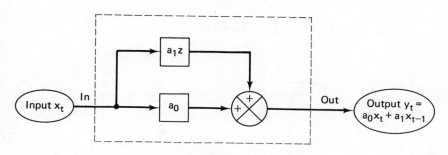

Figure 2-16

43

The output of the subcomponent a_1z is a_1x_{t-1}. These two outputs are then fed as inputs to the mixer, which adds them and hence produces the output $y_t = a_0x_t + a_1x_{t-1}$. Let us illustrate numerically the action of this filter for $a_0 = \frac{1}{2}$ and $a_1 = \frac{1}{4}$, so that our filter is $\frac{1}{2} + \frac{1}{4}z$:

Time Index, t	Input: x_t	a_0x_t	x_{t-1}	a_1x_{t-1}	Output: $y_t = a_0x_t + a_1x_{t-1}$
0	10	5	—	—	5
1	20	10	10	2.5	12.5
2	10	5	20	5	10
3	0	0	10	2.5	2.5
4	−10	−5	0	0	−5
5	0	0	−10	−2.5	−2.5
6	—	—	0	0	0

We see that this filter is causal because its output at time t depends only upon its input at times t and $t-1$.

It is not difficult to verify that the causal filter $a_0 + a_1z$ may also be illustrated by the block diagram shown in Figure 2-17.

As we have seen, the series connection of two unit-delay filters is equivalent to a filter with a two-unit delay (Figure 2-18). The dashed rec-

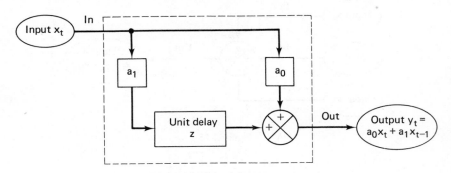

Figure 2-17

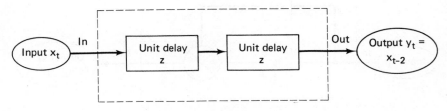

Figure 2-18

tangular box is the filter z^2, which is a two-unit delay filter. We recall that z^n represents an n-unit delay filter (Figure 2-19).

A constant filter a_n connected in series with the n-unit delay filter z^n yields the filter $a_n z^n$, shown by Figure 2-20. For example, suppose that

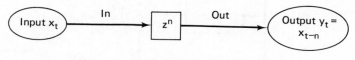

Figure 2-19

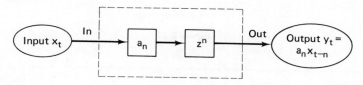

Figure 2-20

$n = 2$ and $a_2 = \frac{3}{4}$. Then the filter $a_2 z^2$ is illustrated by the table

Time Index, t	Input: x_t	x_{t-2}	Output: $y_t = a_2 x_{t-2}$
0	10	Time lag is two units	—
1	20	—	—
2	10	10	7.5
3	0	20	15
4	−10	10	7.5
5	0	0	0
6	—	−10	−7.5
7	—	0	0

The most general causal filter with a <u>finite number of delay elements</u> has the form

$$a_0 + a_1 z + a_2 z^2 + a_3 z^3 + a_4 z^4 + \cdots + a_n z^n$$

For example, the filter $a_0 + a_1 z + a_2 z^2$ has the block diagram shown in Figure 2-21. We may numerically illustrate the filter $a_0 + a_1 z + a^2 z^2$ by letting $a_0 = \frac{1}{2}$, $a_1 = \frac{1}{4}$, and $a_2 = \frac{3}{4}$:

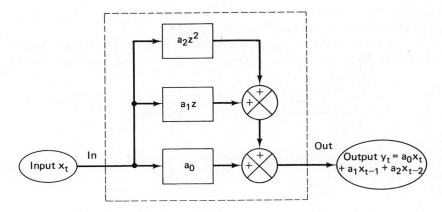

Figure 2-21

Time Index, t	Input: x_t	$a_0 x_t$	$a_1 x_{t-1}$	$a_2 x_{t-1}$	Output: y_t
0	10	5	—	—	5
1	20	10	2.5	—	12.5
2	10	5	5	7.5	17.5
3	0	0	2.5	15	17.5
4	−10	−5	0	7.5	2.5
5	0	0	−2.5	0	−2.5
6	—	—	0	−7.5	−7.5
7	—	—	—	0	0

where

$$y_t = a_0 x_t + a_1 x_{t-1} + a_2 x_{t-2} = \tfrac{1}{2}x_t + \tfrac{1}{4}x_{t-1} + \tfrac{3}{4}x_{t-2} \qquad (2\text{-}1)$$

We see that this filter is <u>causal</u> because its output at time t depends only on its input at <u>times t, $t-1$, and $t-2$.</u>

An equivalent block diagram for the filter $a_0 + a_1 z + a_2 z^2$ is given by Figure 2-22. The nth-order causal filter $a_0 + a_1 z + a_2 z^2 + \cdots + a_n z^n$ may be illustrated by either of the two block diagrams shown in Figure 2-23. In both these diagrams we see that the input is fed forward through the weighting-coefficient boxes and the delay boxes in order to produce the output. As a result we call a causal filter of this type a *causal feedforward filter*. Another name for this type of filter is a *moving-average* (MA) *filter* (see Chapter 16).

As a matter of terminology, the nth-order polynomial in z given by

$$A(z) = a_0 + a_1 z + a_2 z^2 + \cdots + a_n z^n \qquad (2\text{-}2)$$

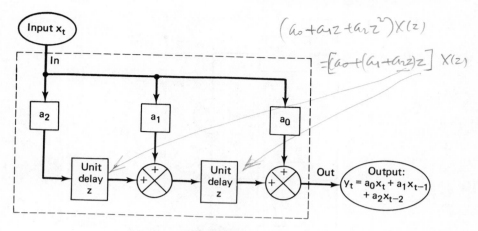

The following handwritten annotations appear:

$$(a_0 + a_1 z + a_2 z^2)X(z)$$

$$= [a_0 + (a_1 + a_2 z)z] X(z)$$

Figure 2-22

is called the *z* transform[1] of the *n*th-order causal feedforward filter with the constant weighting coefficients

$$a_0, a_1, a_2, \ldots, a_n$$

The weighting coefficients of the causal feedforward filter are also called the *memory function* or *impulse response function* of the filter.

Let us consider the action of the *n*th-order causal feedforward filter on an input given by the equally spaced sampled values $x_0, x_1, \ldots, x_m$. Proceeding as in (2-1), we write for the output at time *t*:

$$y_t = a_0 x_t + a_1 x_{t-1} + a_2 x_{t-2} + \cdots + a_n x_{t-n}$$

A more compact notation for this operation is

$$y_t = \sum_{s=0}^{n} a_s x_{t-s} \qquad \text{for } t = 0, 1, 2, \ldots, m + n \tag{2-3}$$

(handwritten: $t = 0 - (m+n)$)

where it is understood that $y_t = 0$ when *t* falls outside the range $t = 0, 1, 2, \ldots, m + n$.

This expression is the discrete representation of the linear operation commonly known as *convolution*. In the literature one often sees this operation represented by an integral rather than by a discrete summation. This is because analog filters operate in continuous time, which calls for an integral

---

[1]This definition of the *z* transform follows mathematical usage as originated by Laplace. Laplace used the term "generating function" instead of *z* transform. Most engineering texts use z^{-1} in place of our *z*. We prefer Laplace's version since it leads to *z* transforms that are polynomials in powers of *z* rather than polynomials in powers of z^{-1}.

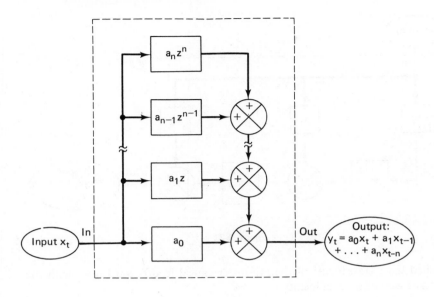

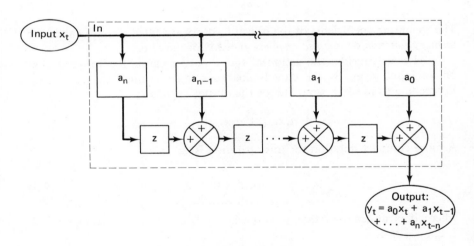

Figure 2-23

representation of the convolution operation. Since we are dealing here with discrete-time data, it is necessary to represent the convolution process by a summation. The output of the nth-order causal feedforward filter is thus obtainable by the discrete convolution of the input $x_0, x_1, \ldots, x_m$ with the filter's weighting coefficients $a_0, a_1, \ldots, a_n$. Convolution will be discussed further in Chapter 3.

The Amplitude and Phase Characteristics of Digital Filters

The filters that we discussed in the previous section all operate in the time domain. Therefore, they may be called _time-domain digital filters._ The numerical examples presented thus far illustrate digital filtering in the time domain. Many people are more accustomed to think about filtering in the frequency domain. One can profitably study the action of filters either in the time or in the frequency domains, or in a combination of both. The choice of a particular approach depends on the nature of the problem at hand. We shall now proceed to sketch the relation that exists between _time-domain and frequency-domain filter_ing. Before doing this, a brief discussion of simple harmonic motion is in order.

    Simple harmonic or sinusoidal motion at a given frequency may be illustrated by a wheel rotating at a constant angular velocity. Figure 2-24

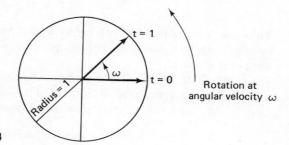

Figure 2-24

shows a wheel of unit radius rotating counterclockwise at an angular velocity of ω radians per unit time. The two vectors show that an angle of ω radians is swept out in one time unit. The lower vector corresponds to time $t = 0$, the upper vector to $t = 1$. Instead of considering a rotating wheel, we may simply think of a vector that rotates at a constant angular velocity ω (Figure 2-25). At time $t = 0$, suppose that the vector lies in the positive direction along the horizontal coordinate axis. Then, at some arbitrary time t, the vector will make an _angle of ωt radians with the horizontal axis._ The projec-

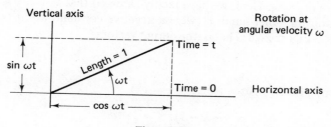

Figure 2-25

tion of this vector on the horizontal axis is seen to be

$$\text{Horizontal component} = \cos \omega t$$

while its projection on the vertical axis is

$$\text{Vertical component} = \sin \omega t$$

Letting the vertical axis be the imaginary axis (i.e., a unit distance on the vertical axis is $i = \sqrt{-1}$), we see that we can represent the vector at time t in terms of its components $\cos \omega t$ and $\sin \omega t$ as follows:

$$\text{Vector (at time } t) = \cos \omega t + i \sin \omega t$$

which, by Euler's formula, is

$$\text{Vector (at time } t) = e^{i\omega t}$$

Thus, the exponential $e^{i\omega t}$ represents a unit vector (or wheel) rotating at constant angular velocity ω. The components of this vector represent simple harmonic motion.

We shall now let simple harmonic motion be the input to the various digital filters that we have considered. First, let us consider the constant filter a_0. We have the diagram shown in Figure 2-26, so that the output is

$$a_0 e^{i\omega t} = a_0(\cos \omega t + i \sin \omega t)$$
$$= a_0 \cos \omega t + ia_0 \sin \omega t$$

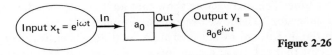

Figure 2-26

Hence, the output is also a rotating vector, but instead of having unit length, the output vector has length a_0. For example, for $a_0 = \frac{1}{2}$, we have Figure 2-27. Since both the input vector at time t and the output vector at time t make the same angle ωt with the horizontal axis, we say that the input and output are *in phase*. Here, we have tacitly assumed that a_0 is a positive constant. If, on the other hand, a_0 were a negative constant, say $a_0 = -\frac{1}{2}$, we would have Figure 2-28. That is, the output vector is

$$-\tfrac{1}{2}e^{i\omega t}$$

which may also be written (since $-1 = e^{i\pi}$)

$$+\tfrac{1}{2}(-1)e^{i\omega t} = \tfrac{1}{2}e^{i\pi}e^{i\omega t}$$

which is

$$\tfrac{1}{2}e^{i(\omega t + \pi)}$$

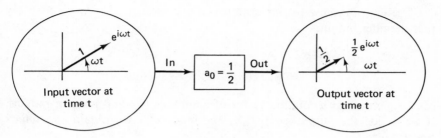

Figure 2-27

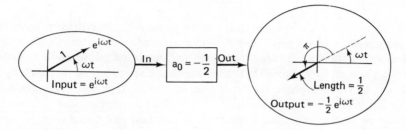

Figure 2-28

We have thus converted the output vector into the product of the positive constant $\frac{1}{2}$ times $e^{i(\omega t + \pi)}$. The positive constant $\frac{1}{2}$ is the length of the output vector, and hence is called the *magnitude* of this output vector. Now the quantity

$$e^{i(\omega t + \pi)} = \cos{(\omega t + \pi)} + i \sin{(\omega t + \pi)}$$

represents the rotating vector of length 1, which may be drawn as shown in Figure 2-29. At time $t = 0$, this vector lies on the horizontal axis in the negative direction and makes an angle of π radians with the positive horizontal axis. This angle π is called the *phase* of the vector:

$$e^{i(\omega t + \pi)}$$

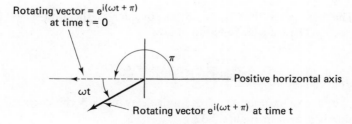

Figure 2-29

Returning now to our example of the filter $a_0 = -\frac{1}{2}$, we may say that the filter output,

$$-\tfrac{1}{2}e^{i\omega t} = \tfrac{1}{2}e^{i(\omega t + \pi)}$$
$$= \tfrac{1}{2}\cos(\omega t + \pi) + i\tfrac{1}{2}\sin(\omega t + \pi)$$

can be pictured as a rotating vector of amplitude $+\frac{1}{2}$ and phase π radians. If we divide the output $\frac{1}{2}e^{i(\omega t + \pi)}$ by the input $e^{i\omega t}$, we obtain the filter's *transfer function*:

$$\frac{\text{Output}}{\text{Input}} = \tfrac{1}{2}e^{i\pi}$$

This quantity, which is in general complex, may be described in terms of its magnitude and its phase angle. The magnitude of the transfer function is known as the filter's *magnitude spectrum*, while its phase angle is called the filter's *phase spectrum*. Thus, the magnitude spectrum of the filter $a_0 = -\frac{1}{2}$ is $+\frac{1}{2}$, while its phase spectrum is π. We notice that both the magnitude spectrum and the phase spectrum of *this* filter are *independent of angular velocity* ω. Because the phase spectrum is constant and equal to π, we may say that this filter's input and output are *out of phase* by π for all ω, or simply that they are 180° out of phase.

In the same way, we may compute the transfer function of any constant filter a_0 (Figure 2-30). The transfer function is

$$\frac{\text{Output}}{\text{Input}} = \frac{a_0 e^{i\omega t}}{e^{i\omega t}} = a_0$$

which is just the constant vector a_0.

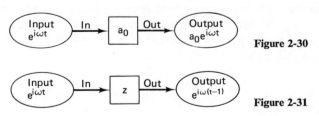

Figure 2-30

Figure 2-31

The next filter that we wish to consider is the unit-delay filter z (Figure 2-31). The transfer function is now

$$\frac{\text{Output}}{\text{Input}} = \frac{e^{i\omega(t-1)}}{e^{i\omega t}} = e^{-i\omega}$$

That is, the transfer function of the unit-delay filter is the constant vector $e^{-i\omega}$, which is shown in Figure 2-32.

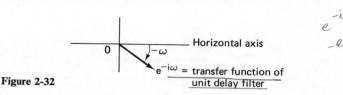

Figure 2-32

Now we recall that the *magnitude spectrum* of a filter is equal to the magnitude of the filter's transfer function. Because the vector $e^{-i\omega}$ has unit magnitude, we see that the magnitude spectrum of the unit-delay filter is 1. We also recall that the *spectrum* of a filter is equal to the phase angle of the transfer function of the filter. Since the vector $e^{-i\omega}$ makes an angle of $-\omega$ with the positive horizontal axis, we see that the phase spectrum of this filter is $-\omega$. Hence, the phase spectrum of this filter is a *function of angular velocity* ω, although its magnitude spectrum is independent of ω. The magnitude and phase spectra are plotted in Figure 2-33.

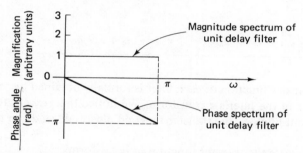

Figure 2-33. Magnitude and phase spectra of the unit delay filter.

Summing up, we see that the transfer function of the filter a_0 is a_0, and that the transfer function of the filter z is $e^{-i\omega}$. For the causal feedforward filter $a_1 z$ we have the transfer function

$$\frac{\text{Output}}{\text{Input}} = \frac{a_1 e^{i\omega(t-1)}}{e^{i\omega t}} = a_1 e^{-i\omega}$$

The causal feedforward filter $a_0 + a_1 z$ has the transfer function

$$\frac{\text{Output}}{\text{Input}} = \frac{a_0 e^{i\omega t} + a_1 e^{i\omega(t-1)}}{e^{i\omega t}} = a_0 + a_1 e^{-i\omega}$$

The transfer function of the nth-order causal feedforward filter, given by Figure 2-34, may now be written down by induction:

$$\frac{\text{Output}}{\text{Input}} = \frac{a_0 e^{i\omega t} + a_1 e^{i\omega(t-1)} + \cdots + a_n e^{i\omega(t-n)}}{e^{i\omega t}}$$

$$= a_0 + a_1 e^{-i\omega} + \cdots + a_n e^{-i\omega n}$$

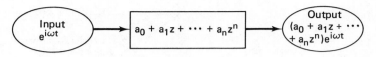

Figure 2-34

Our results may be tabulated in the form

Causal Feedforward Filter	Corresponding Transfer Function
a_0	a_0
z	$e^{-i\omega}$
$a_1 z$	$a_1 e^{-i\omega}$
$a_0 + a_1 z$	$a_0 + a_1 e^{-i\omega}$
.	.
.	.
.	.
$a_0 + a_1 z + \cdots + a_n z^n$	$a_0 + a_1 e^{-i\omega} + \cdots + a_n e^{-i\omega n}$

Thus, the transfer function of each filter is formally obtained by the substitution $e^{-i\omega} = z$ in the filter's z transform. We notice that, except for the case of the constant filter a_0, the transfer function always depends on the angular velocity ω.

If we write the transfer function in polar form,

$$A(\omega) = |A(\omega)| e^{i\psi(\omega)}$$

we see that the magnitude $|A(\omega)|$ and the angle $\psi(\omega)$ represent, respectively, the *magnitude* and *phase spectra* of the filter. For example, the transfer function of the filter

$$a_0 + a_1 z$$

is

$$A(\omega) = a_0 + a_1 e^{-i\omega}$$
$$= (a_0 + a_1 \cos \omega) - i(a_1 \sin \omega)$$

Now $A(\omega)$ (for a fixed value of ω) is the vector that is the sum of the vectors a_0 and $a_1 e^{-i\omega}$ (Figure 2-35). The length $|A(\omega)|$ of the vector $A(\omega)$ is

$$|A(\omega)| = + \sqrt{(a_0 + a_1 \cos \omega)^2 + (a_1 \sin \omega)^2}$$
$$= + \sqrt{a_0^2 + 2a_0 a_1 \cos \omega + a_1^2}$$

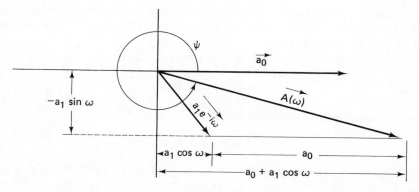

Figure 2-35

This quantity is the *magnitude spectrum* of the filter $a_0 + a_1 z$. The angle ψ, which is a function of ω,

$$\psi(\omega) = \tan^{-1}\left(\frac{-a_1 \sin \omega}{a_0 + a_1 \cos \omega}\right)$$

$$= -\tan^{-1}\left(\frac{a_1 \sin \omega}{a_0 + a_1 \cos \omega}\right)$$

yields the *phase spectrum* of the filter $a_0 + a_1 z$. We see that both the magnitude and phase spectra are functions of angular frequency ω. Instead of dealing with the phase spectrum $\psi(\omega)$, we shall deal with its negative, which we call the *phase-lag spectrum* $\phi(\omega)$. The phase-lag spectrum is defined as

$$\phi(\omega) = -\psi(\omega)$$

where

$$A(\omega) = |A(\omega)|\, e^{-i\phi(\omega)} = |A(\omega)|\, e^{i\psi(\omega)}$$

In the example above, $a_0 + a_1 z$, the phase-lag spectrum, is then

$$\phi(\omega) = \tan^{-1}\left(\frac{a_1 \sin \omega}{a_0 + a_1 \cos \omega}\right)$$

The Minimum-Phase-Lag Spectrum of a Digital Filter

We have seen in the previous section that the transfer function of a digital filter can be conveniently expressed in terms of a magnitude spectrum $A(\omega)$ and a phase-lag spectrum $\phi(\omega)$. It is much easier to visualize the physical significance of the magnitude spectrum of a filter than it is to understand the corresponding phase spectrum. Some people therefore tend to neglect the phase-lag spectra of filters when solving actual problems. It turns out, however, that the phase-lag spectra are of fundamental importance in classifying filters with identical amplitude characteristics. This fact is perhaps best illustrated by a simple example.

Let us consider two filters, one with a z transform

$$A_0(z) = 1 + 0.5z$$

and the other with a z transform

$$A_1(z) = 0.5 + 1z$$

In other words, the weighting coefficients of filter A_0 are $(1, 0.5)$ while the weighting coefficients of filter A_1 are $(0.5, 1)$. We recall that the magnitude spectrum of the filter

$$A(z) = a_0 + a_1 z$$

is

$$|A(\omega)| = +\sqrt{a_0^2 + 2a_0 a_1 \cos \omega + a_1^2}$$

Setting $a_0 = 1$ and $a_1 = 0.5$ in this formula, we find that the magnitude spectrum of the $A_0(z)$ filter is

$$|A_0(\omega)| = +\sqrt{1 + \cos \omega + 0.25}$$
$$= +\sqrt{1.25 + \cos \omega}$$

Setting $a_0 = 0.5$ and $a_1 = 1$, we find that the magnitude spectrum of the $A_1(z)$ filter is

$$|A_1(\omega)| = +\sqrt{0.25 + \cos \omega + 1}$$
$$= +\sqrt{1.25 + \cos \omega}$$

which is the same as the foregoing expression for $|A_0(\omega)|$. Thus, the two filters have the same magnitude spectrum, which is shown in Figure 2-36 for the frequency range $-\pi \le \omega \le \pi$. Negative frequencies are used for mathematical convenience. In our discussion of simple harmonic motion, a positive frequency would indicate that the wheel is rotating in a counter-clockwise direction, whereas a negative frequency would indicate rotation in a clockwise direction.

The question now arises: What is the relation between the phase-lag spectra of the two filters $A_0(z)$ and $A_1(z)$? We recall that the phase-lag spectrum of the filter $a_0 + a_1 z$ is given by

$$\phi(\omega) = \tan^{-1} \left(\frac{a_1 \sin \omega}{a_0 + a_1 \cos \omega} \right)$$

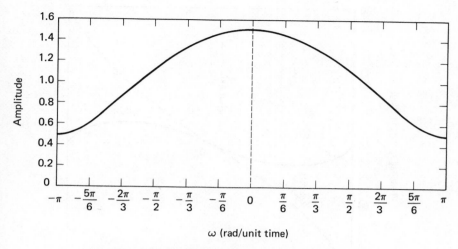

Figure 2-36. Magnitude spectra of the filters $A_0(z)$ and $A_1(z)$.

Thus, for the $A_0(z)$ filter, we have, with $a_0 = 1$ and $a_1 = 0.5$,

$$\phi_0(\omega) = \tan^{-1}\left(\frac{0.5 \sin \omega}{1 + 0.5 \cos \omega}\right)$$

On the other hand, letting $a_0 = 0.5$ and $a_1 = 1$, we have for the $A_1(z)$ filter,

$$\phi_1(\omega) = \tan^{-1}\left(\frac{\sin \omega}{0.5 + \cos \omega}\right)$$

These two-phase-lag spectra are plotted in Figure 2-37 for the range $-\pi \leq \omega \leq \pi$.

Since phase as well as phase-lag curves are odd functions of angular frequency ω (i.e., they are antisymmetrical about the origin), we need only to plot $\phi_0(\omega)$ and $\phi_1(\omega)$ for ω in the range 0 to π. This is done in Figure 2-38. We thus see that the phase-lag spectrum $\phi_0(\omega)$ lies below the phase-lag spectrum $\phi_1(\omega)$. We can now state that the filter $A_0(z)$ has a phase-lag spectrum which is less than the phase-lag spectrum of the filter $A_1(z)$ for the range $0 < \omega \leq \pi$. At $\omega = 0$, the phase-lag spectra of both filters are zero.

Let us restrict ourselves to real weighting coefficients, and let us consider only causal digital filters, for which the output can never precede the input in time. Then we see that the pair of filters $\{A_0(z), A_1(z)\}$ represents a set of digital causal filters each with the same amplitude spectrum. This set is exhaustive in the sense that the two real weighting coefficients a_0 and a_1 can only occur either in the sequence a_0, a_1 or in the sequence a_1, a_0. We can now say that the filter $A_0(z)$ has the *minimum-phase-lag spectrum* of the

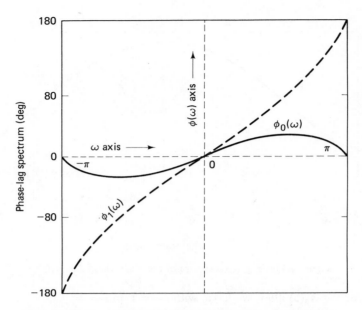

Figure 2-37. Phase-lag spectra of the filters $A_0(z)$ and $A_1(z)$ in the range $-\pi$ to $+\pi$.

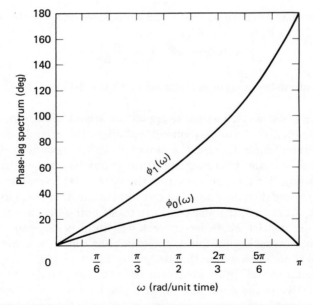

Figure 2-38. Phase-lag spectra of the filters $A_0(z)$ and $A_1(z)$ in the range 0 to $+\pi$.

filter set $\{A_0(z), A_1(z)\}$. The concept of minimum-phase-lag filters is quite general, and can be extended to other sets of causal filters, each filter in the set having the same amplitude spectrum. There is in each such set one filter whose phase-lag spectrum is minimum with respect to the phase-lag spectra of all other members of that set, and this filter is the minimum-phase-lag filter. All filters within the given set by definition have the same amplitude spectrum.

Concluding Remarks

Our present discussion has been primarily concerned with the underlying principles of digital filtering. We have shown in some detail how digital filters operate on a discrete input to yield a discrete output. We found that the mechanics of this process can be visualized with greatest ease in the time domain, but it is usually necessary to think of filtering in terms of both the time and the frequency domains. We have thus attempted to present a thorough, although heuristic description of filter behavior in both these domains. We have introduced the minimum-phase-lag concept for classifying digital filters.

minimum-delay and causal feedback filters

Summary

We consider various types of signals: in particular, power signals, which are represented by time series with finite power, and energy signals, which are represented by time series with finite energy. A wavelet is defined as an energy signal with a definite arrival time. Convolution is described as a folding operation between two signals. The autocorrelation of a signal is defined as the convolution of the signal with its reverse. A minimum-delay wavelet is defined as a wavelet with its energy concentrated near its arrival time, as opposed to a mixed-delay wavelet, which is defined as a wavelet with its energy distributed away from its arrival time. The concept of maximum delay is defined for finite-length wavelets, namely the maximum-delay wavelet is the time reverse of a finite-length minimum-delay wavelet. The first-order causal feedback filter is defined, and its stability is related to the property of minimum delay.

Introduction

In this chapter we wish to present and explain the minimum-delay concept in detail, so that it may be useful to those actively engaged in design. To show how the techniques can be used, we shall frequently illustrate the mathematics by examples of systems with numerical values included.

The concept of minimum delay is developed here for systems operating in discrete time. Examples of such systems are:

1. *Sampled-data systems* in which the variables appear as a train of pulses at discrete, equally spaced sampling instants.
2. *Digital computer systems* in which the variables appear as a sequence of numbers at discrete, equally spaced time instants.

In other words, the values of the variables of discrete-time systems are only given at discrete instants of time.

By contrast, systems operating in continuous time have variables that are functions of continuous time. That is, the values of the variables are given at all instants of time. The concept of minimum delay can also be developed for such systems, although we will not treat the continuous-time case in this book.

It is worth noting that models of both discrete-time systems and continuous-time systems should take into account uncertainties in the amplitudes of the variables. A signal, whether it be a train of pulses or a function of continuous time, always has imperfections in its amplitude, and these make its precise measurement uncertain. Moreover, a signal that appears as a sequence of numbers always has an uncertainty in magnitude, the degree of this uncertainty being at least equal to one unit of measurement.

The reason for exploring discrete-time systems instead of continuous-time systems is twofold:

1. We may present the basic ideas in more simple mathematical terms.
2. We may develop programs for a digital computer, in a universal computer language such as FORTRAN, to carry out the mathematical operations.

Our objective is always to keep the mathematics simple and operational, as far as possible.

Time Series

We have seen that discrete-time systems have variables that may be either trains of pulses or sequences of numbers, each pulse or each number being identified by a fixed time instant. Such a series of data, whether pulses or numbers, in a time sequence is called a *time series*. In other words, a time series x_t is a series of data, each data value x_t being associated with a discrete, equally spaced time instant t.

Time series occur in all branches of science. Economic data always appear in the form of numerical time series. Some meteorological data, such as daily temperatures, are numerical time series; other meteorological data, such as continuous barographic records, are continuous-time functions. Continuous-time functions appear as the rule in the engineering and physical sciences. Nevertheless, such continuous-time functions may be read (or measured, or observed, or sampled) at equal intervals of time, thereby generating time series. Because a time series represents only the sampled values of a continuous-time function, it provides only a limited description of the function. That is, the time series represents sampled data of the continuous time function, the sampling instants being discrete, equally spaced time points. By taking the sampling instants close enough together, the amount of information that is lost by replacing a well-behaved continuous-time function by a time series can be made small. The time spacing must be chosen relative to the frequency characteristics of the continuous-time function being sampled. A time spacing that is too gross would mean substantial information lost in the sampling process. At the other extreme, a time spacing that is too fine would mean substantial redundancy in information produced by the sampling process. Thus, in determining the time interval, we must always balance redundant information versus lost information, with consideration given to their relative costs.

For simplicity, we shall designate the interval of discrete time to be one unit, whether it be 1 μs, or 10 ms, 3 s, or whatever. In this way the time variable t is always a whole number, or integer.

The term *time series* is a generic one. Often the term is used to represent a signal continuing over all time, from the remote past to the distant future. Thus, we may say that such a time series has infinite time duration, or infinite length. Nevertheless, in any actual situation we can only obtain the values of a time series over a finite interval of time. Because a finite number of data values represents a sample of an infinite-length time series, we shall call a finite portion of a time series a *sample time series*. Here we must distinguish between:

1. The sampling process that extracts a time series (over all discrete time) from a continuous-time function (over all continuous time) and
2. The sampling process that extracts a sample time series (over a finite interval of discrete time) from a time series (over all discrete time).

A sample time series, then, is a finite portion of a time series, that is, the portion from some fixed time to a later time. For example, for a given time series, we may have available only the portion from time $t = 1$ to time $t = 15$, which consists of the 15 values

$$(x_1, x_2, \ldots, x_{15})$$

A numerical example of a sample time series with 15 values (or readings, or measurements, or observations, or samples) is

$$(8, 20, 29, 34, 33, 26, 18, 11, 3, -3, 9, 25, 25, 15, 5)$$

In this numerical example, one would have to make a notation that the first value, 8, is the value for time $t = 1$. The following values, 20, 29, ..., would automatically be in sequence, that is, the values for $t = 2, 3, \ldots,$ 14, 15.

Wavelets

Let us now introduce the concept of a wavelet. A *wavelet* is a signal characterized by two properties:

1. The *one-sided property*: A wavelet has a definite origin (or arrival) time; that is, all values of the wavelet before its origin time are zero.
2. The *stability property*: A wavelet has finite energy; that is, it is a transient or dying-out phenomenon as time progresses.

We must always be careful to distinguish between the concepts of *wavelet*, on the one hand, and *sample time series*, on the other hand. The point of difference is that a wavelet is a self-contained or entire entity, and this entity is a transient phenomenon with a definite arrival time. In contrast, a sample time series is only a portion of an entire entity, namely, a time series, over all time, and this entity is a continuing phenomenon.

When we plot a wavelet, we see its entire life history, from the time it arrives until the time it damps out. Its origin time, or arrival time, is called time index 0, which serves as the reference date for the wavelet. At time index 0 we plot the amplitude of the wavelet at its origin. At time index 1 we plot the amplitude of the wavelet when it is 1 unit old. At time index 2 we plot the amplitude of the wavelet when it is 2 units old, and so on.

For example, suppose that the wavelet has amplitude 4 at time index 0, amplitude 2 at time index 1, amplitude 1 at time index 2, and zero amplitude for all succeeding times. The following table summarizes this wavelet:

Time index: 0 1 2 3 4 . . .
Wavelet: 4 2 1 0 0 . . .

More concisely, we may summarize this wavelet by the row vector

$$(4, 2, 1)$$

where it is understood that 4 is the initial amplitude (i.e., at time index 0) so that all preceding amplitudes would be zero, 2 is the amplitude at time index 1, and 1 is the amplitude at time index 2, or the final amplitude (so that all following amplitudes are zero).

Wavelets may also be represented algebraically. Thus we may write the wavelet in the form

$$\mathbf{b} = (b_0, b_1, b_2)$$

This notation means that wavelet $\mathbf{b}$ has amplitude b_0 at time index 0, amplitude b_1 at time index 1, and amplitude b_2 at time index 2. For example, the equation

$$(b_0, b_1, b_2) = (4, 2, 1)$$

would mean that $b_0 = 4$, $b_1 = 2$, $b_2 = 1$.

It is possible to consider wavelets with complex amplitudes, that is, amplitudes of the form $u + iv$, where u and v are real numbers and $i = \sqrt{-1}$. An example is the wavelet $(i, 0.5)$, where i is the amplitude at time index 0 and 0.5 is the amplitude at time index 1. Another example is the wavelet $(2 + i, 1 - i, i)$, where $2 + i$ is the amplitude at time index 0, $1 - i$ is the amplitude at time index 1, and i is the amplitude at time index 2. The amplitudes of the wavelet are also called the coefficients of the wavelet. For example, for the wavelet (b_0, b_1, b_2), we may call b_0, b_1, and b_2 either the amplitudes or the coefficients of the wavelet. That is, b_0 would be the coefficient at time index 0, and so on. Because the wavelet (b_0, b_1, b_2) has three coefficients, we say that it has *length (or time duration)* equal to 3, or that it is a 3-length wavelet.

Up to now we have considered only finite wavelets, that is, wavelets with a finite number of coefficients, or, in other words, wavelets with finite length. These wavelets die out completely (i.e., become zero) after a certain age. For example, the wavelet $(4, 2, 1)$ dies out (i.e., becomes zero) at time index 3.

It is possible to have infinite-length wavelets, but for stability we must require that they have *finite energy*. Thus, an infinite-length wavelet may be represented by

$$(b_0, b_1, b_2, \ldots)$$

where b_0 is the coefficient for time index 0, b_1 is the coefficient for time index 1, b_2 is the coefficient for time index 2, and the three dots indicate that the coefficients extend on for all positive times to form an infinite sequence. For stability, it is required that the wavelet's energy, given by

$$b_0^2 + b_1^2 + b_2^2 + \cdots$$

be finite. (*Note:* If the coefficients of the wavelet are complex, the energy is given by

$$b_0 b_0^* + b_1 b_1^* + b_2 b_2^* + \cdots$$

where the asterisk in the superscript position indicates the complex conjugate of the quantity to which it is attached. For example, if $b_0 = u + iv$, then $b_0^* = u - iv$ and $b_0 b_0^* = u^2 + v^2$.)

An example of an infinite-length wavelet is

$$\mathbf{b} = (1, \tfrac{1}{2}, \tfrac{1}{4}, \tfrac{1}{8}, \tfrac{1}{16}, \tfrac{1}{32}, \tfrac{1}{64}, \ldots)$$

where the first coefficient, 1, is the coefficient for time index 0, the next coefficient, $\tfrac{1}{2}$, is the coefficient for time index 1, and so on. An alternative way of representing this wavelet would be

$$\mathbf{b} = (b_0, b_1, b_2, \ldots)$$

where $b_t = (\tfrac{1}{2})^t$ for $t = 0, 1, 2, \ldots$. Because of the stability property, we see that the magnitudes of the coefficients asymptotically approach zero as time increases.

The energy distribution of a wavelet is displayed by its partial energy. Consider the (real) wavelet

$$(b_0, b_1, b_2, b_3)$$

The partial energy for time 0, denoted by p_0, is b_0^2. Since partial energy is cumulative, the partial energy p_1 at time 1 is $b_0^2 + b_1^2$, and so on. That is,

$$
\begin{aligned}
p_0 &= b_0^2 \\
p_1 &= b_0^2 + b_1^2 & &= p_0 + b_1^2 \\
p_2 &= b_0^2 + b_1^2 + b_2^2 & &= p_1 + b_2^2 \\
p_3 &= b_0^2 + b_1^2 + b_2^2 + b_3^2 &&= p_2 + b_3^2
\end{aligned}
$$

The last value of partial energy, in this case p_3, is the total energy of the wavelet. (If the wavelet is complex, we would use $b_0 b_0^*$ instead of b_0^2, $b_1 b_1^*$ instead of b_1^2 and so on.)

Convolution

Let us now describe the process of convolution in more detail. The *convolution* of the two wavelets

$$\mathbf{a} = (a_0 \ a_1, a_2, \ldots)$$

and

$$\mathbf{b} = (b_0, b_1, b_2, \ldots)$$

is defined to be

$$\mathbf{c} = \mathbf{a} * \mathbf{b} = (c_0, c_1, c_2, \ldots)$$

where the coefficients of $\mathbf{c}$ are given by the formula

$$c_t = \sum_{s=0}^{t} a_s b_{t-s}$$

(*Note:* The asterisk $*$ used in this way, that is, as a binary operation between two time functions, denotes convolution.)

To illustrate the use of this formula, suppose that

$$\mathbf{a} = (a_0, a_1) = (2, 1)$$

and

$$\mathbf{b} = (b_0, b_1) = (3, 4)$$

Each of these wavelets has two coefficients (i.e., it has length 2). Using the convolution formula, we see that the convolution of $\mathbf{a}$ and $\mathbf{b}$ is

$$\mathbf{c} = \mathbf{a} * \mathbf{b} = (c_1, c_2, c_3)$$

where

$$
\begin{aligned}
c_0 &= a_0 b_0 & &= 2 \cdot 3 & &= 6 \\
c_1 &= a_0 b_1 + a_1 b_0 & &= 2 \cdot 4 + 1 \cdot 3 &&= 11 \\
c_2 &= & a_1 b_1 &= & 1 \cdot 4 &= 4
\end{aligned}
$$

Hence,

$$\mathbf{c} = \mathbf{a} * \mathbf{b} = (6, 11, 4)$$

Convolution is a *folding operation.* In fact, the German word for convolution is *faltung,* which means folding. To see how convolution may be performed by folding, let us construct the table shown in Figure 3-1, whose entries are the products of the wavelets $\mathbf{a}$ and $\mathbf{b}$ (which are on the margins). Thus, we have the table of entries (without the margins) given in Figure 3-2, which we are going to fold successively on the dashed lines. The element in the upper left-hand corner, that is, $a_0 b_0$, or 6, is c_0. Now fold this entry over,

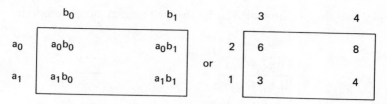

Figure 3-1

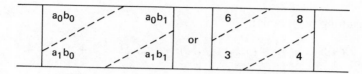

Figure 3-2

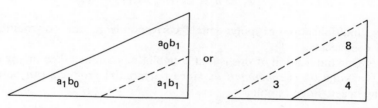

Figure 3-3

thus obtaining Figure 3-3. The sum of the two entries before the next fold is

$$a_0b_1 + a_1b_0 \quad \text{or} \quad 8 + 3 = 11$$

which is c_1. Now fold these entries over, thus obtaining Figure 3-4. This last entry is c_2. Thus, we have found the convolution series

$$(c_0, c_1, c_2) = (a_0b_0, a_0b_1 + a_1b_0, a_1b_1) = (6, 11, 4)$$

Figure 3-4

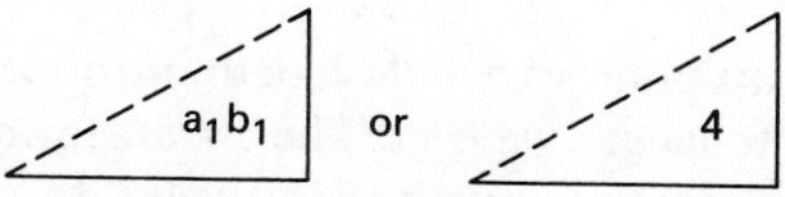

Convolution may also be performed by *multiplication of polynomials*. Thus, we write the z transforms

$$A(z) = a_0 + a_1z \qquad A(z) = 2 + z$$
$$\text{or}$$
$$B(z) = b_0 + b_1z \qquad B(z) = 3 + 4z$$

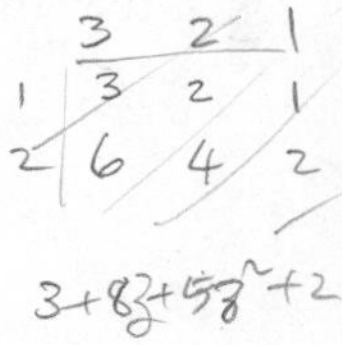

Multiplying the polynomial $A(z)$ by the polynomial $B(z)$, we obtain

$$
\begin{array}{r}
b_0 + b_1 z \\
a_0 + a_1 z \\
\hline
a_0 b_0 + a_0 b_1 z \\
a_1 b_0 z + a_1 b_1 z^2 \\
\hline
a_0 b_0 + (a_0 b_1 + a_1 b_0)z + a_1 b_1 z^2
\end{array}
\qquad \text{or} \qquad
\begin{array}{r}
3 + 4z \\
2 + z \\
\hline
6 + 8z \\
3z + 4z^2 \\
\hline
6 + 11z + 4z^2
\end{array}
$$

The resulting polynomial,

$$C(z) = a_0 b_0 + (a_0 b_1 + a_1 b_0)z + a_1 b_1 z^2 = 6 + 11z + 4z^2$$

has coefficients that are equal to the convolution

$$c_0 = a_0 b_0 = 6, \qquad c_1 = a_0 b_1 + a_1 b_0 = 11, \qquad c_2 = a_1 b_1 = 4$$

Thus, multiplication of polynomials corresponds to the convolution of their coefficients.

One implication of this is that convolutions can be taken in any order (i.e., convolution is *commutative*), since polynomial products can be taken in any order. For example,

$$\mathbf{a} * \mathbf{b} = \mathbf{b} * \mathbf{a}$$

since

$$A(z)B(z) = B(z)A(z)$$

By the same reasoning, we see that convolution is *associative*; that is,

$$(\mathbf{a} * \mathbf{b}) * \mathbf{c} = \mathbf{a} * (\mathbf{b} * \mathbf{c})$$

and convolution is *distributive* with respect to addition; that is,

$$\mathbf{a} * (\mathbf{b} + \mathbf{c}) = (\mathbf{a} * \mathbf{b}) + (\mathbf{a} * \mathbf{c})$$

At this point we would like to introduce the concept of the *z transform* of an infinite-length wavelet. The *z* transform of a finite wavelet is the polynomial in *z* associated with the wavelet. In other words, the *z* transform of the wavelet

$$\mathbf{b} = (b_0, b_1, b_2, \ldots, b_n)$$

is the *n*th-degree polynomial in *z*,

$$B(z) = b_0 + b_1 z + b_2 z^2 + \cdots + b_n z^n$$

While the z transform of a finite-length wavelet is a polynomial, the z transform of an infinite-length wavelet is a power series in z. In other words, the z transform of the infinite-length wavelet

$$\mathbf{b} = (b_0, b_1, b_2, \ldots)$$

is the power series

$$B(z) = b_0 + b_1 z + b_2 z^2 + \cdots$$

Linear Systems

In the preceding section we defined the convolution of two wavelets. There is no reason, however, why this definition cannot be extended to the convolution of an arbitrary time series with a wavelet. Let the wavelet be

$$\mathbf{a} = (a_0, a_1, a_2, \ldots)$$

and let the arbitrary time series be

$$\mathbf{x} = (\ldots, x_{-2}, x_{-1}, x_0, x_1, x_2, x_3, \ldots)$$

which we may think as being infinitely long in both the positive and negative directions. Then their convolution is the time series

$$\mathbf{y} = (\ldots, y_{-2}, y_{-1}, y_0, y_1, y_2, y_3, \ldots)$$

where y_t is given by the formula

$$y_t = \sum_{s=0}^{\infty} a_s x_{t-s}$$

Like the $\mathbf{x}$ time series, the $\mathbf{y}$ time series is also infinitely long in both directions.

Now let us illustrate this convolution process schematically. We suppose that the numbers x_t are going into a filter; that is,

$$x_{-1} \text{ enters at time } t = -1$$
$$x_0 \text{ enters at time } t = 0$$
$$x_1 \text{ enters at time } t = 1$$
$$x_2 \text{ enters at time } t = 2$$
$$x_3 \text{ enters at time } t = 3$$

etc.

so that the time series **x** is the *input* to the filter. Coming out of the filter are the numbers y_t; that is,

$$y_{-1} \text{ emerges at time } t = -1$$
$$y_0 \text{ emerges at time } t = 0$$
$$y_1 \text{ emerges at time } t = 1$$
$$y_2 \text{ emerges at time } t = 2$$

etc.

so that the time series **y** is the *output* from the filter. We see that the wavelet **a** inside the filter is its *memory function.* Schematically, we have Figure 3-5.

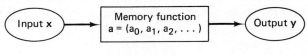

Figure 3-5

Whenever we illustrate such an input/output relationship, we mean the following: The output **y** is equal to the convolution of the input **x** and the memory function **a**, or

$$\mathbf{y} = \mathbf{a} * \mathbf{x}$$

If the input to a causal filter is the Kronecker delta function, or unit impulse function, given by

$$\mathbf{x} = (1, 0, 0, 0, \ldots)$$
$$\uparrow$$
$$t = 0$$

then the output is seen to be the memory function **a**; that is,

$$\mathbf{y} = \mathbf{a} * (1, 0, 0, 0, \ldots) = \mathbf{a}$$

Thus the memory function **a** is the response of a causal filter to a unit impulse. For this reason the memory function of a causal filter may also be called its *impulse response function.*

Consider the situation shown in Figure 3-6, where two filters are cas-

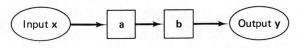

Figure 3-6

caded, which we have already treated in Chapter 2. The output of the first filter is $\mathbf{a} * \mathbf{x}$, which is the input to the second filter. Hence, the output of the second filter is

$$\mathbf{y} = \mathbf{b} * (\mathbf{a} * \mathbf{x})$$

Hence, the two filters may be replaced with one filter, the memory function of which is the convolution $\mathbf{c}$ of the memory functions $\mathbf{a}$ and $\mathbf{b}$ of the two filters; that is,

$$\mathbf{c} = \mathbf{b} * \mathbf{a}$$

Thus, Figure 3-6 is equivalent to Figure 3-7.

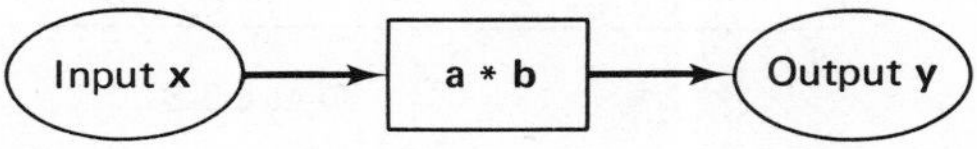

Figure 3-7

Input x ⟶ a * b ⟶ Output y

Autocorrelation

The autocorrelation of a wavelet $(b_0, b_1, b_2, \ldots)$ is defined to be

$$r_k = \sum_{j=0}^{\infty} b_{j+k} b_j^*$$

This formula gives the autocorrelation coefficient r_k for each integer k. For a finite-length wavelet $(b_1, b_1, \ldots, b_n)$, this equation becomes

$$r_k = \begin{cases} \sum_{j=0}^{n-k} b_{j+k} b_j^* & \text{for } k = 0, 1, 2, \ldots, n-1, n \\ r_{-k} & \text{for } k = -n, -n+1, \ldots, -2, -1 \\ 0 & \text{for } k < -n \text{ and } k > n \end{cases}$$

The superscript asterisk indicates the complex conjugate; the superscript asterisk can be omitted except in those cases when the wavelet coefficients are complex.

Let us now define the reverse wavelet. If $(b_0, b_1, \ldots, b_n)$ is a finite wavelet, then $(b_n^*, b_{n-1}^*, \ldots, b_1^*, b_0^*)$ is called the *reverse wavelet*. For example, $(3, 0, 1)$ is the reverse of the wavelet $(1, 0, 3)$. Another example is $(i, 0.5)$, which is the reverse wavelet of $(0.5, -i)$, since $-i^* = i$. We may also say that the wavelets $(1, 0, 3)$ and $(3, 0, 1)$ are the reverses of each other; the same is true for $(0.5, -i)$ and $(i, 0.5)$.

To autocorrelate a finite-length wavelet, we merely convolve the wavelet with its reverse. For example, let us find the autocorrelation of the wavelet (b_0, b_1, b_2). It reverse wavelet is (b_2^*, b_1^*, b_0^*). The folding table is shown

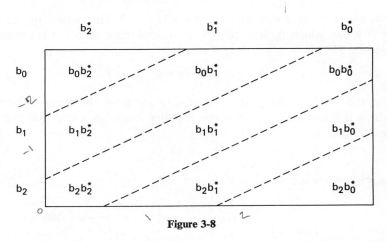

Figure 3-8

in Figure 3-8. We see that the autocorrelation r_0 for time index 0 is equal to the sum on the main southwest–northeast diagonal:

$$r_0 = b_0 b_0^* + b_1 b_1^* + b_2 b_2^* \qquad \text{(for each summand, the first index minus the second index is 0)}$$

The value r_0 gives the energy of the wavelet. Working up from the diagonal, we always get the r_k's for the negative k's; that is,

$$r_{-1} = b_0 b_1^* + b_1 b_2^* \qquad \text{(for each summand, the first index minus the second index is } -1)$$

and

$$r_{-2} = b_0 b_2^* \qquad \text{(the first index minus the second index is } -2)$$

Working down from the diagonal, we always get the r_k's for positive k's; that is,

$$r_1 = b_2 b_1^* + b_1 b_0^* \qquad \text{(for each summand, the first index minus the second is 1)}$$

and

$$r_2 = b_2 b_0^* \qquad \text{(the first index minus the second index is 2)}$$

Let us multiply each coefficient of a wavelet $(b_0, b_1, b_2, \ldots)$ by a constant

$$c = e^{ia} = \cos a + i \sin a$$

of unit magnitude

$$|c| = \sqrt{cc^*} = \sqrt{\cos^2 a + \sin^2 a} = 1$$

Examples of such constants c are $1, i, -1, -i$, and $(1 + i)/\sqrt{2}$. We thereby obtain a new wavelet,

$$(cb_0,\ cb_1,\ cb_2, \ldots)$$

the respective coefficients of which are equal to the coefficients of the original wavelet, except for the constant factor c of magnitude $cc^* = 1$. It is easy to see that the new wavelet has the same autocorrelation as the original wavelet. Also, because $cc^* = 1$, the partial energies of both wavelets are the same; that is,

$$p_0 = b_0 b_0^* = (cb_0)(cb_0)^*$$
$$p_1 = p_0 + b_1 b_1^* = p_0 + (cb_1)(cb_1)^*$$
$$p_2 = p_1 + b_2 b_2^* = p_0 + (cb_2)(cb_2)^*$$
$$\vdots$$

As a result, it is convenient to introduce the convention that two wavelets are equivalent whenever their corresponding coefficients are equal, except for a constant factor of magnitude 1. Thus, the two wavelets $(2, 1)$ and $(-2, -1)$ are equivalent, since their corresponding coefficients are equal, except for the factor $c = -1$; that is,

$$-2 = c \cdot 2 = (-1) \cdot 2$$
$$-1 = c \cdot 1 = (-1) \cdot 1$$

Similarly, the two wavelets $(2, 1)$ and $(2i, i)$ are equivalent, since their corresponding coefficients are equal, except for the factor $c = i$; that is,

$$2i = c \cdot 2 = (i)(2)$$
$$i = c \cdot 1 = (i)(1)$$

We can let any wavelet in such an equivalence represent the entire class. For example, we may let the wavelet $(2, 1)$ represent the equivalence class that contains the wavelets $(2, 1), (-2, -1), (2i, i)$.

Minimum Delay, Mixed Delay, and Maximum Delay

Let us now consider the wavelet $(2, 1)$. Its reverse is $(1, 2)$, so its autocorrelation table is that shown in Figure 3-9, and its autocorrelation is

$$r_{-1} = 2$$
$$r_0 = 4 + 1 = 5$$
$$r_1 = 2$$

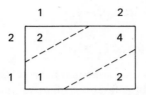

Figure 3-9

On the other hand, consider the wavelet (1, 2). Its reverse is (2, 1), so its autocorrelation table is that given in Figure 3-10, and its autocorrelation is

$$r_{-1} = 2$$
$$r_0 = 1 + 4 = 5$$
$$r_1 = 2$$

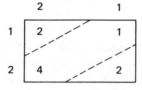

Figure 3-10

Thus, we see that the wavelet (2, 1) has the *same* autocorrelation as its reverse (1, 2). This is always so. That is, any wavelet has the same autocorrelation as its reverse.

But the foregoing example illustrates something further. Let (b_0, b_1) be a 2-length wavelet. Then the *only* other 2-length wavelet with the *same* autocorrelation as (b_0, b_1) is its reverse, (b_1^*, b_0^*). [Of course, when we speak about the wavelet (b_0, b_1) we mean the equivalence class of wavelets (cb_0, cb_1), where $cc^* = 1$. Likewise, by (b_1^*, b_0^*) we mean the equivalence class $(c^*b_1^*, c^*b_0^*)$, where $cc^* = 1$.] Thus, the only other 2-length wavelet with the same autocorrelation as (2, 1) is (1, 2). It turns out that there are wavelets of greater length with the same autocorrelation, but there is no other wavelet of length 2 with the same autocorrelation.

Consequently, we may pair every 2-length wavelet (b_0, b_1) with another 2-length wavelet, namely its reverse, (b_1^*, b_0^*). Such a pair is called a *dipole* and is of the form

$$(b_0, b_1) \quad \text{and} \quad (b_1^*, b_0^*)$$

An example of a dipole is

$$(2, 1) \quad \text{and} \quad (1, 2)$$

Another example is

$$(0.5, -i) \quad \text{and} \quad (i, 0.5)$$

where $i = \sqrt{-1}$. It is convenient to name each member of the dipole. Let us call one 2-length wavelet of the dipole the minimum-delay wavelet and the other the maximum-delay wavelet. The *minimum-delay wavelet* is the one that has the largest coefficient (in magnitude) at the front, whereas the *maximum-delay wavelet* is the one that has the largest coefficient (in magnitude) at the end. Thus, $(2, 1)$ is minimum-delay, and $(1, 2)$ is maximum-delay. Also, $(i, 0.5)$ is minimum-delay (because $|i| = 1 > 0.5$), whereas $(0.5, -i)$ is maximum-delay (because $|-i| = 1 > 0.5$).

The concepts of minimum delay and maximum delay have been defined for 2-length wavelets. Now let us define these concepts for wavelets of greater length. To do so, take a group of dipoles, each with the minimum-delay wavelet on the left. For example,

	Minimum Delay		*Maximum Delay*
Dipole 1:	$(2, 1)$	and	$(1, 2)$
Dipole 2:	$(i, 0.5)$	and	$(0.5, -i)$
Dipole 3:	$(-i, 0.5)$	and	$(0.5, i)$

The convolution of the minimum-delay wavelets in the group (i.e., the convolution of the wavelets on the left) yields a *minimum-delay wavelet*. Thus,

$$(2, 1) * (i, 0.5) * (-i, 0.5) = (2, 1, 0.5, 0.25)$$

is a minimum-delay wavelet. [We have evaluated its coefficients as follows: First, we construct the table shown in Figure 3-11, so

$$(i, 0.5) * (-i, 0.5) = (1, 0, 0.25)$$

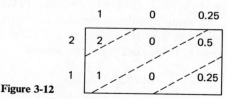

Figure 3-11

Next we construct the table given in Figure 3-12, so

$$(2, 1) * (1, 0, 0.25) = (2, 1, 0.5, 0.25)$$

which is the minimum-delay wavelet given above.]

Figure 3-12

On the other hand, the convolution of the maximum-delay wavelets in the group of dipoles (i.e., the convolution of the wavelets on the right) yields a *maximum-delay wavelet*. Thus,

$$(1, 2) * (0.5, -i) * (0.5, i) = (0.25, 0.5, 1, 2)$$

is a maximum-delay wavelet. [We see that the maximum-delay wavelet is the reverse of the minimum-delay wavelet $(2, 1, 0.5, 0.25)$ found above, as we would expect.]

Furthermore, the convolution of one wavelet from each dipole in the group, chosen so that we have a mixture of minimum-delay and maximum-delay wavelets, gives a *mixed-delay wavelet*. Thus,

$$(1, 2) * (i, 0.5) * (-i, 0.5) = (1, 2) * (1, 0, 0.25) = (1, 2, 0.25, 0.5)$$

is mixed-delay. (Here we have used the table shown in Figure 3-13 to perform the convolution.)

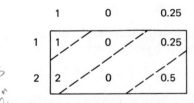

Figure 3-13

If every 2-length wavelet in a group of n dipoles is distinct, there are 2^n possible $(n + 1)$-length wavelets that can be generated by taking one 2-length wavelet from each dipole. If, of course, some of the 2-length wavelets are not distinct, some of the $(n + 1)$-length wavelets will not be distinct either. In any case, all the $(n + 1)$-length wavelets generated in this way from one group of dipoles will have the *same* autocorrelation. These $(n + 1)$-length wavelets, so generated, are said to belong to a *suite of wavelets* all *with the same autocorrelation*.

Of the 2^n possible $(n + 1)$-length wavelets in the suite, one is minimum-delay, another is maximum-delay, and the others are mixed-delay. However, these 2^n $(n + 1)$-length wavelets are not the only wavelets in the suite, because there are other wavelets which have greater length but which have the same autocorrelation; hence these other wavelets are also members of the suite.

If we consider all the members of the suite, there is only one minimum-delay wavelet, which is the $(n + 1)$-length minimum-delay wavelet which we have found. Thus, the minimum-delay concept is unique. In contrast, the maximum-delay concept is a relative one, as a wavelet is maximum-delay with respect to its length. In our example, the wavelet

$$(0.25, 0.5, 1, 2)$$

is the maximum-delay 4-length wavelet of the suite. By delaying this wavelet by one time unit, we obtain the wavelet

$$(0, 0.25, 0.5, 1, 2)$$

which is the maximum-delay 5-length wavelet of the suite. Similarly,

$$(0, 0, 0.25, 0.5, 1, 2)$$

is the maximum-delay 6-length wavelet of the suite, and so on. All the other wavelets of the suite, that is, the wavelets other than the minimum-delay wavelet and the maximum-delay wavelets, are mixed-delay wavelets.

Partial Energy

Another way of describing the delay properties of wavelets is by means of the *partial energy*, which we introduced earlier in this chapter. For example, the partial energy of the minimum-delay wavelet $(2, 1, 0.5, 0.25)$ is

$$p_0 = 4$$
$$p_1 = 4 + 1 = 5$$
$$p_2 = 4 + 1 + 0.25 = 5.25$$
$$p_3 = 4 + 1 + 0.25 + 0.0625 = 5.3125$$

On the other hand, the partial energy of the maximum-delay 4-length wavelet $(0.25, 0.5, 1, 2)$ is

$$p_0 = 0.0625$$
$$p_1 = 0.0625 + 0.25 = 0.3125$$
$$p_2 = 0.0625 + 0.25 + 1 = 1.3125$$
$$p_3 = 0.0625 + 0.25 + 1 + 4 = 5.3125$$

By comparing the two partial energy curves, we see that the partial energy of the maximum-delay wavelet never exceeds that of the minimum-delay wavelet. We would expect this behavior from the way we constructed the two wavelets: the minimum-delay wavelet being the one with the energy concentrated at the front, and the maximum-delay wavelet being the one with the energy concentrated at the end.

The partial energy curves of the mixed-delay $(n + 1)$-length wavelets in the suite lie between the partial energy curves of the minimum-delay wavelet and the maximum-delay $(n + 1)$-length wavelet of the suite. That is, the mixed-delay $(n + 1)$-length wavelets have their energy concentrated between the two extremes. Thus, the mixed-delay wavelet $(1, 2, 0.25, 0.5)$ has partial energy

$$p_0 = 1$$
$$p_1 = 1 + 4 = 5$$
$$p_2 = 1 + 4 + 0.0625 = 5.0625$$
$$p_3 = 1 + 4 + 0.0625 + 0.25 = 5.3125$$

and this curve lies between the partial energy curves of the minimum-delay wavelet and the maximum-delay 4-length wavelet of the suite, as follows:

	Partial Energy			
	p_0	0_1	p_2	p_3
Minimum-delay wavelet	4	5	5.25	5.3125
Above mixed-delay 4-length wavelet	1	5	5.0625	5.3125
Maximum-delay 4-length wavelet	0.0625	0.3125	1.3125	5.3125

Our development up to this point may be summarized as follows. A dipole consists of a 2-length wavelet and its reverse, that is,

$$(a, b) \quad \text{and} \quad (b^*, a^*)$$

One of these two wavelets is called minimum delay, and the other maximum delay. If $|a| > |b|$, then (a, b) is the minimum-delay one and (b^*, a^*) is the maximum-delay one. On the other hand, if $|b| > |a|$, then (a, b) is the maximum-delay one and (b^*, a^*) is the minimum-delay one. Next, we consider a group of n dipoles, say

	Minimum Delay	*Maximum Delay*
Dipole 1	(α_0, α_1)	(α_1^*, α_0^*)
Dipole 2	(β_0, β_1)	(β_1^*, β_0^*)
.	.	.
.	.	.
.	.	.
Dipole n	(ω_0, ω_1)	(ω_1^*, ω_0^*)

where

$$|\alpha_0| > |\alpha_1|$$
$$|\beta_0| > |\beta_1|$$
$$\vdots$$
$$|\omega_0| > |\omega_1|$$

(That is, the arrangement is such that the 2-length wavelets on the left-hand side are minimum-delay and those on the right-hand side are maximum-delay.) We then form a $(n + 1)$-length wavelet by convolving n of the 2-length wavelets, one from each dipole. Since there are n dipoles, there are 2^n possible $(n + 1)$-length wavelets. These 2^n possible wavelets, it turns out, all have the same autocorrelation, so they belong to a suite of wavelets all with the same autocorrelation. The wavelet formed by the convolution of all the minimum-delay 2-length wavelets is given by

$$(\alpha_0, \alpha_1) * (\beta_0, \beta_1) * \cdots * (\omega_0, \omega_1)$$

and is called the *minimum-delay wavelet* of the suite. The wavelet formed by the convolution of all the maximum-delay 2-length wavelets is given by

$$(\alpha_1^*, \alpha_0^*) * (\beta_1^*, \beta_0^*) * \cdots * (\omega_1^*, \omega_0^*)$$

and is called the *maximum-delay $(n + 1)$-length wavelet* of the suite. Clearly, the maximum-delay $(n + 1)$-length wavelet is the reverse of the minimum-delay wavelet. Any other wavelet of these 2^n $(n + 1)$-length wavelets that is, any wavelet formed by a mixture of 2-length minimum-delay and maximum-delay wavelets is called *mixed delay*.

Maximum delay is a property that also depends on the length of the wavelet, so that, to be more explicit, we speak of a maximum-delay $(n + 1)$-length wavelet. Clearly, the maximum-delay $(n + 1)$-length wavelet of the suite is the reverse of the minimum-delay wavelet of the suite.

Let us note that the coefficient for time index 0 of the minimum-delay wavelet of the suite is

$$\alpha_0 \beta_0 \ldots \omega_0$$

whereas the coefficient for time index 0 of the maximum-delay $(n + 1)$-length wavelet is

$$\alpha_1^* \beta_1^* \ldots \omega_1^*$$

Because

$$|\alpha_0| > |\alpha_1|$$
$$|\beta_0| > |\beta_1|$$
$$\cdot$$
$$\cdot$$
$$\cdot$$
$$|\omega_0| > |\omega_1|$$

it follows that

$$|\alpha_0 \beta_0 \ldots \omega_0| > |\alpha_1^* \beta_1^* \ldots \omega_1^*|$$

which says that the coefficient for time index 0 of the minimum-delay wavelet is greater in magnitude than the coefficient for time index 0 of the maximum-delay $(n + 1)$-length wavelet of the same suite. Consider the mixed-delay $(n + 1)$-length wavelet of the suite given by

$$(\alpha_1^*, \alpha_0^*) * (\beta_0, \beta_1) * \cdots * (\omega_0, \omega_1)$$

Its coefficient for time index 0 is

$$\alpha_1^* \beta_0 \ldots \omega_0$$

Since

$$|\alpha_0| > |\alpha_1|$$

it follows that

$$|\alpha_0 \beta_0 \ldots \omega_0| > |\alpha_1^* \beta_0 \ldots \omega_0|$$

which says that the coefficient for time index 0 of the minimum-delay wavelet is greater in magnitude than the coefficient for time index 0 of this mixed-delay wavelet of the suite. By the same argument, it is clear that *the coefficient for time index 0 of the minimum-delay wavelet of a suite is greater in magnitude than the coefficient for time index 0 of any other wavelet in the suite.*

Feedback Stability and Minimum Delay

Let us now look at the simple *feedback filter* shown in Figure 3-14. What can we learn about the concept of minimum delay from this system? The input to the system is the time series x_t, where t denotes time. According to our

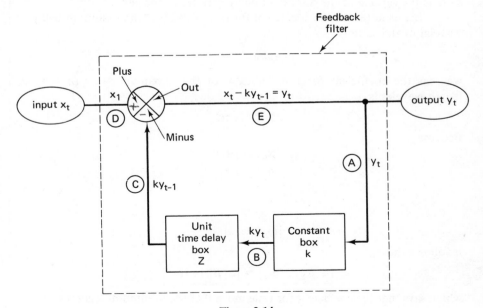

Figure 3-14

convention, the time index t is taken at discrete-time instants spaced one unit apart. Let us now analyze the feedback filter by tracing the path around the loop. Let us start at the beginning of the feedback loop (point A in the diagram). We see that the filter output y_t is fed back through the loop and enters the *constant box* as its input. The constant box produces a constant multiplication. That is, the output of the constant box is equal to k times its input, where k is a fixed number. Thus, the memory function of the constant box is the wavelet $(k, 0, 0, 0, \ldots)$, so its z transform is

$$k + 0 \cdot z + 0 \cdot z^2 + 0 \cdot z^3 + \cdots = k$$

attenuation
amplification

If the magnitude of k is less than 1, that is, if $|k| < 1$, this box produces a constant attenuation. On the other hand, if the magnitude of k is greater than 1, that is, if $|k| > 1$, this box produces a constant amplification. Since the input to the constant box is y_t, the output from the constant box is ky_t (as shown at point B in the diagram). The action of the constant box in terms of z transforms is

Input y_t to constant box has z transform $Y(z)$.

Constant box has z transform k.

Hence, output ky_t from constant box has z transform $kY(z)$.

The output ky_t from the constant box enters the *unit-time-delay* box (point B in the diagram). The unit-time-delay box produces a delay of one time unit from input to output. That is, the output of this box is delayed one time unit with respect to its input. The memory function of the unit-time-delay box is the wavelet $(0, 1, 0, 0, 0, \ldots)$, so its z transform is

$$0 + 1 \cdot z + 0 \cdot z^2 + 0 \cdot z^3 + 0 \cdot z^4 + \cdots = z$$

As we have just seen, the input to the unit-time-delay box is ky_t; thus, the output from this box is ky_{t-1} (as shown at point C in the diagram). The action of the unit-time-delay box in terms of z transforms is

Input ky_t to delay box has z transform $kY(z)$.

Delay box has z transform z.

Hence, output ky_{t-1} from delay box has z transform $zkY(z)$.

The output ky_{t-1} from the unit-time-delay box enters one input channel of the mixer (point C in the diagram). This input channel is the minus channel. At the same time, the closed-loop feedback system input x_t enters the other input channel of the mixer (point D in the diagram). This input channel is the plus channel. The mixer produces a subtraction. That is, the input ky_{t-1} to the minus channel is subtracted from the input x_t to the plus channel. Thus, the output of the mixer is x_t-ky_{t-1} (point E in the diagram).

$y_t = x_t - ky_{t+1}$

But the output of the mixer is also the output y_t of the closed-loop feedback system. Hence, we have the *basic equation*

$$y_t = x_t - ky_{t-1}$$

for the feedback filter.

In terms of z transforms, this equation becomes

$$Y(z) = X(z) - kzY(z)$$

Transposing terms, we obtain

$$X(z) = Y(z) + kzY(z)$$

or

$$X(z) = Y(z)(1 + kz)$$

The *transfer function* $A(z)$ of the feedback filter is defined to be the ratio of the z transform $Y(z)$ of the system output y_t to the z transform $X(z)$ of the system input x_t. Thus, the transfer function is

$$A(z) = \frac{Y(z)}{X(z)} = \frac{1}{1 + kz}$$

This is a key equation, which we shall now interpret. The z transform of the 2-length wavelet $(1, k)$ is

$$1 + kz$$

Therefore, the transfer function $A(z)$ of the feedback filter is the reciprocal of the z transform of the wavelet $(1, k)$.

We recall that the wavelet $(1, k)$ is *minimum-delay* provided that $|k| < 1$, and maximum delay provided that $|k| > 1$. We now state that the feedback filter is *stable* provided that the wavelet $(1, k)$ is minimum-delay, and is *unstable* provided that the wavelet $(1, k)$ is maximum-delay.

For, if $|k| < 1$, the constant box produces an attenuation, and the effect of the feedback damps out with respect to time, thereby making the feedback filter stable. For example, the system output at the time index $t = 3$ is

$$y_3 = x_3 - ky_2$$

Let us develop an expression for y_3 by *looking backward in time*. In this equation we may substitute for the system output y_2, which is

$$y_2 = x_2 - ky_1$$

Thus, the system output y_3 may be written

$$y_3 = x_3 - k(x_2 - ky_1) = x_3 - kx_2 + k^2y_1$$

Now in this equation we may substitute for the system output y_1, which is

$$y_1 = x_1 - ky_0$$

Thus, the system output y_3 may be written

$$y_3 = x_3 - kx_2 + k^2(x_1 - ky_0) = x_3 - kx_2 + k^2x_1 - k^3y_0$$

Continuing such substitutions backward into time, we finally obtain the expression

$$y_3 = x_3 - kx_2 + k^2x_1 - k^3x_0 + k^4x_{-1} - k^5x_{-2} + \cdots$$

for the feedback filter's output y_3 at time index $t = 3$.

When $|k| < 1$, the terms of the sequence

$$1, -k, k^2, -k^3, k^4, -k^5, \ldots$$

tend to zero, thereby making this expression for the system output y_3 converge, and hence the feedback filter stable. As a numerical example, suppose that k is equal to $\frac{1}{2}$. Then the system output at time index 3 is

$$y_3 = x_3 - 0.5x_2 + 0.25x_1 - 0.125x_0 - 0.0625x_{-1} - 0.03125x_{-2} + \cdots$$

which is convergent.

But if $|k| > 1$, the constant box produces an amplification, and the feedback effect grows unboundedly with respect to time. For example, by the same reasoning as before, the feedback filter's output at time index 3 is given by the expression

$$y_3 = x_3 - kx_2 + k^2x_1 - k^3x_0 + k^4x_{-1} - k^5x_{-2} + \cdots$$

and when $|k| > 1$, the terms of the sequence

$$1, -k, k^2, -k^3, k^4, -k^5, \ldots$$

tend to infinity, thereby making the expression for the system output y_3 diverge, and hence the feedback filter unstable. As a numerical example, suppose that k is equal to 2. Then the system output at time index 3 is

$$y_3 = x_3 - 2x_2 + 4x_1 - 8x_0 + 16x_{-1} - 32x_{-2} + \cdots$$

which is divergent.

Feedback Loop Amplification: Stability versus Delay

The foregoing stability analysis is all well and good, but what if we demand a feedback filter with feedback loop amplification, that is, $|k| > 1$. What weapon do we have to make such a filter stable? The only weapon we have is *time delay*. We must delay the filter's output in order to obtain a semblance of stability in our system. Thus, by demanding a feedback loop amplification $|k| > 1$, the factors of *stability* and *output delay* must be balanced against each other. With no delay in the output, the feedback filter is necessarily unstable. But the longer we are willing to delay the filter's output, the greater the chance we have to construct a stable filter that does approximately the same job as the unstable one.

Let us now illustrate this point, although for the moment we shall speak metaphorically. We live in a world in which we only know a one-way flow of time. Everything we do is subject to this unidirectional flow. Now let us look at the two-faced Roman god Janus (after whom the month of January is named). Janus can see both forward and backward in time. We can only see backward in time, that is, we can see only the history of a time series up to the present time *t*, and cannot see the future development of the time series no matter how hard we try. But Janus can see that future, just as well as he can see the past. To him time is not a unidirectional flow. He has a vantage point denied to us.

How can we reach this vantage point? We can reach it only by waiting, that is, by delay, until a finite amount of time has passed, during which period some or most of the information required for stability has become past history. Thus, in order to reach for stability in our feedback filter we must delay our output; the greater the delay, the greater are our chances.

In the previous section we derived the basic equation

$$y_t = x_t - ky_{t-1}$$

which describes the feedback filter which we are using for illustrative purposes (see Figure 3-14). Let us now examine this equation by *looking forward in time* instead of *looking backward in time* as we did in the last section. We rewrite the equation as

looking forward — look for future

$$y_{t-1} = \frac{1}{k}x_t - \frac{1}{k}y_t$$

Thus, the output at time $t - 1 = 3$ is

$$y_3 = \frac{1}{k}x_4 - \frac{1}{k}y_4$$

In this equation we may substitute for the output y_4, which is

$$y_4 = \frac{1}{k}x_5 - \frac{1}{k}y_5$$

Thus, the output y_3 may be written

$$y_3 = \frac{1}{k}x_4 - \frac{1}{k}\left(\frac{1}{k}x_5 - \frac{1}{k}y_5\right)$$

$$= \frac{1}{k}x_4 - \frac{1}{k^2}x_5 + \frac{1}{k^2}y_5$$

Now in this equation we may substitute for the output y_5, which is

$$y_5 = \frac{1}{k}x_6 - \frac{1}{k}y_6$$

Thus, the output y_3 may be written

$$y_3 = \frac{1}{k}x_4 - \frac{1}{k^2}x_5 + \frac{1}{k^2}\left(\frac{1}{k}x_6 - \frac{1}{k}y_6\right)$$

$$= \frac{1}{k}x_4 - \frac{1}{k^2}x_5 + \frac{1}{k^3}x_6 - \frac{1}{k^3}y_6$$

Continuing such substitutions forward into time, we finally obtain the expression

$$y_3 = \frac{1}{k}x_4 - \frac{1}{k^2}x_5 + \frac{1}{k^3}x_6 - \frac{1}{k^4}x_7 + \frac{1}{k^5}x_8 - \cdots$$

for the feedback filter's output y_3 at time index 3. Here y_3 is expressed in terms of the future values

$$x_4, x_5, x_6, x_7, x_8, \ldots$$

of the input. When $|k| > 1$, the terms of the sequence

$$\frac{1}{k}, -\frac{1}{k^2}, \frac{1}{k^3}, -\frac{1}{k^4}, \frac{1}{k^5}, \ldots$$

tend to zero, thereby making the expression for the output y_3 converge, and hence the closed-loop feedback system *stable in terms of future time.* As a numerical example, suppose that k is equal to 2. Then the system output at time index 3 is

$$y_3 = 0.5x_4 - 0.25x_5 + 0.125x_6 - 0.0625x_7 + 0.03125x_8 - \cdots$$

which is convergent.

We have only one problem. At time index 3, the future values

$$x_4, x_5, x_6, x_7, x_8, \ldots$$

that are needed for the foregoing expression for y_3 are not available. There-fore, to make our expression for y_3 meaningful, we must wait until enough of these future values are available.

As an approximation, then, let us truncate our expression for y_3 so that we use the expression

$$0.5x_4 - 0.25x_5 + 0.125x_6 - 0.0625x_7 + 0.03125x_8$$

This truncated expression serves as an approximation for the desired output y_3. Rearranging the order of the terms in this expression, we obtain

$$y_3 = 0.03125x_8 - 0.0625x_7 + 0.125x_6 - 0.25x_5 + 0.5x_4$$

Suppose that we build a causal feedforward filter with the 5-length memory function

$$(0.03125, -0.0625, 0.125, -0.25, 0.5)$$

Since this filter is a feedforward filter with a finite-length memory function, it is necessarily stable. If we let the time series x_t be the input to this filter, the output at time index $t = 8$ will be an approximation to the desired y_3 at time index $t = 3$. In other words, to approximate the desired output y_3, we have introduced a time delay of $8 - 3 = 5$ time units, thereby obtaining a stable causal feedforward filter with a 5-length memory function. By increasing this time delay, stable causal feedforward filters that yield better approximations can be obtained. A perfect approximation is only reached by introducing an infinite time delay.

Concluding Remarks

We have described some of the basic concepts required for digital filter implementation. The stability of a filter is determined by its delay properties. In particular, the feedback filter is stable if and only if its transfer function is minimum delay. Feedback filters which are unstable can be stabilized approxi-mately by the introduction of output delay. The approximation becomes better and better as this delay becomes larger and larger.

the stability of inverse filters

memory function
anticipation function

Summary

A fundamental problem in the analysis of signals is increasing the resolution of overlapping events. One method of attack is the use of inverse digital filters. An inverse digital filter can have the undesirable characteristic that its memory function grows without limit with increasing time. Such an inverse filter is said to be unstable. In the present chapter, filter stability is investigated in some detail. A digital filter of the minimum-delay type is shown to have an inverse that consists only of a stable memory function. On the other hand, a filter of the maximum-delay type has an inverse that consists only of a stable anticipation function. A filter that is neither minimum-delay nor maximum-delay, but is of the mixed-delay type, has a stable inverse, and this inverse consists of both a memory component and an anticipation component. Filters with (nonvanishing anticipation components) are not causal) if one wishes them to work in real time; that is, filters functioning in real time cannot operate on future values of the input. This difficulty may be overcome by making a complete and permanent record of the entire data sequence to be analyzed prior to any filtering operation. In this manner, filters need only operate on already available input values. Any filter design allowing one to delay the output until the entire input has been recorded will thus satisfy the requirement of being computationally realizable.

The geometry of the zeros of a filter in the complex z plane establishes the stability of the inverse filter. A filter, the zeros of which all lie outside the unit circle $|z| = 1$, is called a minimum-phase-lag filter. The minimum-delay filter and the minimum-phase-lag filter are shown to be identical.

Introduction

In Chapters 2 and 3, we have attempted to outline some of the basic concepts of the theory of digital causal filters. One of the main objectives of signal analysis is to increase the resolution of overlapping events, that is, to diminish the width of a given pulse on a record to the point that it can be clearly distinguished from neighboring pulses. A way to accomplish this goal is through the use of *inverse digital filters*. It is well known that inverse filters very often "blow up"—that is, their response functions grow without limit with increasing time. Such inverse filters are called *unstable*. The purpose of the present chapter is to treat the subject of filter stability in considerable detail. The stability problem is not as serious as might appear to be the case at first sight. Indeed, in most cases this stability problem can be solved by making use of inverse filters that have response functions with an anticipation component as well as a memory component. In other words, these inverse filters are two-sided; that is, their weighting function is defined for negative as well as for positive values of the time variable. The synthesis of such filters in real time may always be achieved by incorporating a sufficiently long time delay in the filter. Of course, the output of such a filter is delayed with respect to the input by the amount of the time delay, but this delay is harmless in many applications.

This chapter treats the problem of finding the exact inverse of a digital filter, An approximate inverse may be obtained by truncating the memory function and/or the anticipation function of the exact inverse. Still another kind of approximate inverse may be found by the use of least-squares methods, which are described in Chapter 6.

Although our analysis is carried out in terms of discrete time, it is of interest to examine the relationship of the discrete-time theory to the corresponding continuous-time theory. This is done in Appendix 4-1. Finally, inverse filters can, of course, also be computed in the frequency domain; we present a brief description of inverse filtering in the frequency domain in Appendix 4-2.

The Stability of Inverse Filters

Let us assume that we are given an input described by the sampled values $b_0, b_1, b_2, \ldots, b_n$, and that we desire to find a digital causal feedforward filter a_t which transforms this input into a spike of unit magnitude at $t = 0$. To do this, we allow the desired filter a_t to have a memory function with an infinite number of coefficients, but we make no statement as to whether this memory function is stable or not. We wish to find the coefficients a_s, $s = 0, 1, 2, \ldots$ such that

$$c_t = \sum_{s=0}^{t} b_s a_{t-s} \qquad (t = 0, 1, 2, \ldots) \qquad (4\text{-}1)$$

where

$$c_t = 1 \qquad \text{for } t = 0$$
$$c_t = 0 \qquad \text{for } t = 1, 2, 3, \ldots$$

We recognize equation (4-1) to represent the convolution between the filter coefficients a_t and the input series b_t.

The desired filter with an infinite-length memory function $(a_0, a_1, a_2, \ldots)$ is called the *inverse* to the input time series $(b_0, b_1, b_2, \ldots, b_n)$. We recall that an infinite-length memory function

$$(a_0, a_1, a_2, \ldots)$$

is called *stable* provided that its energy, defined by

$$|a_0|^2 + |a_1|^2 + |a_2|^2 + \cdots \qquad \text{is finite}$$

is finite. In particular, we see that the magnitudes of the coefficients of a stable, infinitely long memory function asymptotically approach zero as the time index increases; that is,

$$|a_t| \longrightarrow 0 \qquad \text{as } t \longrightarrow \infty$$

Although equation (4-1) may be solved for the coefficients a_s directly, we shall find it much more instructive to consider this problem in the complex z plane. For this purpose we need the z transforms $B(z)$, $A(z)$, and $C(z)$ of the input, filter, and output, respectively. For z transform of the output is

$$C(z) = c_0 + c_1 z + c_2 z^2 + \cdots$$

but since $c_t = 1$ for $t = 0$ and $c_t = 0$ for $t > 0$, we have simply that

$$C(z) = 1$$

Because

$$C(z) = B(z)A(z)$$

we have

$$A(z) = \frac{1}{B(z)}$$

or, in expanded form,

$$a_0 + a_1 z + a_2 z^2 + \cdots = \frac{1}{b_0 + b_1 z + b_2 z^2 + \cdots + b_n z^n}$$

If we actually perform the polynomial division indicated by the right-hand member of this equation, we can obtain the values of the filter coefficients in terms of the input coefficients b_t. Let us illustrate such a calculation by considering an input that consists of the two sampled values (b_0, b_1). Without loss of generality, we may suppose that $b_0 = 1$ and $b_1 = k$, so the input series becomes $(b_0, b_1) = (1, k)$. (*Note*: k may be complex.) We thus have the block diagram of Figure 4-1. In other words, the output is a unit spike at time $t = 0$. Hence, we wish to find the coefficients $(a_0, a_1, a_2, \ldots)$ such that when they are convolved with $(1, k)$, the spike $(1, 0, 0, \ldots)$ is produced,

$$(a_0, a_1, a_2, \ldots) * (1, k) = (1, 0, 0, \ldots)$$

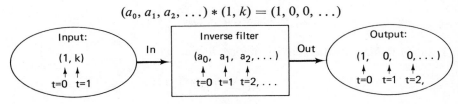

Figure 4-1

Let us use the z-transform representation of this process, so that the preceding equation becomes

$$(a_0 + a_1 z + a_2 z^2 + \cdots)(1 + kz) = 1$$

Dividing each side by $(1 + kz)$, we obtain

$$a_0 + a_1 z + a_2 z^2 + \cdots = \frac{1}{1 + kz}$$

Expanding by the binomial theorem, we have

$$\frac{1}{1 + kz} = 1 - kz + k^2 z^2 - k^3 z^3 + k^4 z^4 - \cdots$$

Identifying coefficients of like powers of z, we find that the memory function of the desired inverse filter is

$$(a_0, a_1, a_2, a_3, \ldots) = (1, -k, +k^2, -k^3, \ldots)$$

or

$$a_t = \begin{cases} 0 & t = -1, -2, -3, \ldots \\ (-k)^t & t = 0, 1, 2, \ldots \end{cases} \tag{4-2}$$

For example, suppose that $k = 0.5$, so the input series

$$(1, k) = (1, 0.5)$$

is minimum-delay. Let us check the property that

$$(a_0, a_1, a_2, \ldots) * (1, k) = (1, 0, 0, \ldots)$$

We have

$$(1, -0.5, 0.25, -0.125, \ldots) * (1, 0.5)$$

which gives

$$(1, 0, 0, 0, \ldots)$$

as desired.

Another example would be if $k = 2$. Then the input series $(1, k) = (1, 2)$ is maximum-delay. Using our formula, we obtain

$$(a_0, a_1, a_2, a_3, \ldots) = (1, -k, k^2, -k^3, \ldots)$$
$$= (1, -2, 4, -8, 16, -32, 64, -128, \ldots)$$

It is easy to check that a spike is produced; that is,

$$(1, -2, 4, -8, 16, -32, \ldots) * (1, 2) = (1, 0, 0, 0, 0, \ldots)$$

but it is disquieting to see that the coefficients of the inverse filter

$$(1, -2, 4, -8, 16, -32, 64, -128, \ldots)$$

are growing in magnitude. This unfortunate condition is called *instability;* that is, the filter with this memory function is *unstable.*

If fact, we see that

$$(a_0, a_1, a_2, \ldots) = (1, -k, k^2, -k^3, \ldots)$$

will be stable whenever $|k| < 1$, and will be unstable whenever $|k| > 1$. in other words, if $(1, k)$ is *minimum-delay*, its *inverse* $(1, -k, k^2, -k^3, \ldots)$

is *stable*. If $(1, k)$ is *maximum-delay*, its *inverse* $(1, -k, k^2, -k^3, \ldots)$ is *unstable*.

Query: Can we avoid this unfortunate condition of instability?

The answer is no, if we require the inverse filter to have a pure memory function, that is, to have its weighting function be a one-sided sequence of the form

$$(a_0, a_1, a_2, \ldots)$$

where a_0 is the coefficient for the time index 0, a_1 for time index 1, a_2 for time index 2, and so on, and where it is understood that all coefficients before time index 0 are automatically zero.

Query: Is there any simple way that we can circumvent the inconvenience of instability?

The answer is yes, provided that we introduce the notion of the *anti-memory function*, or *anticipation function*. An anticipation function has the form

$$(\ldots, a_{-3}, a_{-2}, a_{-1}, a_0)$$

where a_0 is the coefficient at the time index 0, a_{-1} is the coefficient at time index -1, a_{-2} at time index -2, and so on, and where it is understood that the coefficient at time index 1 and all following coefficients (at time indices 2, 3, 4, $\ldots$) are automatically zero.

An infinite-length anticipation function is called *stable* provided that its energy

$$|a_0|^2 + |a_{-1}|^2 + |a_{-2}|^2 + \cdots$$

is finite. This means that the magnitudes of the coefficients of a stable, infinitely long anticipation function asymptotically approach zero as the time index increases in the negative time direction; that is,

$$|a_t| \longrightarrow 0 \qquad \text{as } t \longrightarrow -\infty$$

A *causal filter* is a filter with a memory function as impulse response function. A *purely noncausal filter* is a filter with an anticipation function as impulse response function. A *noncausal filter* is a filter whose impulse response function includes an anticipation component. Thus a causal filter is called one-sided to the past, a purely noncausal filter one-sided to the future, and a noncausal filter two-sided.

Now let us consider the problem of the inverse for a maximum-delay input series. Given the maximum-delay input series

$$(1, k) \qquad \text{with } |k| > 1$$

can we find an inverse that is stable? The answer is yes, provided that we allow the inverse to be the anticipation funtion

$$(\ldots, a_{-3}, a_{-2}, a_{-1})$$

where the above notation indicates that $a_t = 0$ for $t \geq 0$.
That is, we wish to find $(\ldots, a_{-3}, a_{-2}, a_{-1})$ so that the equation

$$(\ldots, a_{-3}, a_{-2}, a_{-1}) * (1, k) = (1, 0, 0, 0, \ldots)$$

is satisfied. Using the z transform representation (where the anticipation function requires negative powers of z), the equation above becomes

$$(\cdots + a_{-3}z^{-3} + a_{-2}z^{-2} + a_{-1}z^{-1})(1 + kz) = 1$$

Dividing each side by $(1 + kz)$, we obtain

$$(\cdots a_{-3}z^{-3} + a_{-2}z^{-2} + a_{-1}z^{-1}) = \frac{1}{1 + kz}$$

Now the right-hand side may be expressed in powers of z^{-1} by the binomial expansion

$$\frac{1}{kz + 1} = k^{-1}z^{-1} - k^{-2}z^{-2} + k^{-3}z^{-3} - k^{-4}z^{-4} + k^{-5}z^{-5} - \cdots$$

Identifying coefficients in like powers of z, we obtain the inverse

$$(\ldots, a_{-5}, a_{-4}, a_{-3}, a_{-2}, a_{-1}) = (\ldots, k^{-5}, -k^{-4}, k^{-3}, -k^{-2}, k^{-1})$$

or

$$a_t = \begin{cases} -(-k)^t & t = -1, -2, -3, \ldots \\ 0 & t = 0, 1, 2, \ldots \end{cases}$$

Returning to our example where $k = 2$, so that $(1, 2)$ is maximum-delay, we see that the inverse is

$$(\ldots a_{-4}, a_{-3}, a_{-2}, a_{-1}) = (\ldots, -0.0625, 0.125, -0.25, 0.5) \qquad (4\text{-}3)$$

which is stable, as expected. Also, it is easy to check that the desired inverse property

$$(\ldots, -0.0625, 0.125, -0.25, 0.5) * (1, 2) = (1, 0, 0, 0, \ldots)$$

is true; that is, a spike at time index $t = 0$ is produced.

Up to now we have seen that the input series $(1, k)$ has as its inverse the stable memory function $(1, -k, k^2, -k^3 \ldots)$ when $|k| < 1$ [i.e., when $(1, k)$

is mimimum-delay]. On the other hand, the input $(1, k)$ has as its inverse the stable anticipation function $(\ldots, -k^{-4}, k^{-3}, -k^{-2}, k^{-1})$ when $|k| > 1$ [i.e., when $(1, k)$ is maximum-delay].

A convenient way to find the *causal* inverse of a 2-length wavelet (b_0, b_1) is to remember *Newton's binomial expansion* in the form

$$\frac{1}{b_0 + b_1 z} = \frac{1}{b_0} \frac{1}{1 + kz}$$

$$= \frac{1}{b_0}(1 - kz + k^2 z^2 - k^3 z^3 + k^4 z^4 - \cdots)$$

where $k = b_1/b_0$. Likewise, to find the *anticausal* inverse of a 2-length wavelet (b_0, b_1), we use *Newton's binomial expansion* in the form

$$\frac{1}{b_0 + b_1 z} = \frac{1}{b_1 z(1 + k^{-1} z^{-1})}$$

$$= \frac{1}{b_1 z}(1 - k^{-1} z^{-1} + k^{-2} z^{-2} - k^{-3} z^{-3} + k^{-4} z^{-4} - \cdots)$$

$$= \frac{1}{b_1}(z^{-1} - k^{-1} z^{-2} + k^{-2} z^{-3} - k^{-3} z^{-4} + k^{-4} z^{-5} - \cdots)$$

where $k^{-1} = b_0/b_1$.

Let us consider the inverse filter for an arbitrary $(n + 1)$-length input

$$(b_0, b_1, \ldots, b_n)$$

Its z transform is

$$B(z) = b_0 + b_1 z + b_2 z^2 + \cdots + b_n z^n$$

This polynomial can be factored into a product of n factors

$$B(z) = (\alpha_0 + \alpha_1 z)(\beta_0 + \beta_1 z)(\gamma_0 + \gamma_1 z) \cdots (\omega_0 + \omega_1 z)$$

Each factor is the z transform of a 2-length memory function, where (α_0, α_1) is the first and (ω_0, ω_1) is the last of these factors. Because multiplying polynomials is equivalent to convolving their coefficients, we have

$$(b_0, b_1, \ldots, b_n) = (\alpha_0, \alpha_1) * (\beta_0, \beta_1,) * \cdots * (\omega_0, \omega_1)$$

As an example, consider the input series $(6, 11, 4)$. It has z transform

$$6 + 11z + 4z^2$$

which may be factored into

$$(2 + z)(3 + 4z)$$

Hence,

$$(6, 11, 4) = (2, 1) * (3, 4)$$

We know that we must be prepared to admit factors that have complex coefficients when we factor polynomials. However, if the original polynomial has real coefficients, all the complex factors occur in complex-conjugate pairs. For example, the polynomial

$$2 + z + 0.5z^2 + 0.25z^3$$

may be factored as

$$(2 + z)(i + 0.5z)(-i + 0.5z)$$

where the corresponding coefficients of the two complex factors are conjugates of each other.

Now to find the stable inverse to the input series $(b_0, b_1, b_2, \ldots, b_n)$, we need apply only the results that we have already developed. First, we represent the input series in factored form as a cascaded combination of 2-length series, that is,

$$(b_0, b_1, \ldots, b_n) = (\alpha_0, \alpha_1) * (\beta_0, \beta_1) * \cdots * (\omega_0, \omega_1)$$

or, in terms of z transforms,

$$b_0 + b_1 z + \cdots + b_n z^n = (\alpha_0 + \alpha_1 z)(\beta_0 + \beta_1 z) \ldots (\omega_0 + \omega_1 z)$$

Second, we see that the inverse has the z transform

$$\frac{1}{b_0 + b_1 z + \cdots + b_n z^n} = \left(\frac{1}{\alpha_0 + \alpha_1 z}\right)\left(\frac{1}{\beta_0 + \beta_1 z}\right) \cdots \left(\frac{1}{\omega_0 + \omega_1 z}\right)$$

Third, we must distinguish three possible cases. The input series $(b_0, b_1, \ldots, b_n)$ may be (1) minimum-delay, (2) maximum-delay, or (3) mixed-delay. In case (1), the stable inverse is the stable memory function $(a_0, a_1, a_2, \ldots)$, the coefficients of which are found by the polynomial division

min. delay

$$\frac{1}{b_0 + b_1 z + \cdots + b_n z^n} = a_0 + a_1 z + a_2 z^2 + \cdots$$

Alternatively, in case (1) we may find the inverse of each of the n 2-length minimum-delay series (α_0, α_1), (β_0, β_1), $\ldots$, (ω_0, ω_1), and then convolve these inverses to yield the overall memory function $(a_0, a_1, a_2, \ldots)$.

In case (2), the *stable* inverse is the stable anticipation function

$\ldots, a_{-n-2},$	$a_{-n-1},$	$a_{-n},$	$0,$	$0, \ldots,$	0
$\uparrow$	$\uparrow$	$\uparrow$	$\uparrow$	$\uparrow$	$\uparrow$
$t = -n-2$	$t = -n-1$	$t = -n$	$t = -n+1$	$t = -n+2$	$t = -1$

whose coefficients may be found by the polynomial long division

$$\frac{1}{b_n z^n + b_{n-1} z^{n-1} + \cdots + b_1 z + b_0}$$
$$= a_{-n} z^{-n} + a_{-n-1} z^{-n-1} + a_{-n-2} z^{-n-2} + \cdots$$

We note that the coefficients $a_{-n+1}, a_{-n+2}, \ldots, a_{-1}$ of this stable anticipation function are zero. Furthermore, we see that the coefficients $a_0, a_1, a_2, \ldots$ vanish automatically since this stable inverse is a pure anticipation function. This may be seen by considering, for example, the case when $n = 2$, which gives

$$
\begin{array}{r}
\frac{1}{b_2} z^{-2} - \frac{b_1}{b_2^2} z^{-3} + \cdots \\[4pt]
b_2 z^2 + b_1 z + b_0 \overline{) 1 } \\[4pt]
1 + \frac{b_1}{b_2} z^{-1} + \frac{b_0}{b_2} z^{-2} \\[4pt]
\hline
\frac{-b_1}{b_2} z^{-1} - \frac{b_0}{b_2} z^{-2} \quad \text{etc.}
\end{array}
$$

We see that the long division must be performed as shown above in order that we obtain the desired expansion in *negative* powers of z. Alternatively, we may find the stable inverses of each of the n 2-length maximum-delay series

$$(\alpha_0, \alpha_1), (\beta_0 \ \beta_1), \ldots, (\omega_0, \omega_1)$$

and then convolve the inverses so obtained to yield

$$
\begin{array}{ccccccc}
(\ldots, a_{-n-2}, & a_{-n-1}, & a_{-n}, & 0, & 0, & \ldots, & 0) \\
\uparrow & \uparrow & \uparrow & \uparrow & \uparrow & & \uparrow \\
\text{time:} \ (-n-2) & (-n-1) & (-n) & (-n+1) & (-n+2) & & (-1)
\end{array}
$$

In case (3), the component 2-length series are a mixture of minimum-delay and maximum-delay series. The stable inverse of each component can be found separately. Each stable inverse will be a stable *memory* function if the 2-length series is minimum-delay, and will be a stable *anticipation* function if the 2-length series is maximum-delay. The overall stable inverse is the convolution of all these stable inverses, and it will have the form of a two-sided series

$$(\ldots, a_{-3}, a_{-2}, a_{-1}, a_0, a_1, a_2, a_3, \ldots)$$

That is, *it has both a memory component and an anticipation component.*

For example, let us consider the mixed-delay memory function given by

$$(1, 2.2, 0.4) = (1, 2) * (1, 0.2)$$

In terms of z transforms, we have that

$$1 + 2.2z + 0.4z^2 = (1 + 2z)(1 + 0.2z)$$

The first component is maximum-delay, while the second is minimum-delay. Since $(1, 2)$ is maximum-delay, its inverse is the stable anticipation function

$$(\ldots, a_{-4}, a_{-3}, a_{-2}, a_{-1}) = (\ldots, -0.0625, 0.125, -0.25, 0.5) \quad \text{max-delay}$$

given by equation (4-3). Since $(1, 0.2)$ is minimum-delay, its inverse is the stable memory function

$$(a_0, a_1, a_2, a_3, \ldots) = (1, -0.2, +0.04, -0.008, \ldots)$$

given with the aid of equation (4-2). The overall stable inverse of $(1, 2.2, 0.4)$ is thus

$$(\ldots, -0.0625, 0.125, -0.25, 0.5) * (1, -0.2, +0.04, -0.008, \ldots)$$

This convolution is most conveniently evaluated with the aid of z transforms. Because convolution of two time functions is equivalent to multiplication of their z transforms, we have

$$(\ldots, -0.0625z^{-4} + 0.125z^{-3} - 0.25z^{-2} + 0.5z^{-1})$$
$$\cdot(1z_0 - 0.2z^1 + 0.04z^2 - 0.008z^3 + \cdots) \quad (4\text{-}4)$$
$$= (\ldots, -0.0625z^{-4} + 0.138z^{-3} - 0.278z^{-2} + 0.556z^{-1} \quad \text{inverse filter}$$
$$- 0.111z^0 + 0.022z^1 - 0.004z^2 + \cdots) \quad (4\text{-}5)$$

Since both factors of (4-4) are infinitely long series, the first extending from $-\infty \le t \le -1$ and the second from $0 \le t \le \infty$, the product (4-5) is only approximately correct, because both component factors were "chopped" to the first four terms each. We notice, however, that the successive terms in both factors decrease rapidly, so that we may expect our approximation (4-5) to be satisfactory. This is found to be the case when we perform the convolution

$$(\cdots -0.0625, +0.138, -0.278, +0.556, -0.111,$$
$$+ 0.022, -0.004, \cdots) * (1, 2.2, 0.4)$$

or, in terms of z transforms,

$$(\cdots - 0.0625z^{-4} + 0.138z^{-3} - 0.278z^{-2} + 0.556z^{-1}$$
$$- 0.111z^0 + 0.022z^1 - 0.004z^2 + \cdots) \cdot (1z^0 + 2.2z^1 + 0.4z^2)$$
$$= (\cdots - 0.0625z^{-4} - 0.014z^{-3} - 0.002z^{-2} - 0.002z^{-1}$$
$$+ 1z^0 + 0z^1 + 0z^2 + 0z^3 - 0.002z^4 + \cdots)$$

We see that we do indeed obtain the desired unit spike at $t = 0$. The non-vanishing coefficients on either side of the term $1z^0$ arise because of the truncation of series performed in (4-4). Nevertheless, the retention of only four terms in the component inverses allows us to calculate a very good approximation to the "perfect" stable inverse filter. We notice also that the inverse filter (4-5) is stable and two-sided.

Real-Time versus Nominal-Time Filtering

Let us now look a little more closely at what we have done so far. First, our introduction of the concept of the anticipation function of a filter required the use of z transforms with *negative* powers of z. The filter $z = z^1$ is the unit-delay operator, that is, the filter z *delays* the output relative to the input by one time unit (see Figure 4-2). By symmetry, we may then conclude that the filter z^{-1} advances the output relative to the input by one time unit (see Figure 4-3). We may thus call the filter z^{-1} the *unit advance operator*. Inductive reasoning leads us to the n-unit advance filter z^{-n}, where the exponent $-n$ represents a delay of $-n$, or equivalently an advance of n (see Figure 4-4).

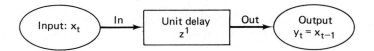

Figure 4-2. Block diagram of the unit-delay filter z^1.

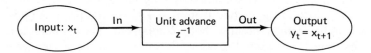

Figure 4-3. Block diagram of the unit-advance filter z^{-1}.

Figure 4-4. Block diagram of the n-unit advance filter z^{-n}.

the z transform to right

The n-unit-*delay* filter shifts an input x_t n time units to the right (later in time). For example, the input $(1, 2)$ with time origin at $t = 0$ has the z transform

$$1 + 2z$$

If we operate on this input with the unit-delay operator z, we obtain

$$z(1 + 2z) = 1z + 2z^2$$

which represents the wavelet $(0, 1, 2)$, namely the input $(1, 2)$ shifted one unit of time to the *right* of the origin $t = 0$. Now if we operate on $1 + 2z$ with the unit advance operator z^{-1}, we obtain

$$z^{-1}(1 + 2z) = 1z^{-1} + 2$$

which represents the sequence

$$
\begin{array}{cc}
(1, & 2) \\
\uparrow & \uparrow \\
t = -1 & t = 0
\end{array}
$$

namely the input $(1, 2)$ shifted one unit of time to the *left* of the origin $t = 0$. These matters are illustrated in Figure 4-5. The nth-order causal feedforward filter, as given in Chapter 2, has the z transform

$$a_0 + a_1 z + a_2 z^2 + \cdots + a_n z^n$$

We may similarly define the nth-order purely noncausal feedforward filter as the filter with z transform

$$a_{-n} z^{-n} + \cdots + a_{-1} z^{-1} + a_0$$

The nth-order causal filter thus has the *memory* function

$$(a_0, a_1, a_2, \ldots, a_n)$$

while the nth-order purely noncausal filter has the *anticipation* function

$$(a_{-n}, \ldots, a_{-1}, a_0)$$

An example of a third-order causal filter is given by the truncated memory component of (4-4):

$$(1z^0 - 0.2z^1 + 0.04z^2 - 0.008z^3)$$

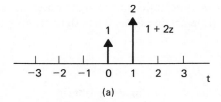

(a)

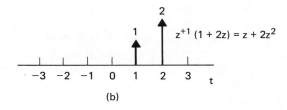

(b)

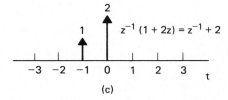

(c)

Figure 4-5. (a) Input $(1, 2)$ in its original position. (b) Input $(1, 2)$ shifted one time unit to the right. (c) Input $(1, 2)$ shifted one time unit to the left.

while an example of a fourth-order purely noncausal filter is given by the truncated anticipation component of (4-4),

$$(-0.0625z^{-4} + 0.125z^{-3} - 0.25z^{-2} + 0.5z^{-1})$$

The causal filter has a memory component only; its impulse response is therefore nonzero for $t \geq 0$, and vanishes for $t < 0$. On the other hand, the purely noncausal filter has an anticipation component only; its impulse response is therefore nonzero for $t \leq 0$ and vanishes for $t > 0$ (see Figure 4-6).

By definition, a noncausal filter has a nonzero weighting function for negative time. An interpretation of this phenomenon frequently encountered in the literature is that this filter "responds before the input arrives at the filter," and that it is therefore *physically nonrealizable*. On the other hand, the causal filter has a nonzero weighting function only for $t \geq 0$, and is thus said to be *physically realizable.*

Consider now the problem of filtering a signal. If this operation is performed in the field, and the arrivals are filtered in real time as they reach the filter, it is certainly true that the filters cannot respond to energy that has not yet arrived, and in this sense the noncausal filter is physically nonrealizable. However, if the entire signal is first recorded prior to further analysis,

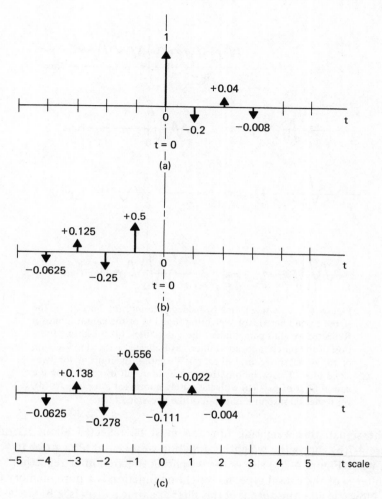

Figure 4-6. (a) Causal filter $(1, -0.2, 0.04, -0.008)$. (b) Purely non-causal filter $(-0.0625, +0.125, -0.25, +0.5)$ (c) Two-sided filter which is the result of the convolution of the filter in part (a) with the filter in part (b).

there is no need to worry about filters which operate on energy that has not arrived, because by the time of filtering the entire signal is at our disposal. These matters are illustrated in Figure 4-7. Figure 4-7(a) shows a given signal that is to be filtered. If we view the action of a filter in the time domain, we know that the filter output is obtained by <u>convolving the input signal with the filter's weighting function.</u> As a result of the convolution formula [see equation (4-1)], we see that before a filter's weighting function can operate

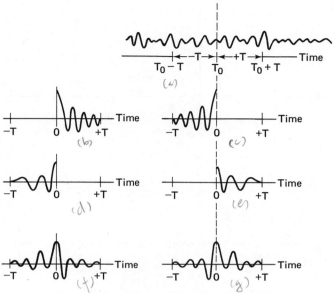

Figure 4-7. One-sided and two-sided filtering. (a) Portion of the signal to be filtered. (b) Weighting function of the causal filter. (c) Reflected weighting function of the causal filter. (d) Weighting function of the purely noncausal filter. (e) Reflected weighting function of the purely noncausal filter. (f) Weighting function of the two-sided filter. This weighting function is the result of convolving the causal filter of part (b) with the purely noncausal filter of part (d). (g) Reflected response function of the two-sided filter.

on the signal, this weighting function must be reflected about its origin. Figure 4-7(b) shows the weighting function of a causal filter prior to reflection, while Figure 4-7(c) shows the weighting function after reflection. Since this filter is of the causal type, its weighting function is a pure memory function. Now consider the output of this filter at a time $t = T_0$ [see Figure 4-7(a)]. If the memory function of the causal filter is $T + 1$ time units in length, the output of the filter at $t = T_0$ depends on the events on the input lying between $t = T_0 - T$ and $t = T_0$. In other words, the output of the causal filter at $t = T_0$ depends on what has happened on the trace in an interval T prior to $t = T_0$.

Next, let us consider the action of the purely noncausal filter pictured in Figure 4-7(d). Let us assume that the weighting function of this filter is also $T + 1$ time units in length. Since this filter is of the purely noncausal type, its weighting function is a pure anticipation function. In Figure 4-7(e) we show the reflected purely noncausal filter. We again examine the action of this filter at an input time $t = T_0$. It is apparent from our figure that the output of the purely noncausal filter at $t = T_0$ depends on the events on the input lying between $t = T_0$ and $t = T_0 + T$. In other words, the output of the purely noncausal filter at $t = T_0$ depends on what has happened on the signal in an interval T after $t = T_0$. Notice our use of the past tense ("has happened") in the previous sentence. The purely noncausal filter does not act on events that "have not arrived," for by the time we perform this kind of filtering, the entire signal is at our disposal for examination. Thus, the causal filter produces output at time $t = T_0$, while the purely noncausal filter produces output at this same time $t = T_0$, on the basis of the signal after $t = T_0$. We see that there is no reason why we cannot profitably use physically non-realizable filters when we operate on a signal that is available in its entirety at the time we wish to analyze it. When we filter as the signal is recorded, we say that we are filtering in real time; when we filter after having recorded the entire signal, we filter in what is termed nominal time. The words "physically nonrealizable" are indeed meaningless when we filter in nominal time!

Let us return now to the filters of Figure 4-7. If the input wavelet that we wish to convert into a unit spike at $t = 0$ is minimum-delay, its *stable* inverse is given by a causal filter; if this input wavelet is maximum-delay, its *stable* inverse is given by a purely noncausal filter. Finally, if the input is mixed-delay, its *stable* inverse is given by a two-sided filter such as illustrated in Figure 4-7(f). As we have shown previously, this filter consists of a memory component and an anticipation component. The stable memory component arises from the minimum-delay portion of the input, while the stable anticipation component arises from the maximum-delay portion of this input. We see that the two-sided filter produces output at time $t = T_0$ on the basis of events *prior to* AND *after* $t = T_0$.

Before leaving the subject of stability, we will find it useful to treat a still further property of digital filters in the complex z plane.

The z-Plane Singularities of a Digital Filter

Consider the pair of 2-length filters $(1, 0.5)$ and $(0.5, 1)$. The filter $(1, 0.5)$ has a memory function that is minimum-delay. Then the filter with the reverse memory function $(0.5, 1)$ is maximum-delay. The z transform of $(1, 0.5)$ is

$$A_0(z) = 1 + 0.5z \qquad \text{min. delay}$$

Now $A_0(z)$ is a first-degree polynomial in z. Its single root, or zero, may be found by solving

$$A_0(z) = 1 + 0.5z = 0$$

so that $z = -2$. Similarly, $(0.5, 1)$ has the z transform

$$A_1(z) = 0.5 + 1z$$

whose single zero is located at $z = -0.5$.

The locus of points $|z| = 1$ in the complex z plane is a circle of unit radius with origin at $z = 0$. This curve is called the *unit circle*. We notice that the zero at $z = -2$ of the minimum-delay memory function $(1, 0.5)$ falls *outside* the unit circle, while the zero at $z = -0.5$ of the maximum-delay memory function $(0.5, 1)$ falls inside the unit circle. The position of these two zeros is indicated in Figure 4-8.

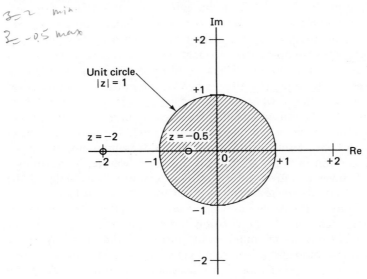

Figure 4-8. Zeros of the memory functions $(1, 0.5)$ and $(0.5, 1)$ in the complex z-plane.

The filter $(1, 0.5)$ is minimum-delay. It therefore has a stable inverse whose *memory* function is given by Newton's binomial expansion as

$$(1, -0.5, +0.25, -0.125, \ldots)$$

On the other hand, the filter $(0.5, 1)$ is maximum-delay. It therefore has a stable inverse whose *anticipation* function is given by Newton's binomial expansion as

$$(\ldots, \underset{\underset{t=-4}{\uparrow}}{-0.125,} \quad \underset{\underset{t=-3}{\uparrow}}{+0.25,} \quad \underset{\underset{t=-2}{\uparrow}}{-0.5,} \quad \underset{\underset{t=-1}{\uparrow}}{1)}$$

In Chapter 2 we discussed the pair of filters,

$$A_0(z) = 1 + 0.5z$$

and

$$A_1(z) = 0.5 + 1z$$

We showed there that the phase-lag characteristic of the filter $A_0(z)$ is less than the phase-lag characteristic of the filter $A_1(z)$ in the angular frequency range $0 \leq \omega \leq \pi$. We were thus led to call $A_0(z)$ the minimum-phase-lag filter. But now we see that the minimum-delay filter $(1, 0.5)$ is also the minimum-phase-lag filter $A_0(z)$, and that the maximum-delay filter $(0.5, 1)$ is also the nonminimum-phase-lag filter $A_1(z)$. On this basis, we may conclude the following:

1. The minimum-delay filter $(1, 0.5)$ is also minimum-phase-lag. Its zero lies *outside* the unit circle. It has a stable inverse which consists of a stable, pure *memory* component.

2. The maximum-delay filter $(0.5, 1)$ is also nonminimum-phase-lag. Its zero lies *inside* the unit circle. It has a stable inverse that consists of a stable, pure *anticipation* component.

More generally, the nth-order digital causal feedforward filter with the overall memory function

$$(a_0, a_1, a_2, \ldots, a_n)$$

is factorable in the z plane in the form

$$
\begin{aligned}
A(z) &= a_0 + a_1 z + a_2 z^2 + \cdots + a_n z^n \\
&= a_n(z - z_1)(z - z_2) \cdots (z - z_n) \\
&= a_n \prod_{i=1}^{n} (z - z_i) \qquad (a_n \neq 0)
\end{aligned}
\tag{4-6}
$$

where the roots $z = z_i$ may be complex. $A(z)$ is thus the product of n 2-length filters of the form $(-z_i + z)$, which can also be written as $(-z_i, 1)$.

Those 2-length filters for which $|z_i| > 1$ have zeros lying *outside* the unit circle. These components are thus both minimum-delay and minimum-phase-lag. On the other hand, those 2-length filters for which $|z_i| < 1$ have zeros lying *inside* the unit circle. These latter components are then both maximum-delay and nonminimum-phase-lag. The location of the zeros of the z transform of a filter thus yields information about the stability of its inverse. Zeros outside the unit circle contribute to the stable memory component of this inverse, while zeros inside the unit circle contribute to the stable anticipation component of this same inverse.[1]

For values of $n > 4$, the factorization (4-6) is no longer amenable to hand calculation, so that one must solve these higher-degree polynomials on a digital computer. A total of n roots will be obtained in this manner. Each

---

[1]The case of zeros *on* the unit circle give rise to equi-delay components; by definition, such components are also both minimum-delay and maximum-delay.

such root will give rise to a 2-length filter, which will be either minimum-delay or maximum-delay. These could all be inverted separately, and the resulting overall two-sided inverse could then be obtained by convolution of the individual inverses. [Complex roots present no difficulty, since our discussion of the inverse of the wavelets $(1, k)$ and $(k, 1)$ was carried through for complex values of the coefficient k. A multiple root of order n would be treated as n simple roots.] However, this procedure is unnecessarily tedious. Having found all the roots of $A(z)$, we combine those roots with moduli less than unity into a maximum-delay z-transform component, while we combine those roots with moduli greater than unity into a minimum-delay z-transform component. Then we have seen that the stable inverse of the maximum-delay component of $A(z)$ is a pure anticipation function, while the stable inverse of the minimum-delay component of $A(z)$ is a pure memory function. The overall response function of this inverse is therefore also stable. It is two-sided, and can be obtained simply by convolving the stable inverse of the maximum-delay component of $A(z)$ with the stable inverse of the minimum-delay component of $A(z)$. The inversion of the z transforms is carried out by synthetic polynomial division. Since the coefficients of $A(z)$ will be real for uses of physical interest, complex roots, if any, will occur in complex-conjugate pairs. Finally, we mention here that a conceptually simple inversion technique exists in the frequency domain. This method is briefly treated in Appendix 4-2.

Concluding Remarks

Inverse digital filters have many practical applications. Their proper use often allows us to shrink information smeared over a broad portion of a signal into a narrower region, with the resulting overall increase in resolution. The present chapter has, however, been primarily concerned with the serious problem of inverse-filter instability. We have seen how this difficulty can be overcome in most cases of physical interest by the use of inverse filters whose response functions are *two-sided*. Both the memory and the anticipation components of such two-sided filters can be made *stable* in most actual situations. Because these inverses are exact, they have an infinite number of weighting coefficients. Very often, however, suitable approximations of finite length can be found by the simple expedient of truncating the exact inverse to an arbitrary but finite number of terms. In those cases where truncation does not lead to a sufficiently short inverse-filter response function, more satisfactory results are usually achieved by the use of the least-squares criterion to find approximate inverses with minimum mean square error. This subject will be treated in Chapter 6.

formulation of the filtering process in discrete and in continuous time

In this book, we follow the notation and conventions generally used by mathematicians who deal with the theory of stochastic processes in discrete time. Because this discrete theory differs in part with the corresponding continuous theory, the purpose of this appendix is to point out some of the important differences between the two approaches.

Many people think primarily in terms of continuous time. However, there is a well-developed mathematical literature geared to handle problems directly in discrete time, with no reliance on the continuous-time concept, although there are, of course, the mappings between the discrete and the continuous-time formulations. In any case, both the concept of discrete time and the concept of continuous time are only two of many possible models of the real physical situation. In principle, one cannot say that one is better than another, and in practice the particular model chosen must depend upon individual judgment. A nice feature of the discrete-time formulation is that all frequency functions can be regarded as band-limited, and this feature is in accord with experience. On the other hand, in continuous time, a one-sided transient (i.e., a transient that has a definite beginning in time) must have a frequency spectrum that is not band-limited, and this feature is not in accord with experience. An interesting discussion of this problem has recently been given by Slepian (1976).

Under the continuous-time formulation, a sampled-time function is written as

$$a(t) = \sum_n a_n \delta(t - n\tau)$$

where $\delta(t)$ is the Dirac delta function, t continuous time, n an integer, and τ the uniform time spacing between sampled values. Under the discrete-time model, a sampled-time function is written simply as a_n, where n is again an integer. The time unit is thus chosen to equal the time spacing τ.

The practical advantage of writing a sampled-time function in the form $a(t)$ shown above is that it, in effect, converts all Riemann integrals into Stieltjes integrals, with the net result that integrals associated with the continuous-time theory can be used for sampled-time functions. For example, the Fourier transform of $a(t)$ is given by the integral

$$A(\omega) = \int_{-\infty}^{+\infty} a(t)e^{-i\omega t}\, dt$$

which is

$$A(\omega) = \int_{-\infty}^{+\infty} \sum_n a_n \delta(t - n\tau)e^{-i\omega t}\, dt$$

This integral is indeed one way of writing a Stieltjes integral, and it reduces to the sum

$$A(\omega) = \sum_n a_n \int_{-\infty}^{+\infty} \delta(t - n\tau)e^{-i\omega t}\, dt$$

$$= \sum_n a_n e^{-i\omega n\tau}$$

Rather than arriving at this result by the foregoing route, the mathematical theory of discrete time *directly* defines $A(\omega)$ as the sum

$$A(\omega) = \sum_n a_n e^{-i\omega n}$$

where, according to our convention, τ has been taken to be one unit of time. It is this latter approach that we have chosen to follow in the present book. In the main text we thus let t be a discrete variable, where $t = 0, 1, 2, \ldots$.

Let us next consider the z transform. There the mathematical theory of discrete time makes use of Laplace's original definition; namely, that the z transform of the discrete-time function a_n is written

$$A(z) = \sum_n a_n z^n$$

Many engineers prefer to use another definition of the z transform, namely

$$A(z) = \sum_n a_n z^{-n}$$

so care must always be exercised to note which definition a particular writer is using. The transformation from one to the other is quite simple, and is accomplished by the replacement of z by z^{-1}.

The z transform

$$A(z) = \sum_n a_n z^n$$

is a function of the complex variable z. The unit circle in the z plane, namely, those values of the complex variable z with unit mangitude $|z| = 1$, can be represented by the equation

$$z = e^{-i\omega}$$

By making this substitution in $A(z)$, we obtain the Fourier transform

$$A(\omega) = \sum_n a_n e^{-i\omega n} \qquad (2)$$

In other words, we use the same symbol A to denote the Fourier transform $A(\omega)$ and to denote the z transform $A(z)$ of the discrete-time function a_n. This convention emphasizes that the Fourier transform is nothing but the z transform evaluated on the unit circle of the z plane.

APPENDIX 4-2

inverse filtering in the frequency domain

Let us assume that we are given the sequence b_t, $t = 0, 1, 2, ..., n$. Its discrete Fourier transform $B(\omega)$ is

$$B(\omega) = \sum_{t=0}^{n} b_t e^{-i\omega t} \qquad (-\pi \le \omega \le \pi)$$

Now suppose that we define the reciprocal transform $A(\omega)$ to be

$$A(\omega) = \frac{1}{B(\omega)} \qquad (4\text{-}7)$$

where we assume that $B(\omega) \ne 0$ for any real value of ω. We can then write

$$A(\omega) \cdot B(\omega) = C(\omega) = 1 \qquad (-\pi \le \omega \le +\pi)$$

But the function $C(\omega) = 1,\ -\pi \leq \omega \leq \pi$ is the Fourier transform of the Kronecker delta δ_t, where

$$\delta_t = \begin{cases} 1 & t = 0 \\ 0 & t \neq 0 \end{cases}$$

Therefore, the frequency-domain calculation (4-7) formally permits us to obtain the inverse of the sequence b_t. The desired response function a_t of this inverse filter can then be found by an inverse Fourier transformation of $A(\omega)$:

$$a_t = \frac{1}{2\pi} \int_{-\infty}^{+\infty} A(\omega) e^{i\omega t}\, d\omega = \frac{1}{2\pi} \int_{-\pi}^{+\pi} \frac{e^{i\omega t}}{B(\omega)}\, d\omega \qquad (4\text{-}8)$$

This procedure is perfectly feasible in practice if one has access to a good direct and inverse Fourier-transform computing scheme, such as the fast Fourier transform (FFT) algorithm (Cooley and Tukey, 1965).

all-pass filters

Summary

Suites of wavelets having identical magnitude spectra are of interest because of the insight they can yield into the fundamental nature of signal processing. Wavelets of a given suite are obtainable from each other by passing them through an appropriate set of all-pass filters. Such filters are of the *pure* phase-shift kind; that is, their magnitude characteristic is constant and independent of frequency, whereas their phase-lag characteristic is a prescribed function of frequency. In general, the finite-length wavelets of a given suite can be distinguished from each other on the basis of the partial energy curve. This curve describes how the energy of a wavelet is distributed over the time range of its duration. The minimum-delay wavelet of a suite is the one whose energy is concentrated at its front, whereas the maximum-delay wavelet of the same suite is the one whose energy is concentrated at its end. Wavelets whose energies are concentrated at intermediate positions between these extreme cases are of the mixed-delay type. Of the wavelet members of a given suite, the minimum-delay wavelet has a phase-lag characteristic that lies as near as possible to the frequency axis, whereas the maximum-delay wavelet has a phase-lag characteristic that lies as far away as possible from the frequency axis. The phase-lag characteristics of the mixed-delay wavelets of a given suite all fall between the phase-lag characteristics of the minimum- and maximum-delay wavelet members of this suite.

Introduction

In Chapters 2 and 3, we have seen that there is generally more than one wavelet having a given magnitude spectrum. To specify a wavelet uniquely in the frequency domain, it is necessary to know both its phase spectrum and its magnitude spectrum. The members of a suite of wavelets with identical magnitude spectra have a number of very interesting properties, some of which we shall discuss in detail in the present chapter. Thus, we wish to concern ourselves with the properties of suites of wavelets, where each wavelet member of a given suite has the same magnitude spectrum but a different phase spectrum. A good deal of insight into the behavior of digital filters can be obtained by studying such wavelet suites.

Energy Characteristics of Wavelets

The wavelet concept is basic in signal processing, and scientists have a good physical picture of it. Roughly speaking, a wavelet is a transient signal having a definite arrival time. In Chapter 3, we defined a wavelet as a one-sided, stable time function. By "one-sided" we mean that the wavelet has a definite origin time (or, in other words, a definite arrival time), so that the values of the wavelet *before* this original time are automatically zero. Thus, a discrete-time function w_t is a wavelet provided that:

1. All the coefficients before the origin time are zero; that is,

$$w_t = 0 \text{ for } t < 0 \quad \text{(the one-sidedness condition)}$$

2. The energy, which is given by the sum of squares of the magnitudes of the coefficients of the wavelet, is finite; that is,

$$|w_0|^2 + |w_1|^2 + |w_2|^2 + \cdots < \infty \quad \text{(the stability condition)}$$

Now in our work we face the problem that we wish to use wavelets with origin times other than $t = 0$. We can avoid this difficulty by saying that w_{t-s} is the wavelet with origin time s, where s can be either positive or negative but where we require that nonzero values of the coefficients only occur for $t - s \geq 0$, so that $w_{t-s} = 0$ whenever $t - s < 0$. The foregoing specification is for a wavelet w_t in discrete time (i.e., $t =$ an integer). A corresponding specification can also be made for a wavelet in continuous time.

In Chapter 3, we introduced the concept of a minimum-delay 2-length wavelet (α_0, α_1), with its leading coefficient α_0 greater in magnitude than its end coefficient α_1, that is, $|\alpha_0| > |\alpha_1|$. We also introduced the concept of the reverse of a wavelet, and said that the reverse (α_1^*, α_0^*) of the minimum-delay 2-length wavelet is called the maximum-delay 2-length wavelet. (The superscript asterisk indicates the complex conjugate, applicable in those cases when the coefficients α_0 and α_1 are complex numbers.) Thus, the leading coefficient α_1^* of the above 2-length maximum-delay wavelet is smaller in magnitude than its end coefficient α_0^*; that is, $|\alpha_1^*| < |\alpha_0^*|$. These two wavelets, (α_0, α_1) and (α_1^*, α_0^*), are said to form a dipole [e.g., $(1, 0.5)$ and $(0.5, 1)$ is a dipole].

Let us now find the magnitude spectrum of the minimum-delay wavelet of the dipole. Its Fourier transform is

$$A_0(\omega) = \alpha_0 + \alpha_1 e^{-i\omega}$$

[handwritten: $z = e^{-i\omega}$]

The squared modulus of this Fourier transform is

$$|A_0(\omega)|^2 = A_0(\omega)A_0^*(\omega) = (\alpha_0 + \alpha_1 e^{-i\omega})(\alpha_0 + \alpha_1 e^{-i\omega})^*$$

where the superscript asterisk indicates the complex conjugate. Hence,

$$|A_0(\omega)|^2 = (\alpha_0 + \alpha_1 e^{-i\omega})(\alpha_0^* + \alpha_1^* e^{+i\omega})$$
$$= (\alpha_0\alpha_0^* + \alpha_0\alpha_1^* e^{+i\omega} + \alpha_1\alpha_0^* e^{-i\omega} + \alpha_1\alpha_1^*)$$

The positive square root of $|A_0(\omega)|^2$ is the magnitude spectrum $|A(\omega)|$; therefore, the magnitude spectrum of the minimum-delay wavelet (α_0, α_1) is

$$|A_0(\omega)| = (\alpha_0\alpha_0^* + \alpha_0\alpha_1^* e^{+i\omega} + \alpha_1\alpha_0^* e^{-i\omega} + \alpha_1\alpha_1^*)^{1/2}$$

For example, the magnitude spectrum of $(1, 0.5)$ is

$$|A_0(\omega)| = (1 + 0.5e^{+i\omega} + 0.5e^{-i\omega} + 0.25)^{1/2}$$
$$= (1.25 + \cos \omega)^{1/2}$$

[handwritten: $\alpha_0 = 1$ min-delay, $\alpha_1 = 0.5$, $\frac{1}{2}(e^{i\omega} + e^{-i\omega})$]

Next, let us find the magnitude spectrum of the maximum-delay wavelet of the dipole. Its Fourier transform is

[handwritten: $z = e^{-i\omega}$]

$$A_1(\omega) = \alpha_1^* + \alpha_0^* e^{-i\omega}$$

The squared modulus is

[handwritten: $\cos\omega + i\sin\omega + (\cos\omega - i\sin\omega)$, (α_1^*, α_0^*)]

$$|A_1(\omega)|^2 = A_1(\omega)A_1^*(\omega) = (\alpha_1^* + \alpha_0^* e^{-i\omega})(\alpha_1^* + \alpha_0^* e^{-i\omega})^*$$
$$= (\alpha_1^* + \alpha_0^* e^{-i\omega})(\alpha_1 + \alpha_0 e^{+i\omega})$$
$$= \alpha_1^*\alpha_1 + \alpha_1^*\alpha_0 e^{+i\omega} + \alpha_0^*\alpha_1 e^{-i\omega} + \alpha_0^*\alpha_0$$

Thus, the magnitude spectrum of the maximum-delay wavelet of the dipole is

$$|A_1(\omega)| = (\alpha_1^*\alpha_1 + \alpha_1^*\alpha_0 e^{+i\omega} + \alpha_0^*\alpha_1 e^{-i\omega} + \alpha_0^*\alpha_0)^{1/2}$$

For example, the magnitude spectrum of $(0.5, 1)$ is

$$|A_1(\omega)| = (0.25 + 0.5e^{+i\omega} + 0.5e^{-i\omega} + 1)^{1/2} = (1.25 + \cos \omega)^{1/2}$$

When compared, the magnitude spectra of the minimum-delay and maximum-delay wavelets of a dipole are seen to be the same; that is, the wavelets of the dipole have the common magnitude spectrum $|A_0(\omega)| = |A_1(\omega)|$.

Now let us consider the group of n dipoles already encountered in Chapter 3. Since both 2-length wavelets of any dipole have the same magnitude spectrum, we may denote the common magnitude spectrum of any dipole as follows:

$$\begin{array}{ll} \text{dipole 1} & |A(\omega)| \\ \text{dipole 2} & |B(\omega)| \\ \quad\cdot & \quad\cdot \\ \quad\cdot & \quad\cdot \\ \quad\cdot & \quad\cdot \\ \text{dipole } n & |\Omega(\omega)| \end{array}$$

For example, the two dipoles

	Minimum-Delay Wavelet	*Maximum-Delay Wavelet*
dipole 1	$(1, 0.5)$	$(0.5, 1)$
dipole 2	$(3, 2)$	$(2, 3)$

have the values

$$\begin{array}{ll} \text{dipole 1} & (1.25 + \cos \omega)^{1/2} \\ \text{dipole 2} & (13 + 12 \cos \omega)^{1/2} \end{array}$$

for the common magnitude spectra of both wavelets in the dipole.

We now wish to form an $(n + 1)$-length wavelet by convolving n of the 2-length wavelets, one from each dipole. Since there are n dipoles, there are 2^n possible $(n + 1)$-length wavelets which are said to form a suite. Returning to our example of two dipoles, we may form the following suite of 2^2 3-length wavelets:

Wavelet Number	*Wavelet*	*Designation*
1	$(1, 0.5) * (3, 2) = (3, 3.5, 1)$	The minimum-delay 3-length wavelet
2	$(1, 0.5) * (2, 3) = (2, 4, 1.5)$	A mixed-delay wavelet
3	$(0.5, 1) * (3, 2) = (1.5, 4, 2)$	A mixed-delay wavelet (reverse of the mixed-delay wavelet above)
4	$(0.5, 1) * (2, 3) = (1, 3.5, 3)$	The maximum-delay 3-length wavelet (reverse of the 3-length minimum-delay wavelet)

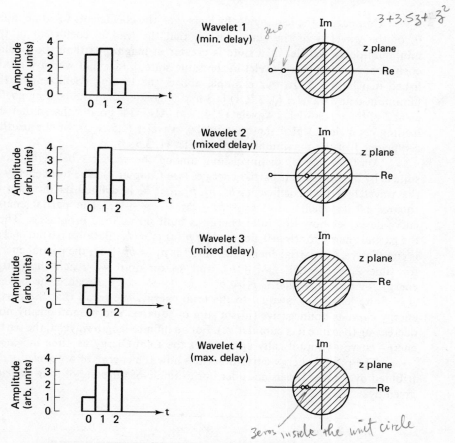

Figure 5-1. Wavelets 1, 2, 3, and 4 and their corresponding z-plane representations. The shaded portions of the z plane correspond to the interior of the unit circle. The small circles indicate the positions of each wavelet's zeros in the z plane.

This suite of wavelets and their corresponding z-plane representations are shown in Figure 5-1.

Now we wish to show that all the wavelets in a suite have the same magnitude spectrum. We recall that the magnitude spectrum of any wavelet formed by the convolution of several wavelets is equal to the product of the magnitude spectra of these several wavelets. Any wavelet in the suite is formed by the convolution of one 2-length wavelet from each dipole. Because both wavelets in any given dipole have the same magnitude spectrum, it follows that the magnitude spectrum of any wavelet in the suite is equal to the product $|A(\omega)||B(\omega)|\cdots|\Omega(\omega)|$. Thus, all wavelets in the suite have a common magnitude spectrum. In our example above, all wavelets in the suite have the magnitude spectrum $(1.25 + \cos \omega)^{1/2} \cdot (13 + 12 \cos \omega)^{1/2}$.

Comparing the leading coefficients (i.e., the coefficients for time index 0) of the wavelets of the suite, we see that the leading coefficient of the minimum-delay wavelet of a suite is greater in magnitude than the leading coefficient of any other wavelet in the same suite, a fact that we have established in Chapter 3. In our example above, the leading coefficient of the minimum-delay wavelet (3, 3.5, 1) is 3 and is greater than the leading coefficient 2 of the mixed-delay wavelet (2, 4, 1.5). Also, 3 is greater than either the leading coefficient 1.5 of the mixed-delay wavelet (1.5, 4, 2), or the leading coefficient 1 of the maximum-delay wavelet (1, 3.5, 3).

Another way of distinguishing among the wavelets belonging to a suite is by means of their partial energies (see Chapter 3). The total energy of the wavelet of real coefficients $b_0, b_1, b_2, \ldots, b_n$ is defined as the sum of squares of its coefficients: $b_0^2 + b_1^2 + b_2^2 + \cdots + b_n^2$. The partial energy curve describes how this total energy is built up as time progresses. Thus, the partial energy for time 0, denoted by p_0, is $p_0 = b_0^2$. Because partial energy is cumulative, the partial energy for time 1 is $p_1 = b_0^2 + b_1^2$, the partial energy for time 2 is $p_2 = b_0^2 + b_1^2 + b_2^2$, and so on until we reach the partial energy for time n, which is $p_n = b_0^2 + b_1^2 + \cdots + b_n^2$. Because the wavelet is of finite length, p_n is equal to the total energy. We recall that the partial energy curve is nonnegative (it is a sum of squares) and monotonically nondecreasing (because it is cumulative). For an infinite-length wavelet, the partial energy curve asymptotically approaches the total energy as time increases.

The partial energy curve describes how the energy of a wavelet is distributed over time. Let us consider our example of the suite of four wavelets given by

Wavelet Number	Wavelet	Designation
1	(3, 3.5, 1)	Minimum delay
2	(2, 4, 1.5)	Mixed delay
3	(1.5, 4, 2)	Mixed delay
4	(1, 3.5, 3)	Maximum delay

The partial energy of the minimum-delay wavelet is

$$p_0 = 9$$
$$p_1 = 9 + 12.25 = 21.25$$
$$p_2 = 9 + 12.25 + 1 = 22.25$$

and the partial energy of the maximum-delay wavelet is

$$p_0 = 1$$
$$p_1 = 1 + 12.25 = 13.25$$
$$p_2 = 1 + 12.25 + 9 = 22.25$$

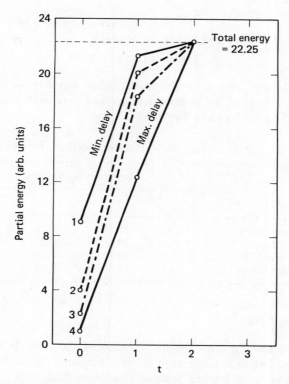

Figure 5-2. Partial energy curves of the wavelets 1, 2, 3, and 4 of Figure 5-1.

By comparing these two partial energy curves (Figure 5-2) we see that the partial energy of the maximum-delay wavelet never exceeds that of the minimum-delay wavelet. This behavior is expected from the construction of the two wavelets: the minimum-delay wavelet has the energy concentrated at the front, and the maximum-delay wavelet has the energy concentrated at the end.

The partial energy curves of the mixed-delay wavelets in the same suite lie between the partial energy curves of the minimum-delay and maximum-delay wavelets of the suite; that is, the mixed-delay wavelets have their energy concentrated between the two extremes. Thus, the mixed-delay wavelet (2, 4, 1.5) has the partial energy

$$p_0 = 4$$
$$p_1 = 4 + 16 = 20$$
$$p_2 = 4 + 16 + 2.25 = 22.25$$

and the mixed-delay wavelet (1.5, 4, 2) has the partial energy

$$p_0 = 2.25$$
$$p_1 = 2.25 + 16 = 18.25$$
$$p_2 = 2.25 + 16 + 4 = 22.25$$

Both partial energy curves of the mixed-delay wavelets lie between the partial energy curves of the minimum-delay and maximum-delay wavelets, as shown in Figure 5-2 and in Table 5-1.

TABLE 5-1.   PARTIAL ENERGY VALUES OF WAVELETS 1 TO 4

Wavelet	*Partial Energy at Time:*			
	$t = 0$	$t = 1$	$t = 2$	$t = 3$
1 (minimum delay)	9	21.25	22.25	22.25
2 (mixed delay)	4	20	22.25	22.25
3 (mixed delay)	2.25	18.25	22.25	22.25
4 (maximum delay)	1	12.25	22.25	22.25

The All-Pass Filter

In this section and the next the membership of the example suite will be enlarged by adding to it more wavelets with the same magnitude spectrum. Such an analysis is of interest because one often deals with signals whose shapes are different but whose magnitude spectra are identical. A phase-shift filter is one that does not alter the magnitude spectrum of the filter input. Such phase-shift filters may be causal (or memory) filters, which operate on only the present and the past of the input; purely non-causal (or anticipation filters), which operate only on the future of the input; or two-sided filters, which operate on the past, present, and future of the input. Because a phase-shift filter does not alter the magnitude spectrum of the input, the magnitude spectrum $|P(\omega)|$ of the filter is perfectly level at unit value; that is, $|P(\omega)| = 1$.

Causal phase-shift filters are also called *all-pass filters* and may be grouped into four types: (a) type 0 all-pass, or trivial filter; (b) type 1 all-pass, or dispersive filter; (c) type 2 all-pass, or pure-delay filter; and (d) type 3 all-pass, or impure-delay filter. The impure-delay filter is a mathematical curiosity, and will not be dealt with here (e.g., Robinson, 1962, p. 53).

The type 0 all-pass filter is a constant filter of which the weighting coefficient may be either 1 or -1, so that its z transform may be either $A(z) = 1$ or $A(z) = -1$. Thus, it passes input to output either with no change or with a change of polarity, as illustrated in Figure 5-3. The membership

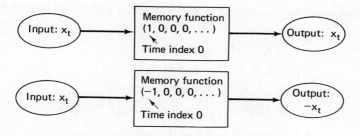

Figure 5-3. Example of a type 0 all-pass filter.

of any suite may therefore be doubled by changing the polarity of each member wavelet. Thus, to the suite of our example, we may add the wavelets

$(-3, -3.5, -1)$ minimum delay

$(-2, -4, -1.5)$ mixed delay

$(-1.5, -4, -2)$ mixed delay

$(-1, -3.5, -3)$ maximum delay 3-length

Multiplying a wavelet by -1 does not change its delay properties, which are based on the partial energy curves and which are thus not affected.

The type 2 all-pass or pure-delay filter has z transform

$$A(z) = z^n \qquad (n > 0)$$

Thus, it passes input to output with no change in shape, but with a delay of n time units, as illustrated in Figure 5-4. Thus, for $n = 1$, we may add to the suite of our example the wavelets

$(0, 3, 3.5, 1)$ mixed delay

$(0, 2, 4, 1.5)$ mixed delay

$(0, 1.5, 4, 2)$ mixed delay

$(0, 1, 3.5, 3)$ maximum delay 4-length

and their negatives, which are 4-length wavelets. Since these new wavelets were formed in effect by convolving the maximum-delay 2-length wavelet $(0, 1)$ with the 3-length wavelets of our original suite, the new wavelets have

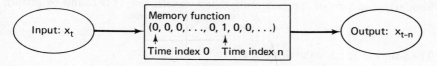

Figure 5-4. Example of a type 2 all-pass filter.

the delay properties indicated. For $n = 2$, we may add to our suite the wavelets

$$(0, 0, 3, 3.5, 1) \qquad \text{mixed delay}$$
$$(0, 0, 2, 4, 1.5) \qquad \text{mixed delay}$$
$$(0, 0, 1.5, 4, 2) \qquad \text{mixed delay}$$
$$(0, 0, 1, 3.5, 3) \qquad \text{maximum delay 5-length} \checkmark$$

and their negatives. We may continue this process for $n = 3, 4, 5, \ldots$.

Note that the 4-length minimum-delay wavelet is $(3, 3.5, 1, 0)$ and that the 5-length minimum-delay wavelet is $(3, 3.5, 1, 0, 0)$. All these minimum-delay wavelets are the same as the 3-length minimum-delay wavelet $(3, 3.5, 1)$ because it is understood that all coefficients after the last nonzero coefficient of a wavelet are zero.

The Type 1 All-Pass or Dispersive Filter

Let us consider the dipole consisting of the minimum-delay wavelet $(1, c)$ with z transform $C_0(z) = 1 + cz$ and the maximum-delay wavelet $(c^*, 1)$ with z transform $C_1(z) = c^* + 1z$. Here c is a real or complex number whose magnitude $|c|$ is less than 1. Whereas this pair of wavelets have identical magnitude spectra, their phase spectra are different. In fact, as we have seen in Chapter 2, the minimum-delay wavelet $(1, c)$ has a phase-lag spectrum which is less than the phase-lag spectrum of the maximum-delay wavelet $(c^*, 1)$ in the frequency range $0 \leq \omega \leq \pi$. For this reason the minimum-delay wavelet is called the minimum-phase-lag wavelet. The zero of the z transform of the minimum-delay wavelet is found by setting

$$1 + cz = 0$$

$(1, z)$
minimum-delay

The solution for z is the desired zero:

$$z_0 = -\frac{1}{c}$$

Since $|c| < 1$, the zero z_0 of the minimum-delay wavelet lies outside the unit circle; that is,

$$|z_0| = \left| -\frac{1}{c} \right| > 1$$

Similarly, the zero of the maximum-delay wavelet $(c^*, 1)$ is found by setting

$$c^* + z = 0$$

The solution for z is the desired zero:

$$z_1 = -c^*$$

C = a+bj
c* = a-bj

Since $|c| < 1$, the zero z_1 of the maximum-delay wavelet lies inside the unit circle; that is,

$$|z_1| = |-c^*| < 1$$

The points z_0 and z_1 are related by

$$z_1 = \frac{1}{z_0^*} \quad \text{or} \quad z_0 = \frac{1}{z_1^*}$$

$z_0 = -\frac{1}{c} \qquad z_0^* = -\frac{1}{c^*} = $

$z_1 = -c^* \qquad = \frac{1}{-c^*} = \frac{1}{z_1}$

   The relationship between the wavelets $(1, c)$ and $(c^*, 1)$ will be examined in greater detail. In particular, we shall attempt to find a digital filter D which will transform the minimum-delay wavelet $(1, c)$ into the maximum-delay wavelet $(c^*, 1)$; that is,

$$(1 + cz)D(z) = (c^* + z)$$

$z_1 = \frac{1}{z_0^*} ; z_0 = \frac{1}{z_1^*}$

Solving for $D(z)$, we obtain

$$D(z) = \frac{c^* + z}{1 + cz} \quad \text{(where } |c| < 1\text{)}$$

min   transfer   max
$(1,c) \rightarrow D(z) \rightarrow (c^*,1)$

Using the equations $z_0 = -1/c$ and $z_1 = -c^*$, we may obtain the following alternative expressions for $D(z)$:

$c^* = -z_1 \quad c = -\frac{1}{z_0}$

$$D(z) = \frac{z - z_1}{(-z/z_0) + 1}$$

$$D(z) = \frac{z - z_1}{-z_1^* z + 1}$$

$\frac{c^* + z}{1 + cz} = \frac{z - z_1}{-(z/z_0) + 1}$

$$D(z) = \frac{-z_0[z - (1/z_0^*)]}{z - z_0}$$

$\frac{1}{z_0} = z_1^*$

The filter $D(z)$ thus does indeed convert the input minimum-delay wavelet $(1, c)$ into the output maximum-delay wavelet $(c^*, 1)$ (Figure 5-5). Since both $(1, c)$ and $(c^*, 1)$ have the same magnitude spectrum, the filter D must have a magnitude spectrum that is unity for all ω. D is a type 1 all-pass or dispersive filter whose action (as for all other phase-shift filters) is contained exclusively in its phase spectrum. A *type 1 (or dispersive) all-pass filter* is defined as a filter whose z transform is a rational function formed by a minimum-delay polynomial as denominator and the corresponding maximum-delay polynomial as numerator.

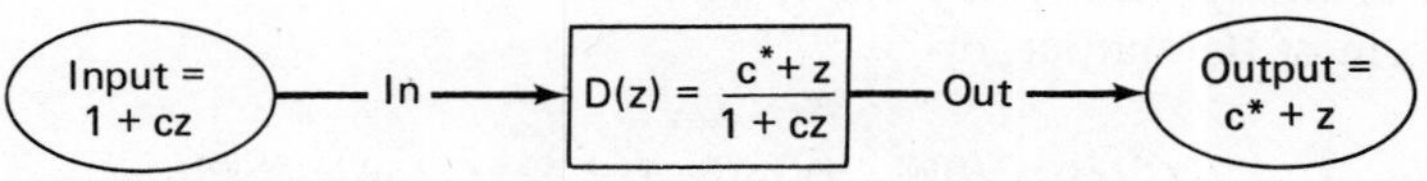

$(1,c)$ min
$(c^*,1)$ max

Input =
$1 + cz$
— In ⟶ $D(z) = \dfrac{c^* + z}{1 + cz}$ — Out ⟶ Output =
$c^* + z$

$\dfrac{1 + cz}{c^* + z}$

Figure 5-5. Dispersive filter $D(z)$.

The z transform

$$D(z) = \frac{c^* + z}{1 + cz} \qquad \text{(where } |c| < 1\text{)}$$

is a ratio of two polynomials (in this case both of the first degree), so $D(z)$ is a rational function. The denominator polynomial is a minimum-delay z transform, whereas the numerator polynomial is the corresponding maximum-delay z transform. Points at which the numerator of a rational function vanishes are called *zeros*, and points at which the denominator vanishes are called *poles*. The dispersive filter $D(z)$ has a zero at $z_1 = -c^*$ and a pole at $z_0 = -1/c$. The dispersive filter $D(z)$ removes (or "divides out") the zero of $(1 + cz)$ at z_0 and replaces it with a new zero located at $z_1 = 1/z_0^*$. Since the point z_0 lies outside and the point $z_1 = 1/z_0^*$ lies inside the unit circle, the filter $D(z)$ has converted the minimum-delay wavelet $(1, c)$ into the maximum-delay wavelet $(c^*, 1)$. It is the pole of $D(z)$ at $z = z_0$ that removes the zero of $(1 + cz)$. The zero at $z = 1/z_0^*$ of the output wavelet $(c^*, 1)$ is provided by the numerator of $D(z)$.

If we assume that $c = 0.5$, the wavelet $(1, c) = (1, 0.5)$ has the z transform $C_0(z) = 1 + 0.5z$, and the wavelet $(c^*, 1) = (0.5, 1)$ has the z transform $C_1(z) = 0.5 + 1z$. The positions of the zeros of $C_0(z)$ and $C_1(z)$ and the positions of the zero and the pole of $D(z)$ are shown in Figure 5-6.

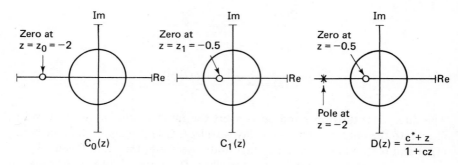

Figure 5-6. Zeros of the wavelets $(1, c)$ and $(c^*, 1)$ and the zeros and poles of the dispersive filter $D(z)$. The circle shown is the unit circle.

Having examined the action of the dispersive filter $D(z)$ in the z plane, let us next look at this filter in the frequency domain. The representation of the input/output relationship is that the product of the Fourier transform of the input multiplied by the Fourier transform of the filter equals the Fourier transform of the output, or

$$|C_0(\omega)|e^{-i\phi_0(\omega)} \cdot |D(\omega)|e^{-i\delta(\omega)} = |C_1(\omega)|e^{-i\phi_1(\omega)}$$

Here $|C_0(\omega)|, |C_1(\omega)|$, and $|D(\omega)|$ are, respectively, the magnitude spectra of the wavelets $(1, c)$ and $(c^*, 1)$ and of the dispersive filter D; $\phi_0(\omega), \phi_1(\omega)$, and $\delta(\omega)$ are the corresponding phase-lag spectra. If we equate the magnitude and phase-lag parts of both sides of the equation above, we obtain the magnitude spectra equation,

$$|C_0(\omega)| \cdot |D(\omega)| = |C_1(\omega)|$$

and the phase-lag spectra equation,

$$\phi_0(\omega) + \delta(\omega) = \phi_1(\omega)$$

Since $|C_0(\omega)| = |C_1(\omega)|$, we have at once from the magnitude spectra equation that $|D(\omega)| = 1$ for all ω, as we have already shown. The phase-lag spectra equation can be written in the form

$$\delta(\omega) = \phi_1(\omega) - \phi_0(\omega) \tag{5-1}$$

This relation allows us to compute $\delta(\omega)$ if we know the phase-lag spectra $\phi_0(\omega)$ and $\phi_1(\omega)$.

For the moment, we shall assume that the constant c is a real number, positive or negative. Then from Chapter 2 the tangent of the phase-lag spectrum of the wavelet $(1, c)$ is

$$\tan \phi_0(\omega) = \frac{c \sin \omega}{1 + c \cos \omega} \tag{5-2a}$$

and the tangent of the phase-lag spectrum of the wavelet $(c, 1)$ is

$$\tan \phi_1(\omega) = \frac{\sin \omega}{c + \cos \omega} \tag{5-2b}$$

However, we have from (5-1) that

$$\tan \delta(\omega) = \tan [\phi_1(\omega) - \phi_0(\omega)]$$

$$= \frac{\tan \phi_1(\omega) - \tan \phi_0(\omega)}{1 + \tan \phi_1(\omega) \tan \phi_0(\omega)} \tag{5-3}$$

Substitution of (5-2a) and (5-2b) into (5-3) yields, after some algebra,

$$\delta(\omega) = \tan^{-1} \left[\frac{(1 - c^2) \sin \omega}{2c + (1 + c^2) \cos \omega} \right] \tag{5-4}$$

This is the phase-lag spectrum of the dispersive filter D. Again returning to our previous numerical example, we set $c = 0.5$ in (5-4), so that

$$\delta(\omega) = \tan^{-1}\left(\frac{3\sin\omega}{4 + 5\cos\omega}\right) \tag{5-5}$$

This phase lag is plotted in Figure 5-7 for the frequency range $0 \le \omega \le \pi$, where we also show the phase lag of the wavelets $(1, 0.5)$ and $(0.5, 1)$. As stated in equation (5-1), the sum of the phase lags $\phi_0(\omega)$ and $\delta(\omega)$ yields the phase lag $\phi_1(\omega)$.

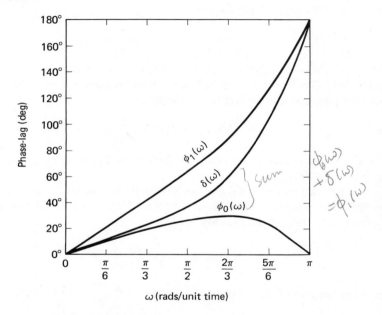

Figure 5-7. Phase lags $\phi_0(\omega)$, $\phi_1(\omega)$, and $\delta(\omega)$ of the wavelets $1 + 0.5z$ and $0.5 + 1z$ and the all-pass filter $D(z)$, respectively.

Let us again allow c to be either a real or a complex number. We recall from Chapter 4 that the 2-length minimum-delay wavelet $(1, c)$, where $|c| < 1$, has an inverse that is a stable memory function (i.e., a stable causal filter). The z transform of this inverse is

$$\frac{1}{1 + cz} = 1 - cz + c^2z^2 - c^3z^3 + c^4z^4 - \cdots$$

Thus, the inverse may be constructed as a feedforward filter, as shown by the block diagram of Figure 5-8. We note that an infinite number of components are needed to construct this feedforward configuration. Let $X(z)$ be the z transform of the input x_t, and let $Y(z)$ be the z transform of the output y_t. Since

$$Y(z) = (1 - cz + c^2z^2 - c^3z^3 + c^4z^4 - \cdots)X(z)$$

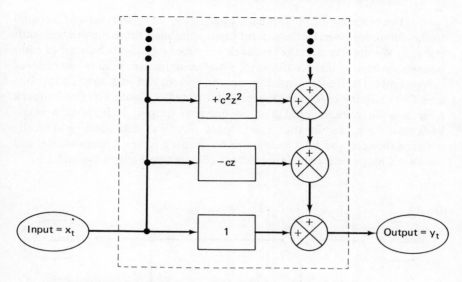

Figure 5-8. Block diagram of the feedforward filter $(1 - cz + c^2z^2 - \ldots)$.

it follows that the input x_t and the output y_t are related by

$$y_t = (1 - cz + c^2z^2 - c^3z^3 + c^4z^4 - \ldots)x_t$$
$$= x_t - cx_{t-1} + c^2x_{t-2} - c^3x_{t-3} + c^4x_{t-4} - \ldots \qquad (5\text{-}6)$$

where we recall that z^n is the n-unit-delay operator. We observe that the feedforward filter

$$1 - cz + c^2z^2 - c^3z^3 + c^4z^4 - \ldots \qquad (5\text{-}7)$$

acts on the input x_t by means of parallel delay lines z^n with multiplications $(-c)^n$.

Now it is possible to exhibit a feedback configuration of the same inverse, which we write in the form

$$\frac{1}{1 + cz} \qquad (5\text{-}8)$$

rather than in the previous form of equation (5-7). Since we know that $Y(z) = [1/(1 + cz)] X(z)$, we may write

$$X(z) = (1 + cz) Y(z)$$

Proceeding as before, we then have

$$x_t = (1 + cz)y_t = y_t + cy_{t-1}$$

We can solve this equation for the output y_t and obtain

$$y_t = x_t - cy_{t-1} \qquad (5\text{-}9)$$

The output y_t of the feedback filter $1/(1 + cz)$ is thus seen to be equal to the difference given by the present input x_t minus c times the previous output y_{t-1}. We observe that the feedback filter contains a finite number of components, whereas the feedforward filter contains an infinite number of components. The feedforward filter acts on present and past input; it involves an infinite number of forward loops, as illustrated in Figure 5-8. The feedback filter acts on present input and on previous output; it involves a single backward loop, as illustrated in Figure 5-9. We distinguish graphically between forward and backward loops by drawing forward loops above, and backward loops below the input/output ellipses (see Figures 5-8 and 5-9).

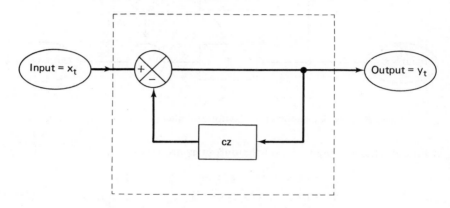

Figure 5-9. Block diagram of the feedback filter $1/(1 + cz)$.

It is not difficult to see that the feedforward filter input/output relation given by equation (5-6) is equivalent to the feedback filter input/output relation given by equation (5-9). Thus, we obtain from (5-6) an expression for y_{t-1},

$$y_{t-1} = x_{t-1} - cx_{t-2} + c^2 x_{t-3} - \ldots$$

Substituting this relation in (5-9), we have

$$y_t = x_t - cy_{t-1} = x_t - cx_{t-1} + c^2 x_{t-2} - c^3 x_{t-3} + \ldots$$

which is exactly equal to equation (5-6).

Let us now consider the maximum-delay wavelet $(c^*, 1)$ where $|c| < 1$; this wavelet is the reverse of the minimum-delay wavelet just considered. The filter with this maximum-delay wavelet as a memory function has the z transform $c^* + z$ and may be constructed as the feedforward filter shown in the block diagram of Figure 5-10. Because $Y(z) = (c^* + z)X(z)$, it follows that input x_t and output y_t are related by the convolution $y_t = c^* x_t + x_{t-1}$.

The question we now wish to consider is: What happens when the filter $c^* + z$ is placed in series with the filter $1/(1 + cz)$? We know that two

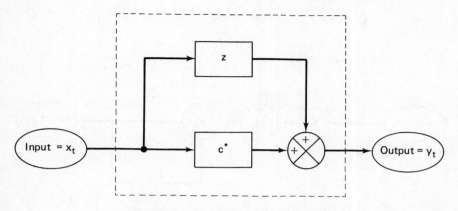

Figure 5-10. Block diagram of the filter $c^* + z$.

filters in series yield an overall filter whose z transform is the product of the z transforms of the two component filters. Hence, the overall filter's z transform is given by

$$D(z) = (c^* + z)\frac{1}{1 + cz}$$

This is the dispersive filter treated earlier in this section, and may be constructed as shown in the block diagram of Figure 5-11.

Because

$$\frac{Y(z)}{X(z)} = \frac{c^* + z}{1 + cz}$$

or

$$Y(z)(1 + cz) = X(z)(c^* + z)$$

it follows that

$$y_t + cy_{t-1} = c^*x_t + x_{t-1}$$

By combining the two adders (each with two inputs) of the foregoing configuration into one adder (with three inputs), we obtain the configuration shown in Figure 5-12.

A configuration that involves no feedback can be found by using

$$D(z) = (c^* + z)\frac{1}{1 + cz} = (c^* + z)(1 - cz + c^2z^2 - c^3z^3 + c^4z^4 - \dots)$$

$$= (c^* - c^*cz + c^*c^2z^2 - c^*c^3z^3 + c^*c^4z^4 - \dots)$$

$$+ (z - cz^2 + c^2z^3 - c^3z^4 + \dots)$$

$$= c^* + (1 - c^*c)z - c(1 - c^*c)z^2 + c^2(1 - c^*c)z^3$$

$$- c^3(1 - c^*c)z^4 + \dots$$

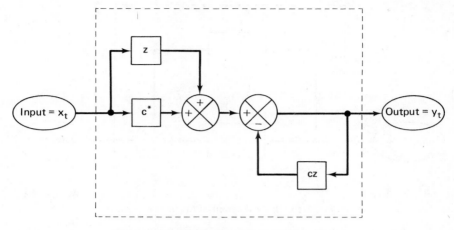

Figure 5-11. Block diagram of the filter $D(z)$.

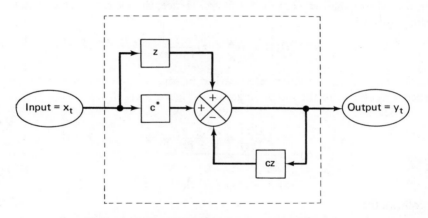

Figure 5-12. Simplified block diagram for the filter $D(z)$.

where $|c| < 1$. Hence, the dispersive filter also has the feedforward configuration shown in the block diagram given in Figure 5-13. If we take $c = \frac{1}{2}$, the z transform of the dispersive filter is

$$\frac{Y(z)}{X(z)} = D(z) = \frac{\frac{1}{2} + z}{1 + \frac{1}{2}z} = \frac{1}{2} + \frac{3}{4}z - \frac{3}{8}z^2 + \frac{3}{16}z^3 - \frac{3}{32}z^4 + \cdots$$

and the input/output relationship is

$$y_t + \tfrac{1}{2}y_{t-1} = \tfrac{1}{2}x_t + x_{t-1}$$

or

$$y_t = \tfrac{1}{2}x_t + \tfrac{3}{4}x_{t-1} - \tfrac{3}{8}x_{t-2} + \tfrac{3}{16}x_{t-3} - \tfrac{3}{32}x_{t-4} + \cdots$$

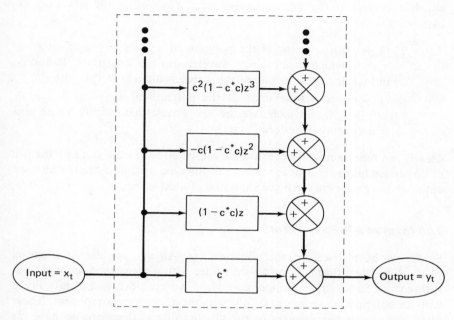

Figure 5-13. Block diagram for the filter $D(z)$ involving feedforward loops only.

Let us now use the(dispersive filter)to enlarge the membership of a suite of wavelets, all with the same magnitude spectrum. Until now, our enlargements of the suite have added only wavelets of finite length. Let $(x_0, x_1, x_2, \ldots, x_n)$ be any $(n + 1)$-length wavelet in the suite. Then a new wavelet may be generated by passing this wavelet through a dispersive filter D. Let

$$X(z) = x_0 + x_1 z + x_2 z^2 + \cdots + x_n z^n$$

be the z transform of the input wavelet, and let

$$D(z) = \frac{c^* + z}{1 + cz} = c^* + (1 - c^*c)z - c(1 - c^*c)z^2 + \ldots$$

(where $|c| < 1$) be the z transform of the dispersive filter. Then the z transform of the output is

$$Y(z) = \frac{c^* + z}{1 + cz}(x_0 + x_1 z + x_2 z^2 + \cdots + x_n z^n)$$

so that the new wavelet of the suite is $(y_0, y_1, y_2, \ldots)$, where $Y(z) = y_0 + y_1 z + y_2 z^2 + \ldots$. This new wavelet has a magnitude spectrum, $|Y(\omega)|$, which is identical to the magnitude spectrum $|X(\omega)|$ of the old wavelet. Two cases may occur:

1. $(1 + cz)$ is a factor of the polynomial $x_0 + x_1z + x_2z^2 + \cdots + x_nz^n$, in which case the new wavelet also has length $n + 1$, and is, in fact, the same as one of the wavelets already in the suite.
2. $(1 + cz)$ is not a factor of the polynomial $x_0 + x_1z + x_2z^2 + \cdots + x_nz^n$, in which case the new wavelet has infinite length and consequently is indeed "new."

Clearly, an infinite number of wavelets can be added to the suite by the use of dispersive filters. Except for the use of the type 3 all-pass filter, which we omit, we have now enlarged the suite to its fullest extent.

The Inverse All-Pass Filter

So far we have described the all-pass filter. An all-pass filter is a causal phase-shift filter; that is, a phase-shift filter with a memory component but without an anticipation component. Now we consider phase-shift filters with an anticipation component, but without a memory component. These filters are simply the inverses to the all-pass filters; therefore we have, as before, four types:

1. Inverse to type 0 all-pass or inverse trivial filter.
2. Inverse to type 1 all-pass or inverse dispersive filter.
3. Inverse to type 2 all-pass, inverse pure-delay, or pure-advance filter.
4. Inverse to type 3 all-pass filter, inverse impure-delay, or impure-advance filter.

Their transfer functions, which are the reciprocals of the transfer functions of the respective all-pass filters, are therefore:

1. $B(z) = \frac{1}{1} = 1$ or $B(z) = 1/(-1) = -1$.
2. $B(z) = 1/D(z) = (1 + cz)/(c^* + z)$, where $|c| < 1$.
3. $B(z) = 1/z^n = z^{-n}$, where $n > 0$.
4. We omit this case, as it is of mathematical interest only.

In case 1 we see that the inverse to a trivial all-pass filter is the same as the all-pass filter; both filters have the same weighting function, which is

$$(\ldots, 0, \quad 0, \pm 1, 0, 0, \ldots)$$
$$\uparrow \quad \uparrow \quad \uparrow \uparrow \uparrow$$
$$\text{time index} = -2, -1, \quad 0, 1\ 2, \ldots$$

In case 2, the transfer function of the inverse dispersive filter may be expressed in the form of its anticipation function

$$B(z) = \frac{1 + cz}{c^* + z} = (1 + cz)(z^{-1} - c^*z^{-2} + c^{*2}z^{-3}\ldots)$$

$$= c + (1 - c^*c)z^{-1} - c^*(1 - cc^*)z^{-2} + \ldots$$

The factor $(c^* + z)^{-1}$ is here expanded in negative powers of z, since $|c| < 1$. In case 3, the pure n-unit advance filter has the anticipation function

$$(\ldots, \quad 0, \quad\quad 1, \quad 0, \quad\quad \ldots, \quad\quad 0, 0)$$
$$\uparrow \quad\quad \uparrow \quad \uparrow \quad\quad\quad\quad \uparrow \uparrow$$
$$\text{time index} = \ldots, -n - 1, -n, -n + 1, \ldots, -1, 0$$

whereas the pure n-unit delay filter has the memory function

$$(0, 0, \ldots, \quad 0, \quad 1, \quad 0, \ldots)$$
$$\uparrow \uparrow \quad\quad\quad \uparrow \quad \uparrow \quad \uparrow$$
$$\text{time index} = 0, 1, \ldots, n - 1, n, n + 1, \ldots$$

so that the convolution of the two filters produces a unit spike at time 0, as is desired.

In all cases the anticipation function of an inverse to an all-pass system can be found by reflecting the memory function of the all-pass system about time index 0 and taking the complex conjugates of the coefficients. For example, reflecting the system

$$[c^*, (1 - c^*c), -c(1 - c^*c), c^2(1 - c^*c), -c^3(1 - c^*c), \ldots]$$

about time index 0, we obtain

$$[\ldots, -c^3(1 - c^*c), c^2(1 - c^*c), -c(1 - c^*c), (1 - c^*c), c^*]$$

Taking complex conjugates, we obtain

$$[\ldots, -c^{*3}(1 - c^*c), c^{*2}(1 - cc^*), -c^*(1 - cc^*), (1 - cc^*), c]$$

which is the memory function of the inverse dispersive filter.

More complicated all-pass filters may be built as series (or cascaded) combinations of basic all-pass filters; also more complicated inverses to all-pass filters may be built as series (or cascaded) combinations of basic inverses to all-pass filters. Arbitrary phase-shift filters P (with both memory and anticipation components) can be built as series (or cascaded) combinations of all-pass filters and inverses to all-pass filters.

The Transfer Function of the Dispersive Filter

We showed that any $(n + 1)$-length wavelet with z transform

$$X(z) = x_0 + x_1 z + x_2 z^2 + \cdots + x_n z^n \qquad (5\text{-}10)$$

can be factored into n 2-length wavelets, some or all of whose zeros may be complex. In the case of real signals, the coefficients $x_0, x_1, x_2, \ldots, x_n$ are real. Then all the nonreal zeros of $X(z)$ must occur in complex-conjugate pairs. We can factor equation (5-10) in the form

$$X(z) = x_n(z - z_1)(z - z_2) \ldots (z - z_n) = x_n \prod_{k=1}^{n} (z - z_k) \qquad (5\text{-}11)$$

where $z = z_k$ is the kth zero of $X(z)$.

Let us now see what happens when we pass the wavelet $X(z)$ through the type 1 all-pass (or dispersive) filter:

$$P_k(z) = -z_k \frac{z - 1/z_k^*}{z - z_k} \qquad (5\text{-}12)$$

As shown previously, the filter $P_k(z)$ removes the zero that the input waveform has at the point $z = z_k$, and replaces it by a new zero at $z = 1/z_k^*$. In the frequency domain, the transfer function of the filter P_k is

$$P_k(\omega) = |P_k(\omega)| \, e^{-i\alpha_k(\omega)} \qquad (5\text{-}13)$$

where $|P_k(\omega)|$ is the magnitude spectrum and $\alpha_k(\omega)$ is the phase-lag spectrum of this dispersive filter. The magnitude spectrum of P_k is unity:

$$|P_k(\omega)| = 1 \qquad (-\pi \leq \omega \leq +\pi) \qquad (5\text{-}14)$$

while the phase lag $\alpha_k(\omega)$ is given by

$$\alpha_k(\omega) = \tan^{-1} \left[\frac{-2 \sin \theta_k + |z_k|^{-1} \sin (\omega + 2\theta_k) - |z_k| \sin \omega}{2 \cos \theta_k - |z_k|^{-1} \cos (\omega + 2\theta_k) - |z_k| \cos \omega} \right] \qquad (5\text{-}15)$$

where θ_k is the phase angle of the complex root z_k, $z_k = |z_k| e^{i\theta_k}$ (see Appendix 5-1). Equation (5-15) is more general than equation (5-4), to which it reduces for the case of a real root $z_k = -1/c$, so that $|z_k| = 1/c$ and $\theta_k = \pi$, when c is positive, and $|z_k| = -1/c$ and $\theta_k = 0$, when c is negative.

Let us now pass the input wavelet $X(z)$, given by (5-11), through the dispersive filter $P_k(z)$,

$$X(z)P_k(z) = [x_n(z - z_1) \ldots (z - z_k) \ldots (z - z_n)]\left(-z_k \frac{z - 1/z_k^*}{z - z_k}\right)$$

$$= x_n(z - z_1) \ldots (-z_k)(z - 1/z_k^*) \ldots (z - z_n) = Y(z) \qquad (5\text{-}16)$$

The output wavelet Y differs from the input wavelet X in that the zero that $X(z)$ had at $z = z_k$ is replaced in $Y(z)$ by a new zero at $z = 1/z_k^*$. Since the magnitude spectrum of P_k is unity, we have

$$|X(\omega)| = |Y(\omega)| \qquad (5\text{-}17)$$

where $|Y(\omega)|$ is the magnitude spectrum of the output wavelet Y. If the root $z = z_k$ is complex, we can see from (5-16) that the weighting coefficients of the output wavelet Y will be complex. Since the input wavelet weighting coefficients $x_0, x_1, \ldots, x_n$ are real, the nonreal roots of $X(z)$ occur in complex-conjugate pairs. If the output wavelet Y is to have real weighting coefficients only, it will be necessary to pass $X(z)$ through two filters of the type P_k, the first of which removes the zero at $z = z_k$, and the second of which removes the zero at $z = z_k^*$. It will be convenient to rewrite (5-11) in the form

$$X(z) = x_n \prod_{p=1}^{r} (z - \hat{z}_p) \prod_{k=1}^{q} (z - z_k)(z - z_k^*) \qquad (5\text{-}18)$$

where the r roots $\hat{z}_p$ are real and the $2q$ roots z_k, z_k^* are complex, so that $r + 2q = n$. The $2q$ complex roots appear in complex conjugate pairs, so that q of them are denoted in equation (5-18) by z_k (where $k = 1, 2, \ldots, q$) and the remaining q of them are the corresponding complex conjugates z_k^* (where $k = 1, 2, \ldots, q$). The dispersive filter P, which will remove the complex-conjugate root pair (z_k, z_k^*) from the input wavelet $X(z)$ and will replace it by the complex-conjugate root pair $(1/z_k^*, 1/z_k)$ in the output wavelet $Y(z)$. is given by

$$P_k(z) = |z_k|^2 \frac{(z - 1/z_k^*)(z - 1/z_k)}{(z - z_k)(z - z_k^*)} \qquad (5\text{-}19)$$

Since the product of two complex-conjugate factors is real, the transfer function $P_k(z)$ is real. This means that if we pass the wavelet $X(z)$ through the filter $P_k(z)$, we obtain

$$
\begin{aligned}
X(z)P_k(z) = {} & x_n \prod_{p=1}^{r} (z - z_p^*) \\
& \cdot \left[(z - z_1)(z - z_1^*) \ldots |z_k|^2 \left(z - \frac{1}{z_k}\right)\left(z - \frac{1}{z_k^*}\right) \ldots \right. \\
& \qquad\qquad\qquad\qquad \left. (z - z_q)(z - z_q^*) \right] \\
= {} & Y(z) \qquad (5\text{-}20)
\end{aligned}
$$

We then see that $Y(z)$ is the z transform of a wavelet with real weighting coefficients only.

$(5.4) \quad \delta(\omega) = \tan^{-1}\left[\dfrac{(1-c^2)\sin\omega}{2c + (1+c^2)\cos\omega}\right]$

From the foregoing discussion, we may write the general expression for the nth-order dispersive filter

$$P_n(z) = \prod_{p=1}^{r} P_p(z) \prod_{k=1}^{q} P_k(z) \qquad (r + 2q = n) \qquad (5\text{-}21)$$

where $P_p(z)$ is the first-order dispersive filter

$$P_p(z) = -\hat{z}_p \frac{z - 1/\hat{z}_p}{z - \hat{z}_p} \qquad (5\text{-}22)$$

and $\hat{z}_p$ is real, whereas $P_k(z)$ is a second-order dispersive filter given by (5-19). By inductive reasoning, we find also that the magnitude spectrum of the nth-order dispersive filter is given by

$$|P_n(\omega)| = 1 \qquad (5\text{-}23)$$

whereas its phase lag can be found by summing n first-order phase lags of the kind given by equation (5-15):

$$\alpha^{[n]}(\omega) = \sum_{k=1}^{n} \alpha_k(\omega) \qquad (5\text{-}24)$$

The Progressive Dispersion of a Wavelet

We shall use some of the results obtained so far in order to present a numerical illustration of our main points. We will use the suite of four 3-length wavelets that we introduced earlier (see Figure 5-1), and which we again list here for convenience:

Wavelet	Wavelet Coefficients	Designation
1	(3, 3.5, 1)	Minimum delay
2	(2, 4, 1.5)	Mixed delay
3	(1.5, 4, 2)	Mixed delay
4	(1, 3.5, 3)	Maximum delay

All wavelets of this suite have the same magnitude spectrum $|A(\omega)| = (1.25 + \cos \omega)^{1/2} (13 + 12 \cos \omega)^{1/2}$.

Our objective is: Given any member wavelet of the suite above, find the other three members of the suite. We first tabulate the factored z transforms of all four wavelets:

Wavelet	Factored z Transform
1	$(1 + 0.5z)(3 + 2z)$
2	$(1 + 0.5z)(2 + 3z)$
3	$(0.5 + z)(3 + 2z)$
4	$(0.5 + z)(2 + 3z)$

Assuming that the minimum-delay wavelet 1 is given, we must first calculate its phase lag. In the present case the simplest way is to add the phase lags of the two component 2-length wavelets $(1, 0.5)$ and $(3, 2)$; the sum is the phase lag of the minimum-delay wavelet, which is shown in Figure 5-14. Let us now attempt to find the remaining three wavelets of this suite by passing the minimum-delay wavelet 1 through a set of type 1 all-pass filters as given by equation (5-12). We shall illustrate this approach by attempting first to find wavelet 2, assuming that wavelet 1 is given.

To find wavelet 2 from wavelet 1, we notice first that the zero at $z = z_1 = -\frac{3}{2}$ (which lies outside the unit circle) of wavelet 1 must be replaced by a zero at $z = -1/z_1 = -\frac{2}{3}$ (which lies inside the unit circle). This may be accomplished by passing wavelet 1 through the type 1 all-pass or dispersive filter

$$P_1(z) = +\frac{3}{2}\left(\frac{z + \frac{2}{3}}{z + \frac{3}{2}}\right)$$

From equation (5-16) we then have

$$X(z)P_1(z) = Y(z)$$

where the input $X(z)$ is the z transform of the minimum-delay wavelet

$$X(z) = (1 + 0.5z)(3 + 2z)$$

The output $Y(z)$ is thus

$$(1 + 0.5z)(3 + 2z)\left[\frac{3}{2}\left(\frac{z + \frac{2}{3}}{z + \frac{3}{2}}\right)\right] = (1 + 0.5z)(2 + 3z) = Y(z)$$

where $Y(z) = (1 + 0.5z)(2 + 3z)$ is the z transform of the mixed-delay wavelet 2. The magnitude spectrum of the dispersive filter $P_1(z)$ is unity [equation (5-14)], and the phase-lag spectrum of this filter is given by equation

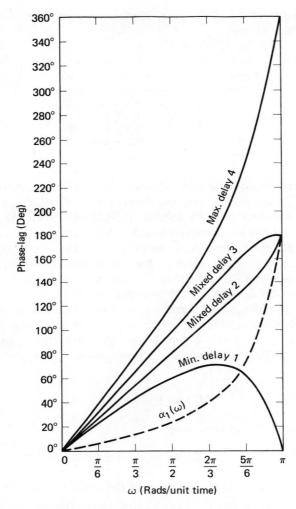

Figure 5-14. Phase-lag spectra of wavelets 1, 2, 3, and 4 and of the dispersive filter $P_1(\omega) = |1| e^{-\alpha_1(\omega)}$.

(5-15). The zero removed from wavelet 1 is $z_1 = -\frac{3}{2} = 3e^{i\pi}/2$, so that in equation (5-15) we must set $|z_1| = \frac{3}{2}$ and $\theta_1 = \pi$. We then obtain

$$\alpha_1(\omega) = \tan^{-1}\left(\frac{5\sin\omega}{12 + 13\cos\omega}\right)$$

where $\alpha_1(\omega)$ is the phase-lag spectrum of the dispersive filter $P_1(\omega)$. Then, we write

$$\phi^{(1)}(\omega) + \alpha_1(\omega) = \phi^{(2)}(\omega)$$

where $\phi^{(1)}(\omega)$ and $\phi^{(2)}(\omega)$ are the phase-lag spectra of wavelets 1 and 2, respectively. These two phase-lag spectra are plotted in Figure 5-14, where we also show the curve for $\alpha_1(\omega)$.

By proceeding in this way, it is possible to generate the phase lags of the remaining wavelets of our suite, the mixed-delay wavelet 3 and the maximum-delay wavelet 4. These phase lags are plotted in Figure 5-14. We observe that the phase lags of the mixed-delay wavelets 2 and 3 lie between the phase lags of the minimum- and maximum-delay wavelets of our suite. We may thus say that the sequence of wavelets 1, 2, 3, and 4 of Figure 5-1 represents the *progressive dispersion* of a wavelet that starts as the minimum-delay wavelet 1 and ends as the maximum-delay wavelet 4. We repeat that it is only the phase-lag spectrum that changes from wavelet to wavelet, whereas the magnitude spectrum remains the same for all members of the suite. We also know that the maximum-delay wavelet is the time reverse of the minimum-delay wavelet; this effect can be seen in the wavelet suite pictured in Figure 5-1.

Concluding Remarks

The theory of phase-shift filters forms a convenient framework within which the peculiarities of wavelet suites having identical magnitude spectra can be studied. We have seen how the phase-lag spectrum of a wavelet is related to the time distribution of the energy of the wavelet and to the position of the singularities of the z transform of the wavelet.

For quick reference, a summary of our classification of the various types of phase-shift filters is given in Table 5-2. For every memory phase-

TABLE 5-2. CLASSIFICATION SCHEME OF THE VARIOUS TYPES OF PHASE-SHIFT FILTERS

All-Pass *(or Causal)[a]*	*Inverse All-Pass* *(or Purely Noncausal)[b]*
Type 0 all-pass or trivial filter	Inverse type 0 all-pass or inverse trivial filter
Type 1 all-pass or dispersive filter	Inverse type 1 all-pass or inverse dispersive filter
Type 2 all-pass or pure-delay filter	Inverse type 2 all-pass or pure-advance filter
Type 3 all-pass or impure-delay filter	Inverse type 3 all-pass or impure-advance filter

[a]One-sided to the past.
[b]One-sided to the future.

shift filter in the first column of Table 5-2, there is a corresponding anticipation phase-shift filter shown in the second column. As we indicated earlier, the filters in the first column are inverses to the corresponding filters in the second column.

---

the phase-lag spectrum of the dispersive filter $P_k(z)$

To derive the phase-lag spectrum of the dispersive filter $P_k(z)$, we start with equation (5-12):

$$P_k(z) = -z_k \frac{z - 1/z_k^*}{z - z_k}$$

$(5\text{-}12)$

Since this dispersive filter consists of the ratio of two first-degree polynomials in z and has one zero and one pole, we say that $P_k(z)$ is a *first-order dispersive filter*. It will be convenient to express the root z_k in the polar form

$$z_k = |z_k| e^{i\theta_k}$$

where $|z_k|$ is the magnitude and θ_k is the phase angle of the root z_k. We then have

$$z_k^* = |z_k| e^{-i\theta_k} \quad \text{and} \quad \frac{1}{z_k^*} = \left(\frac{1}{|z_k|}\right) e^{i\theta_k}$$

Since $z = e^{-i\omega}$, we may write (5-12) in the form

$$P_k(\omega) = \frac{-|z_k| e^{i\theta_k}(e^{-i\omega} - |z_k|^{-1} e^{i\theta_k})}{e^{-i\omega} - |z_k| e^{i\theta_k}}$$

Some tedious algebra allows us to split this expression into real and imaginary parts, from which we can finally obtain the phase-lag spectrum

$$\alpha_k(\omega) = \tan^{-1}\left[\frac{-2\sin\theta_k + |z_k|^{-1}\sin(\omega + 2\theta_k) - |z_k|\sin\omega}{2\cos\theta_k - |z_k|^{-1}\cos(\omega + 2\theta_k) - |z_k|\cos\omega}\right]$$

which is equation (5-15) of the main text. Since

$$z_k = |z_k|\,e^{i\theta_k}$$

we see that $\theta_k = 0$ if z_k is real and positive, whereas $\theta_k = \pi$ if z_k is real and negative.

page 138

principles of
least-squares filtering

Summary

The theory of statistical communication provides an invaluable framework within which it is possible to formulate design criteria and actually obtain solutions for digital filters. These are then applicable in a wide range of geophysical problems. The basic model for the filtering process considered here consists of an input signal, a desired output signal, and an actual output signal. If one minimizes the energy or power existing in the difference between desired and actual filter outputs, it becomes possible to solve for the *optimum*, or *least-squares filter*. In this chapter, we derive from basic principles the theory leading to such filters. The analysis is carried out in the time domain in discrete form. We propose a model of a seismic trace in terms of a statistical communication system. This model trace is the sum of a signal time series plus a noise time series. If we assume that estimates of the signal shape and of the noise autocorrelation are available, we may calculate least-squares filters that will attenuate the noise and sharpen the signal. The net result of these operations can then, in general, be expected to increase seismic resolution. We show a few numerical examples to illustrate the model's applicability to situations one might find in practice.

140

Introduction

The use of digital filters in the analysis of time-series recordings has by now become quite widespread. In this chapter, we give the actual design criteria and computational procedures needed to obtain the filters from the time-series data. One of the most powerful and versatile design criteria for such filters comes from the theory of least squares. The present treatment is an effort to describe in reasonably simple terms how least-squares concepts can be applied in filter design work. Since digital filters are ordinarily applied in the time domain, our development will be carried out with time rather than frequency as the independent variable. In other words, the transfer characteristics of the filters we obtain in this chapter are expressed in terms of their unit impulse response functions, that is, in terms of the memory functions of the filters.

Before dealing with the actual filter design problem, we shall first discuss a few basic concepts from statistical communication theory. These principles will then serve as a basis for our subsequent discussion of least-squares filtering.

Basic Concepts

The word *signal* is a generic term that may denote the time variation of any physical quantity. For example, earth motion in seismic exploration may be regarded as a signal, the voltage that records earth motion is also a signal, and so is the actual recording on digital tape.

Before we can make any analysis or design equipment, we must agree how to characterize the signals. In this book, we are concerned with digital devices, so we shall assume that all signals have been digitized, and thus each signal is available in the form of a *time series*, that is, numbers in a sequence with a unit time spacing between each number of the sequence.

In seismic work there is always a distinction between a wavelet and an arbitrary seismic recording. Qualitatively, a *wavelet* is a signal with negligibly small values except in some finite region of the time scale. Thus, a wavelet is epochal in character; that is, there is a portion of the infinite time scale at which the wavelet occurs.

An important parameter of a wavelet, say with amplitudes

$$b_0, b_1, b_2, \ldots$$

is the sum of squares of its amplitudes

$$b_0^2 + b_1^2 + b_2^2 + \cdots$$

which is called the *energy* of the wavelet. It is always assumed that a wavelet has finite energy. For this reason a wavelet may be called an *energy signal*. Another important type of signal is the stationary time series. A *stationary time series* is a time series with statistical properties that do not change with time. Thus, a segment of such a time series recorded in some past epoch has essentially the same statistics as another segment recorded in some later epoch. A stationary time series is the embodiment of a time phenomenon continuing over all time, from the remote past to the distant future. Thus, we may say that a stationary time series has infinite time duration, or infinite length. Nevertheless, in any actual situation we can only obtain its values over a finite interval of time. Because such a finite segment represents a sample of the infinite length time series, we shall call a finite portion of a stationary time series a *sample* of a stationary time series. In some seismic situations it is permissible to think of a portion of a recorded seismic trace as a sample of a stationary time series, as described in Chapter 10.

We must always be careful to distinguish between the concepts of *wavelet* on the one hand, and *sample time series* on the other hand. The point of difference is that a wavelet is a self-contained or entire entity, and this entity is a transient phenomenon with a definite arrival time. In contrast, a sample time series is only a portion of an entire entity, namely the time series, and this entity is a continuing phenomenon.

Because a stationary time series represents a continuing phenomenon, it may be called a *power signal*. The direct component, or mean value of a power signal can be removed, thereby leaving the alternating component. We shall always assume that this has been done, so that all power signals in our analysis will have mean values of zero.

An important parameter of a power signal, say with amplitudes

$$\ldots, u_{-2}, u_{-1}, u_0, u_1, u_2, u_3, \ldots$$

(where the subscript is the time index) is the average sum of squares of its amplitudes

$$\frac{1}{2T+1}(u_{-T}^2 + u_{-T+1}^2 + \cdots + u_{-1}^2 + u_0^2 + u_1^2 + \cdots + u_T^2)$$

where T tends to infinity. This average is called the *power* of the signal, in conformity with the convention that power represents average energy per unit time.

Let us now introduce the *expected value* symbol E to denote the operation of taking an average. This symbol will have two different meanings, depending upon whether the quantity to be averaged is derived from an energy signal or from a power signal. The reason for this duality of meaning is that it allows us to treat both (energy signals) and (power signals) in a unified manner. As an example of this dual usage, let us again present the concepts , of energy and power, but this time we will make use of the expected value symbol E. In order to clearly distinguish a wavelet, which is an energy signal, from a sample time series, which is a power signal, we use the notation b_t for the equispaced amplitudes of a wavelet and the notation u_t for the equispaced amplitudes of a sample time series.

Now let us correlate a signal with itself; that is, let us correlate a signal with its replica with no time shift. For a wavelet b_t this quantity is its energy, given by

$$E\{b_t^2\} = \sum_{t=0}^{\infty} b_t^2 = b_0^2 + b_1^2 + b_2^2 + \cdots$$

for a wavelet

expected value

be

and for a stationary time series u_t this quantity is its power, given by

$$E\{u_t^2\} = \lim_{T \to \infty} \frac{1}{2T+1}(u_{-T}^2 + u_{-T+1}^2 + \cdots + u_{-1}^2 + u_0^2 + u_1^2 + \cdots + u_T^2)$$

The notation $E\{\cdot\}$ in each case denotes the average of the quantity within the braces; for a wavelet this average is a sum of squares, whereas for a stationary time series it is the limit of the sum of squares divided by the elapsed time as time becomes infinite. In other words, for a wavelet $E\{\cdot\}$ indicates an *energy average*, whereas for a stationary time series $E\{\cdot\}$ indicates a *power average*. This convention is necessary because the *energy* of an infinitely long stationary time series is infinite, whereas its *power* is finite. (*Note:* We have *defined* the energy average of a wavelet to be the sum of squares of its amplitudes, *without* dividing this sum by N, where N is the number of equispaced amplitude values used to describe the wavelet.)

If we correlate a signal not with itself, but with a replica of itself shifted by an amount τ along the time axis, we expect the amount of correlation to be somewhat less. The dependence of correlation upon the shift is an important characteristic, and is called the *autocorrelation* function of the signal. The autocorrelation of a wavelet b_t is

$$\phi_{bb}(\tau) = \sum b_{t+\tau}b_t = E\{b_{t+\tau}b_t\}$$

and the autocorrelation of a stationary time series u_t is

$$\phi_{uu}(\tau) = E\{u_{t+\tau}u_t\} = \lim_{T \to \infty} \frac{1}{2T+1}(u_{t+\tau}u_t \cdots$$

For $\tau = 0$, the autocorrelation gives the energy or the power, as the case may be. The autocorrelation for all values of τ gives information as to the relative amounts of energy or power in the different frequency components of the signal.

The *crosscorrelation* between a signal x_t and a signal y_t is

$$\phi_{xy}(\tau) = E\{x_{t+\tau}y_t\}$$

Here the signal $x_{t+\tau}$ represents a replica of the signal x_t shifted to the left along the time axis by the amount τ. In this case we take into account the algebraic sign of τ, so that if τ were negative, the shift of x_t would actually be to the right. (Following usual practice, t always increases to the right.) The crosscorrelation between y_t and x_t is

$$\phi_{yx}(\tau) = E\{y_{t+\tau}x_t\}$$

Here the signal $y_{t+\tau}$ represents a replica of the signal y_t shifted to the left along the time axis by the amount τ. This is equivalent to shifting the signal x_t to the right by an amount τ, so we see that

$$\phi_{yx}(\tau) = \phi_{xy}(-\tau)$$

If $y = x$, the expression above reduces to

$$\phi_{xx}(\tau) = \phi_{xx}(-\tau)$$

That is, the autocorrelation is an even function.

A fundamental property of a stationary time series is its *wavelet representation*. This principle states that any stationary time series can be expressed as the convolution of a wavelet with white noise. White noise is a stationary time series with zero autocorrelation, except for time shift $\tau = 0$, where the autocorrelation is simply equal to its power. Thus, all frequency components of white noise have the same power, in analogy with "ideal" white light. Other names for *white noise* are "white spike series" and "random orthogonal series."

Suppose that u_t is the stationary time series, b_t is the characteristic wavelet, and ϵ_t is the white noise. Then the wavelet representation is written as

$$u_t = b_0\epsilon_t + b_1\epsilon_{t-1} + b_2\epsilon_{t-2} + \cdots$$
$$= \sum_{s=0}^{\infty} b_s\epsilon_{t-s}$$

Letting an asterisk $*$ denote convolution, the wavelet representation may be written more concisely as

$$u_t = b_t * \epsilon_t$$

Because the autocorrelation of the white noise ϵ_t vanishes except for time shift zero, the white noise has no effect on the shape of the autocorrelation of the time series u_t. The autocorrelation at time shift zero of the white noise, we recall, is its power, so the autocorrelation of the time series u_t is proportional to the power of the white noise. Let us call this power P; that is,

$$P = E\{\epsilon_t^2\}$$

It follows, then, that the shape of the autocorrelation of the time series u_t is given entirely by the shape of the autocorrelation of the characteristic wavelet b_t. Let us denote the autocorrelation of the characteristic wavelet by

$$\phi_{bb}(\tau) = \sum_{t=0}^{\infty} b_{t+\tau}b_t = E\{b_{t+\tau}b_t\}$$

Thus, the autocorrelation of the time series u_t is given by

$$\phi_{uu}(\tau) = E\{u_{t+\tau}u_t\}$$
$$= P\phi_{bb}(\tau) \qquad (6\text{-}1)$$

That is, it is equal to the power P of the white noise multiplied by the autocorrelation $\phi_{bb}(\tau)$ of the characteristic wavelet. The derivation of equation (6-1) is given in Appendix 6-1.

The General Filter Design Model

Before we can design an actual digital filter, we must set up a model. A very useful and general model is shown in Figure 6-1. Here there are three signals:

1. The input signal.
2. The desired output signal.
3. The actual output signal.

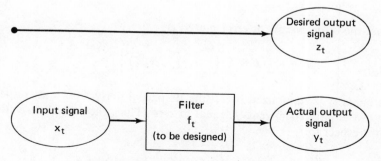

Figure 6-1. General filter design model.

In our analysis all three may be wavelets, or all three may be stationary time series. In the first case we deal with energies and in the second case with powers, but in either case the mathematical analysis is very similar.

Our goal now is to find a technique that allows us to find the filter f_t in terms of the input signal and the desired output signal. Among the various methods at our disposal to accomplish this task, one is particularly outstanding because of the quality of the results obtainable and the simplicity of the concepts involved. The technique we are about to describe is based on the least-squares criterion.

Mathematical Details

The basic principle in filter design is the least-squares criterion: one minimizes the energy or power (as the case may be) existing in the difference between the desired output z_t and the actual output y_t. In other words, we seek the filter coefficients f_t such that the value of

$$J = E\{(z_t - y_t)^2\}$$

is a minimum. The quantity J is the error energy or error power, as the case may be. Its smallest achievable value for a given situation will yield the best, or "optimum" linear filter in the least-squares sense. To make this problem meaningful for calculation on a digital computer, we must restrict ourselves to filters that have a finite number of coefficients,

$$f_0, f_1, f_2, \ldots, f_j, \ldots, f_m$$

We then say that the filter f_t is described by $(m + 1)$ equispaced coefficients, or simply that the filter is $(m + 1)$-length.

The output y_t is the convolution of the filter coefficients f_t with the input x_t; that is,

$$y_t = x_t * f_t = \sum_{\tau=0}^{m} f_\tau x_{t-\tau}$$

so that the *error energy*, or *error power*, is

$$J = E\left\{\left(z_t - \sum_{\tau=0}^{m} f_\tau x_{t-\tau}\right)^2\right\} \tag{6-2}$$

This quantity is minimized by setting its partial derivatives with respect to each of the filter coefficients f_τ equal to zero. Thus, the partial derivative of J with respect to f_1 is

$$\frac{\partial J}{\partial f_1} = E\left\{2\left(z_t - \sum_{\tau=0}^{m} f_\tau x_{t-\tau}\right) \frac{\partial}{\partial f_1}\left(z_t - \sum_{\tau=0}^{m} f_\tau x_{t-\tau}\right)\right\}$$

$$= 2E\left\{\left(z_t - \sum_{\tau=0}^{m} f_\tau x_{t-\tau}\right)(-x_{t-1})\right\}$$

$$= 2E\left\{-z_t x_{t-1} + \sum_{\tau=0}^{m} f_\tau x_{t-\tau} x_{t-1}\right\}$$

$$= 2\left[-E\{z_t x_{t-1}\} + \sum_{\tau=0}^{m} f_\tau E\{x_{t-\tau} x_{t-1}\}\right]$$

$$= 2\left[-\phi_{zx}(1) + \sum_{\tau=0}^{m} f_\tau \phi_{xx}(1 - \tau)\right]$$

Setting this partial derivative equal to zero, we obtain the equation

$$\sum_{\tau=0}^{m} f_\tau \phi_{xx}(1 - \tau) = \phi_{zx}(1)$$

In a similar manner, we may compute the partial derivative of J with respect to f_j ($j = 0, 1, 2, \ldots, m$). In this way we obtain a system of $(m + 1)$ linear simultaneous equations in the unknown filter coefficients f_j, which can be written in the form

$$\sum_{\tau=0}^{m} f_\tau \phi_{xx}(j - \tau) = \phi_{zx}(j) \qquad \text{for } j = 0, 1, 2, \ldots, m \qquad (6\text{-}3)$$

These are the |*normal equations;*| their solution yields the filter coefficients f_j. The known quantities in this system of equations are the autocorrelation $\phi_{xx}(\tau)$ of the input signal and the crosscorrelation $\phi_{zx}(\tau)$ of the desired output signal with the input signal.

If we treat z_t and y_t as vectors, and thus let $z_t = y_t + e_t$, where e_t is the error between desired and actual output, the magnitude of the vector e_t will be minimum when $e_t \perp y_t$; that is, when e_t is normal to y_t. This is why equations (6-3) are called "normal" equations.

Let us next try to find a more convenient expression for the error energy or power J than that given by equation (6-2). Thus,

$$J = E\left\{\left(z_t - \sum_{\tau=0}^{m} f_\tau x_{t-\tau}\right)^2\right\}$$

$$= E\{z_t^2\} - 2E\left\{z_t \sum_{\tau=0}^{m} f_\tau x_{t-\tau}\right\} + E\left\{\sum_{\tau=0}^{m} f_\tau x_{t-\tau} \sum_{\mu=0}^{m} f_\mu x_{t-\mu}\right\}$$

$$= E\{z_t^2\} - 2\sum_{\tau=0}^{m} f_\tau E\{x_t x_{t-\tau}\} + \sum_{\tau=0}^{m} f_\tau \sum_{\mu=0}^{m} f_\mu E\{x_{t-\tau} x_{t-\mu}\} \qquad (6\text{-}4)$$

where μ is a dummy summation index. However, we have that

$$E\{z_t^2\} = \phi_{zz}(0) \tag{6-5a}$$

$$E\{z_t x_{t-\tau}\} = E\{z_{t+\tau} x_t\} = \phi_{zx}(\tau) \tag{6-5b}$$

$$E\{x_{t-\tau} x_{t-\mu}\} = E\{x_{t+\mu-\tau} x_t\} = \phi_{xx}(\mu - \tau) \tag{6-5c}$$

The left member of equation (6-5a) is the energy or power in the desired output z_t, and we see that this quantity is equal to the zeroth lag of the autocorrelation of z_t. Equations (6-5b) and (6-5c) follow from our definitions of crosscorrelation and autocorrelation given previously in this chapter.

If we substitute relations (6-5a) to (6-5c) into (6-4), we have

$$J = \phi_{zz}(0) - 2 \sum_{\tau=0}^{m} f_\tau \phi_{zx}(\tau) + \sum_{\tau=0}^{m} f_\tau \sum_{\mu=0}^{m} f_\mu \phi_{xx}(\mu - \tau) \qquad (6\text{-}6)$$

Now the quantity J will be at a minimum if we substitute into it the normal equations (6-3), since we recall that these equations were obtained by our minimization of J. Keeping in mind that $\phi_{xx}(\mu - \tau) = \phi_{xx}(\tau - \mu)$, we replace the innermost summation of the last term on the right-hand side of equation (6-6) by the right-hand side of equation (6-3). This yields

$$J_{\min} = \phi_{zz}(0) - 2 \sum_{\tau=0}^{m} f_\tau \phi_{zz}(\tau) + \sum_{\tau=0}^{m} f_\tau \phi_{zx}(\tau)$$

$$= \phi_{zz}(0) - \sum_{\tau=0}^{m} f_\tau \phi_{zx}(\tau) \qquad (6\text{-}7)$$

We shall find it convenient to normalize expression (6-7) in such a form that the value of the minimum error energy or error power, as the case may be, will always lie between zero and unity. We do this by dividing both sides of equation (6-7) through by $\phi_{zz}(0)$ (which is never zero except in the trivial case of a desired output of zero for all t) and obtain

$$\frac{J_{\min}}{\phi_{zz}(0)} = 1 - \sum_{\tau=0}^{m} f_\tau \frac{\phi_{zx}(\tau)}{\phi_{zz}(0)}$$

Letting $J_{\min}/\phi_{zz}(0) = E$ and $\phi_{zx}(\tau)/\phi_{zz}(0) = \phi'_{zx}(\tau)$, we can write

$$E = 1 - \sum_{\tau=0}^{m} f_\tau \phi'_{zx}(\tau) \qquad (6\text{-}8)$$

Since E is a sum of squares [see equation (6-4)], it can never be negative. Moreover, E can never be greater than unity, because the value of 1 for E can always be obtained by letting the filter f_τ be identically zero. Hence, we have that

$$0 \leq E \leq 1$$

The quantity E is called the *normalized mean square error*. If it is zero, the filter performs perfectly; that is, there is complete agreement between the desired output z_t and the actual output y_t. On the other hand, if E is unity, there is no agreement between desired output and the actual output, and this situation corresponds to the worst possible case. For fixed values of the parameters required for the specification of its properties, the least-squares filter is that filter for which E is a minimum. However, in the overall design of the filter, we are allowed to vary these parameters subject to constraints imposed by the physical situation. Obviously, then, we shall as a rule seek to find those values of the parameters that yield a least-squares filter f_t for which E is as small as possible.

For many applications it is more convenient to use the one's complement of E, namely, $1 - E$, to measure filter performance. This quantity we call the filter performance parameter, P (see also Chapter 7).

The filter performance for a given desired output z_t always improves as the number of coefficients in the least-squares filter f_t ($t = 0, 1, 2, \ldots, m$) increases. Furthermore, the performance of a given fixed-length filter can in most cases be considerably improved by *delaying* the desired output z_t with respect to the input x_t, as will be described in Chapter 7.

The normal equations (6-3) may be solved by standard techniques. However, this approach becomes cumbersome for the large values of m occurring in practice. For these cases the system (6-3) is solved by the very efficient Toeplitz method, which makes use of the high degree of symmetry that this system possesses. The details of this "Toeplitz recursion" are described in Appendix 6-2.

Consideration of Some Simplified Examples

As an aid to those readers who have found the foregoing mathematics rather difficult, we shall now discuss a simple application of the least-squares filter. Suppose that the input signal is the 2-length wavelet (b_0, b_1), that the filter f_t has two coefficients (f_0, f_1), and that the desired output signal is the 3-length wavelet (d_0, d_1, d_2). The actual output c_t is then obtained by carrying out the convolution $c_t = b_t * f_t$, which is

$$(c_0, c_1, c_2) = (b_0, b_1) * (f_0, f_1)$$
$$= (f_0 b_0, f_0 b_1 + f_1 b_0, f_1 b_1)$$

Now we wish to determine the filter coefficients by minimizing the energy:

$$J = E\{(d_t - c_t)^2\} = (d_0 - c_0)^2 + (d_1 - c_1)^2 + (d_2 - c_2)^2$$
$$= (d_0 - f_0 b_0)^2 + (d_1 - f_0 b_1 - f_1 b_0)^2 + (d_2 - f_1 b_1)^2$$

Since we are here dealing with wavelets, the operation $E\{\cdot\}$ implies a simple sum of squares. Setting the partial derivatives of J with respect to f_0 and f_1

equal to zero, we get the set of simultaneous linear equations

$$f_0(b_0^2 + b_1^2) + f_1(b_0 b_1) = d_0 b_0 + d_1 b_1$$
$$f_0(b_1 b_0) + f_1(b_0^2 + b_1^2) = d_1 b_0 + d_2 b_1$$

which we can also write in the form

$$\phi_{bb}(0) f_0 + \phi_{bb}(1) f_1 = \phi_{ab}(0)$$
$$\phi_{bb}(1) f_0 + \phi_{bb}(0) f_1 = \phi_{ab}(1)$$

These normal equations are a particular case of the more general set given by equation (6-3), if in that expression we set $m = 1$, $b_t = x_t$, $z_t = d_t$, and recall that $\phi_{bb}(-\tau) = \phi_{bb}(\tau)$.

As a numerical example, suppose that we let $b_t = (2, 1)$ and $d_t = (1, 0, 0)$. In other words, we seek to find the 2-length filter (f_0, f_1) which will convert the input wavelet $(2, 1)$ into a unit spike with zero delay with respect to the beginning of the input. Then we have

$$5f_0 + 2f_1 = 2$$
$$2f_0 + 5f_1 = 0$$

We find that the least-squares filter is now given by

$$f_t = (f_0, f_1) = \left(\frac{10}{21}, -\frac{4}{21} \right)$$

The actual output c_t is

$$(c_0, c_1, c_2) = (2, 1) * \left(\frac{10}{21}, -\frac{4}{21} \right) = \left(\frac{20}{21}, \frac{2}{21}, -\frac{4}{21} \right)$$

while the energy of the error between desired output d_t and actual output c_t becomes

$$J_{min} = \left(1 - \frac{20}{21} \right)^2 + \left(0 - \frac{2}{21} \right)^2 + \left(0 - \left[-\frac{4}{21} \right] \right)^2 = \frac{1}{21} \doteq 0.048$$

Since $\phi_{dd}(0) = 1$ in this case (our desired output is a single unit spike), we have from the preceding section that

$$E = \frac{J_{min}}{\phi_{dd}(0)} = \frac{1}{21}$$

This result may also be obtained from the alternative expression (6-8),

$$E = 1 - \sum_{\tau=0}^{1} f_\tau \phi_{ab}(\tau) = 1 - \left(\frac{10}{21} \right) \cdot (2) - \left(-\frac{4}{21} \right) \cdot (0) = \frac{1}{21}$$

We observe that E is quite small, so that the least-squares filter f_t does a very good job in contracting the input wavelet (2, 1) into a unit spike. The results of this simple illustrative calculation are shown in Figure 6-2(a).

The memory function of this spiking filter is only 2-length. We know that the value of E will decrease further as m grows. In fact, because the input

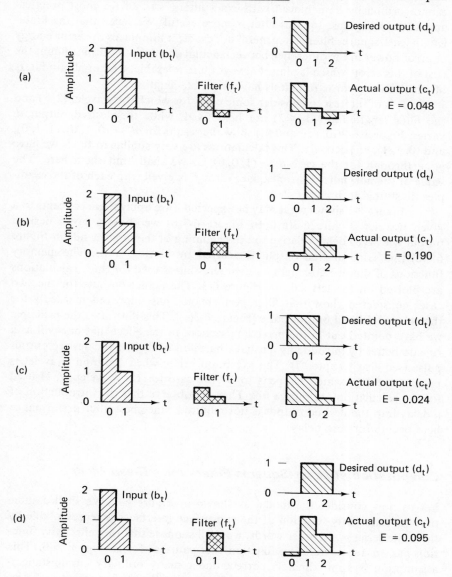

Figure 6-2. Various examples of least-squares filtering for the input pulse (2, 1).

wavelet (2, 1) is minimum-delay, as $m \rightarrow \infty$ the spiking filter will approach the exact inverse filter, which we may obtain by methods described in Chapter 4. One of the great advantages of the present treatment is that we can as a rule obtain very satisfactory filter performance with least-squares filters with memory functions of the order of the duration of the input wavelet. This cannot ordinarily be achieved with exact filters, so that for most practical applications the least-squares filter is more useful. We recall that the finite-length least-squares filter is in general not exact; it minimizes the error energy or error power in the difference between actual and desired output. But at the cost of this error, which is almost always quite tolerable, we can obtain filters of which the response functions have tractable lengths.

It is of interest to consider some cases for which the wavelet (2, 1) and the filter length $(m + 1 = 2)$ are held fixed, while the desired output d_t varies. Figures 6-2(b), (c), and (d) show the results for $d_t = (0, 1, 0)$, $(1, 1, 0)$, and $(0, 1, 1)$, respectively. The calculations are very similar to those we have gone through for the case $d_t = (1, 0, 0)$, so we shall omit them here. The value of the normalized mean square error E is given with each of the examples illustrated in Figure 6-2.

Figure 6-3 shows what may be expected if we apply these concepts to a given wavelet [Figure 6-3(a)]. In Figure 6-3(b) we show a given desired output with zero delay relative to the beginning of the input, while in Figure 6-3(c) this same desired output is delayed by 11 time units. The memory functions of the two 20-length least-squares filters used for these calculations are plotted on the left side of Figure 6-3. The actual outputs for the two cases considered show that filter performance has improved markedly for the delayed desired output pulse [Figure 6-3(c)]. This illustrates the principle we have pointed out in the preceding section, to the effect that one will as a rule do better by introducing a usually harmless delay into the desired output pulse (see also Chapter 7). The exception arises when the input wavelet is minimum-delay, which happens to be the case for the wavelet (2, 1) used for the calculations of Figure 6-2. Then we observe that the introduction of a delay into the desired output is not required, since best filter performance here occurs for zero delay.

Computation of Least-Squares Filters for a Given Model

Let us now consider the model as shown in Figure 6-4. We shall assume that the reflectivity function of the subsurface interfaces is representable by the white series ϵ_t. In other words, we shall suppose that the reflectivity function has an autocorrelation function that vanishes for all lags $\tau \neq 0$. This assumption is approximately correct under many ordinary circumstances. Following usual practice, we suppose further that the ideal seismic time series

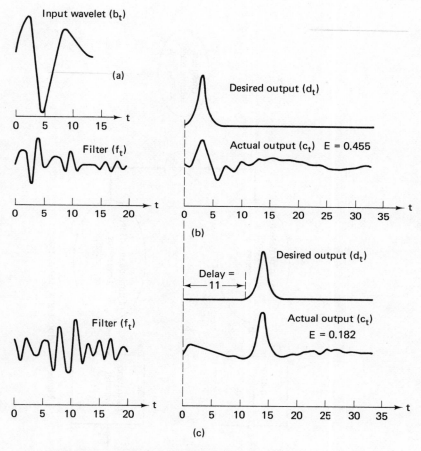

Figure 6-3. Two examples of least-squares filtering of a given wavelet.

u_t is obtainable by convolving ϵ_t with a wavelet of known shape, which we call the characteristic wavelet b_t. Thus, we assume that all individual isolated seismic events have the shape b_t. Although the characteristic wavelet b_t is time-varying, we make the usual assumption that it is approximately time-invariant for the section of seismic trace under consideration. The seismic trace u_t is then given by the wavelet representation

$$u_t = b_t * \epsilon_t$$

We next assume that a further white series β_t exists, which represents disturbances in the observed seismic trace stemming from other causes, such as the effects of buried scatterers, wind and instrument noise, and so on.

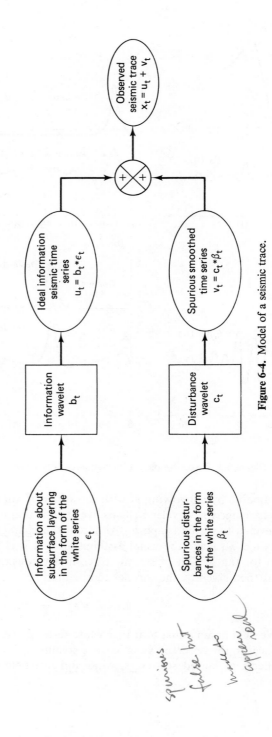

Figure 6-4. Model of a seismic trace.

spurious
false but
made to
appear real

Since pure white noise is not ordinarily present on actual field recordings, we suppose also that the spurious white series β_t is smoothed, or filtered, by convolving it with a second wavelet, which we shall call the disturbance wavelet c_t. We thus obtain the spurious smoothed time series v_t with wavelet representation

$$v_t = c_t * \beta_t$$

The time series u_t and v_t are finally added, and this result yields our model of the observed seismic trace x_t, given by

$$x_t = u_t + v_t = (\epsilon_t * b_t) + (\beta_t * c_t)$$

Here we must stress the fact that our representation of a seismic trace is of course not unique, as will become evident in subsequent chapters. The present model's usefulness is primarily in its ability to further clarify the concepts we have presented above.

Now we wish to design a filter to operate on a trace x_t to give us some desired output. For example, suppose that we know the characteristic wavelet b_t, the power P of the subsurface layering white series ϵ_t, and the autocorrelation of the spurious time series v_t. Then we might want to shape the characteristic wavelet b_t into some other wavelet d_t. The shape of d_t will depend on the purpose we have in mind. Thus, we might want d_t to have less breadth than b_t, which is something we will want to do if we wish to increase the resolution of overlapping events, such as those due to stratigraphic pinchouts. At the same time, we do not want the filter to respond very much to the spurious time series v_t.

Let us now attempt to set up the normal equations (6-3), which we need to solve for the least-squares filter f_t. The input is given by $x_t = u_t + v_t$. Its autocorrelation is

$$\phi_{xx}(\tau) = E\{x_{t+\tau}x_t\} = E\{(u_{t+\tau} + v_{t+\tau})(u_t + v_t)\}$$
$$= E\{u_{t+\tau}u_t\} + E\{v_{t+\tau}u_t\} + E\{u_{t+\tau}v_t\} + E\{v_{t+\tau}v_t\}$$

We now assume that the white series ϵ_t used in the construction of u_t and the white series β_t used in the construction of v_t are uncorrelated. This means that v_t and u_t are themselves uncorrelated; that is, the crosscorrelations $\phi_{vu}(\tau)$ and $\phi_{uv}(\tau)$ vanish for all τ, so that we have

$$\phi_{xx}(\tau) = \phi_{uu}(\tau) + \phi_{vv}(\tau)$$

But from Appendix 6-1 we have that

$$\phi_{uu}(\tau) = P\phi_{bb}(\tau)$$

where P is the power of the white series ϵ_t and $\phi_{bb}(\tau)$ is the autocorrelation of the characteristic wavelet b_t. Thus, the autocorrelation of the trace x_t is

$$\phi_{xx}(\tau) = P\phi_{bb}(\tau) + \phi_{vv}(\tau) \tag{6-9}$$

which may be computed from the known quantities P, $\phi_{bb}(\tau)$, and $\phi_{vv}(\tau)$.

Next, we must compute the crosscorrelation of the desired output with the input. We let the desired output z_t be

$$z_t = \epsilon_t * d_t$$

where we recall that the white series ϵ_t represents the reflectivity function of the subsurface interfaces. If we are concerned with increased resolution, then d_t represents a wavelet sharper than the original, known characteristic wavelet b_t. At this point, however, d_t is more general, and its shape can be specified later for any particular application. The crosscorrelation of the desired output with the input, $\phi_{zx}(\tau)$, is then

$$
\begin{aligned}
\phi_{zx}(\tau) = E\{z_{t+\tau}x_t\} &= E\{z_{t+\tau}(u_t + v_t)\} \\
&= E\{z_{t+\tau}u_t\} + E\{z_{t+\tau}v_t\} \\
&= E\{z_{t+\tau}u_t\}
\end{aligned}
$$

ϵ_t, v_t uncorrelated

The last step above follows from the fact that the white series ϵ_t, used in the construction of z_t and the white series β_t, used in the constsuction of v_t, are uncorrelated. But since $z_t = \epsilon_t * d_t$ and $u_t = \epsilon_t * b_t$, we can write

$$
\begin{aligned}
\phi_{zx}(\tau) &= E\{(\epsilon_{t+\tau} * d_{t+\tau})(\epsilon_t * b_t)\} \\
&= PE\{d_{t+\tau}b_t\} \\
&= P\phi_{db}(\tau) \tag{6-10}
\end{aligned}
$$

where $\phi_{db}(\tau)$ is the crosscorrelation of the wavelet d_t with the characteristic wavelet b_t. We see that $\phi_{zx}(\tau)$ may be computed from the *known* quantities P, d_t, and b_t. The last step in equation (6-10) follows by arguments similar to those used in Appendix 6-1

We have thus been able to derive expressions for $\phi_{xx}(\tau)$ and $\phi_{zx}(\tau)$ in terms of known quantities. These we can insert into the system of normal equations (6-3) and solve for the least-squares filter f_t. Substituting equations (6-9) and (6-10) into equations (6-3), we have

$$\sum_{\tau=0}^{m} f_\tau[P\phi_{bb}(j - \tau) + \phi_{vv}(j - \tau)] = P\phi_{db}(j)$$

$$\text{for } j = 0, 1, 2, \ldots, m \tag{6-11}$$

The numbers f_t obtained by solving this set of $(m + 1)$ linear simultaneous equations are the <u>weighting coefficients of the least-squares filter we seek.</u>

We shall now consider a few numerical applications of the normal equations (6-11). For this purpose we have generated a particular example of the trace x_t according to the specifications given in Figure 6-4. We recall that our model seismic trace is given by

$$x_t = u_t + v_t = (\epsilon_t * b_t) + (\beta_t * c_t)$$

As we have stated earlier, the white series ϵ_t represents the reflectivity function of the layered earth, and $u_t = \epsilon_t * b_t$ is then the familiar, noise-free seismic trace. The white series ϵ_t and the ideal seismic time series u_t are shown in Figure 6-5, where the series ϵ_t is represented by a set <u>of 10 impulses</u> of various strengths and arrival times. The characteristic wavelet b_t is the one we used in Figure 6-3(a). The spurious series v_t was generated by smoothing a set of random numbers of zero mean with an assumed 3-length disturbance wavelet c_t. In practice we will rarely know the shapes of c_t or of v_t, but only the autocorrelation $\phi_{vv}(\tau)$, which is sometimes estimable by means of field seismic noise studies. For our present artificial example $\phi_{vv}(\tau)$, which we need in order to set up the normal equations (6-11), was computed directly from the assumed series v_t.

Figure 6-6 shows the model seismic trace x_t, where we display again the impulses ϵ_t. A comparison of Figures 6-5 and 6-6 permits us to see to

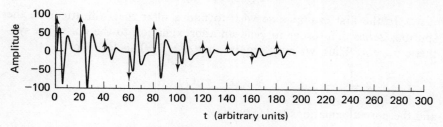

Figure 6-5. White series ϵ_t (represented by impulses), and the noise-free synthetic trace y_t, drawn as a smooth curve.

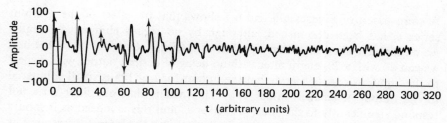

Figure 6-6. Model of the observed seismic trace x_t, shown together with the impulse white series ϵ_t.

what extent the noise-free synthetic trace u_t has become perturbed by the spurious series v_t. We notice that, although the last value of the u_t series occurs at $t = 195$, the x_t series extends through $t = 302$. Thus, the last 107 values of the x_t series are actually points from the spurious series v_t. The v_t series was purposely made longer than the u_t series to permit us to study the response of our least-squares filters to pure spurious energy.

We have selected three examples to illustrate the action of certain least-squares filters on the trace x_t. These three cases are summarized in Table 6-1. A 20-length filter was computed for each case (so that $m = 19$), and each filter so obtained was convolved with the input x_t. The actual outputs are shown in Figures 6-7, 6-8, and 6-9, where we also display the white series, or noise-free impulsive response ϵ_t.

TABLE 6-1. SUMMARY OF THE THREE EXAMPLES OF LEAST-SQUARES FILTERING OF THE TRACE x_t

Example Number	Input	Desired Output	Illustrated by Figure Number
1	$x_t = u_t + v_t$	$z_t = u_t$	6-7
2	$x_t = u_t + v_t$	$z_t = u_{t-k}, k > 0$	6-8
3	$x_t = u_t + v_t$	$z_t = w_{t-k}, k > 0$	6-9

In the first example we wish to find a filter that will attenuate the spurious series v_t in order to yield an approximation to the noise-free synthetic trace u_t. Thus, we must require that $d_t = b_t$. This leads to

$$\phi_{db}(j) = \phi_{bb}(j)$$

and the normal equations (6-11) become

$$\sum_{\tau=0}^{m} f_\tau [P\phi_{bb}(j - \tau) + \phi_{vv}(j - \tau)] = P\phi_{bb}(j) \qquad \text{for } j = 0, 1, 2, \ldots, m$$

A comparison of Figures 6-6 and 6-7 shows that the effect of the spurious series v_t has been diminished, although by no means eliminated. Ideally, we would have liked to obtain a replica of u_t (Figure 6-5), but this is the best we can do with a 20-length filter with no delay of desired output with respect to the input. We also observe that the output for $t > 195$ is small compared to the output at smaller values of t. This means that our filter does not respond significantly to the spurious series v_t, and this is indeed as it should be, since we recall that our design criteria required the filter f_t to respond as little as possible to spurious noise.

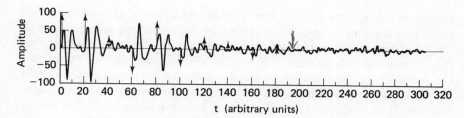

Figure 6-7. Actual output for the filter of Example 1, shown together with the impulse white series ϵ_t.

Let us next investigate the effect of delaying the desired output with respect to the input (Example 2). We again wish to obtain an approximation to the noise-free seismic trace u_t. These requirements lead to

$$d_t = b_{t-k}, \qquad k > 0$$

where k is the delay of the desired output with respect to the input. We may substitute this relation in equation (6-10) and obtain

$$\phi_{zx}(\tau) = PE\{d_{t+\tau}b_t\}$$
$$= PE\{b_{t-k+\tau}b_t\}$$

which can be written

$$\phi_{zx}(\tau) = P\phi_{bb}(\tau - k)$$

For this case, then, the normal equations (6-11) become

$$\sum_{\tau=0}^{m} f_\tau[P\phi_{bb}(j-\tau) + \phi_{vv}(j-\tau)] = P\phi_{bb}(j-k)$$
$$\text{for } j = 0, 1, 2, \ldots, m; \quad k > 0$$

The actual output shown in Figure 6-8 was computed with a 20-length filter for a delay $k = 11$ time units. To facilitate comparison with this actual

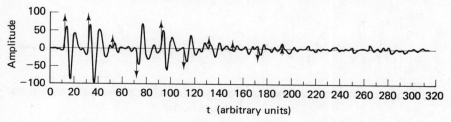

Figure 6-8. Actual output for the filter of Example 2, shown together with the impulse white series ϵ_t. The series ϵ_t has been delayed by 11 units with respect to the time origin.

output, the white series ϵ_t has been plotted at a delay of 11 time units with respect to the time origin. We observe that the agreement with the desired output u_t (Figure 6-5) is now much better than in the case with the 20-length filter computed with no delay (see Figure 6-7). In particular, we notice that the shape of the large-amplitude events in the early portion of the output trace of Figure 6-8 has been quite well restored. However, the smaller-amplitude events occurring at $t \geq 150$ are much less clearly resolved. This simply means that the S/N ratio for these events is too small for the least-squares filter to be effective.

Finally, we shall see what happens if we not only delay the desired output, but require in addition that the wavelet d_t in this desired output be sharper than the wavelet b_t of the input (Example 3). These requirements lead to

$$z_t = \epsilon_t * d_{t-k} = w_{t-k}$$

Therefore, we must replace d_t by d_{t-k} in equation (6-10). We then obtain

$$\phi_{zx}(\tau) = PE\{d_{t-k+\tau}b_t\}$$
$$= P\phi_{db}(\tau - k)$$

Thus, the normal equations (6-11) become

$$\sum_{\tau=0}^{m} f_\tau[P\phi_{bb}(j - \tau) + \phi_{vv}(j - \tau)] = P\phi_{db}(j - k)$$

$$\text{for } j = 0, 1, 2, \ldots, m; \quad k > 0$$

In this example we assume again that $k = 11$. For the desired wavelet d_t we take the pulse shown in Figure 6-3(b). The result of convolving the 20-length least-squares filter calculated for this case with the input x_t is shown in Figure 6-9. Here we have again plotted the white series ϵ_t at a delay of 11 time units with respect to the time origin. We observe that the larger-

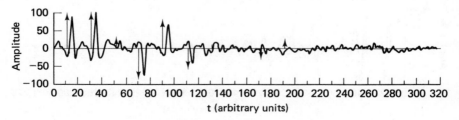

Figure 6-9. Actual output for the filter of Example 3, shown together with the impulse white series ϵ_t. The series ϵ_t has been delayed by 11 units with respect to the time origin.

amplitude events for $t \leq 150$ are sharpened in comparison to corresponding events of the noise-free synthetic trace u_t (see Figure 6-5). This sharpening has been achieved in spite of the presence of the spurious series v_t in the input (see Figure 6-6). Again we note that the smaller-amplitude events for $t \geq 130$ are beyond the resolution capacity of our present filter.

The three examples we have discussed here demonstrate the versatility of the least-squares filter. We note in particular the usefulness of a delay introduced into the desired output z_t, which, as we can appreciate from Figures 6-8 and 6-9, is quite harmless as far as the output is concerned; yet filter performance is improved considerably by this artifice.

It is of some interest to consider the values of the normalized mean square error, E, for the examples that we have just treated. This quantity can be calculated with the aid of equation (6-8). The results are displayed in Table 6-2.

TABLE 6-2. VALUES OF THE NORMALIZED
MEAN SQUARE ERROR E FOR THE THREE
EXAMPLES TREATED IN THIS SECTION

Example Number	E
1	0.158
2	0.122
3	0.427

A comparison of the values of E for Examples 1 and 2 confirms our earlier findings, namely that the introduction of a delay into z_t reduces the error power. In the case of Example 3, we had required shaping of the characteristic wavelet b_t into a new, sharper wavelet d_t. This need was stipulated in addition to the introduction of a delay into z_t identical to that of Example 2. Since this means that we are now asking more of the filter, we may expect that the value of E will increase. Table 6-2 indicates that this is indeed the case.

Concluding Remarks

Our main concern in the present chapter has been to demonstrate both analytically and numerically that the digital least-squares filter and related concepts drawn from time-series theory constitute an interesting and valuable approach to the problem of seismic analysis. The design criteria for a given filter are usually simple to formulate, and the calculation of the corresponding filter unit impulse response functions can in general be carried out rapidly.

the autocorrelation function of
a stationary time series

Let the stationary time series u_t be given by

$$u_t = \sum_{s=0}^{\infty} b_s \epsilon_{t-s}$$

where b_t is the characteristic wavelet and ϵ_t is white noise of zero mean and power P; that is,

$$E\{\epsilon_t\} = 0 \qquad E\{\epsilon_s \epsilon_t\} = \begin{cases} P & s = t \\ 0 & s \neq t \end{cases} \tag{6-12}$$

The autocorrelation of the stationary time series u_t is

$$\phi_{uu}(\tau) = E\{u_{t+\tau} u_t\}$$

If we write

$$u_t = \sum_{s=0}^{\infty} b_s \epsilon_{t-s} = b_0 \epsilon_t + b_1 \epsilon_{t-1} + b_2 \epsilon_{t-2} + \cdots$$

$$u_{t+\tau} = \sum_{s=0}^{\infty} b_s \epsilon_{t+\tau-s} = b_0 \epsilon_{t+\tau} + \cdots + b_\tau \epsilon_t + b_{\tau+1} \epsilon_{t-1} + b_{\tau+2} \epsilon_{t-2} + \cdots$$

we find that

$$\phi_{uu}(\tau) = E\{b_0 b_\tau \epsilon_t^2 + b_1 b_{\tau+1} \epsilon_{t-1}^2 + \cdots + \text{cross-product terms in } \epsilon_t\}$$

Since the characteristic wavelet coefficients b_t are constant, they can be taken outside of the averaging symbol E. Then we see by equation (6-12) that all the cross-product terms involving the white noise vanish, and all the auto-product terms give the constant value P; that is,

$$\phi_{uu}(\tau) = b_0 b_\tau E\{\epsilon_t^2\} + b_1 b_{\tau+1} E\epsilon\{_{t-1}^2\} + \cdots$$
$$= P(b_0 b_\tau + b_1 b_{\tau+1} + \cdots)$$
$$= P \sum_{t=0}^{\infty} b_t b_{t+\tau}$$

We recognize that the summation above yields the autocorrelation of the wavelet. Hence, this expression may be written

$$\phi_{uu}(\tau) = P\phi_{bb}(\tau) \tag{6-1}$$

which is equation (6-1).

---

the Toeplitz recursion

The system of normal equations (6-3)

$$\sum_{\tau=0}^{m} f_\tau \phi_{xx}(j-\tau) = \phi_{zx}(j) \qquad (j=0,1,\ldots,m)$$

can be solved by a very efficient recursive procedure, which we shall now describe. The method is based on the classic work of Levinson (1947), although it differs from his approach in several respects. At this point, it will be convenient to introduce new notation for both the autocorrelation ϕ_{xx} and for the crosscorrelation ϕ_{zx}, which we replace by the symbols r and g, respectively, so that equations (6-3) become

$$\sum_{\tau=0}^{m} f_\tau r_{j-\tau} = g_j \qquad (j=0,1,\ldots,m)$$

These normal equations can also be written as a set of $(m+1)$ linear simultaneous equations,

$$
\begin{aligned}
f_0 r_0 + f_1 r_1 + \cdots + f_m r_m &= g_0 \\
f_0 r_1 + f_1 r_0 + \cdots + f_m r_{m-1} &= g_1 \\
&\cdots \\
f_0 r_m + f_1 r_{m-1} + \cdots + f_m r_0 &= g_m
\end{aligned}
\tag{6-13}
$$

where the autocorrelation coefficients $r_0, r_1, \ldots, r_m$, as well as the right-hand-side crosscorrelation coefficients $g_0, g_1, \ldots, g_m$, represent the known quantities, and the filter coefficients $f_0, f_1, \ldots, f_m$ represent the unknown quantities.

These relations can be expressed in still another form, namely as the matrix equation

$$\mathbf{R}_m \mathbf{f} = \mathbf{g}$$

where $\mathbf{R}_m$ is the $(m + 1)$ by $(m + 1)$ autocorrelation or Toeplitz matrix,

$$\mathbf{R}_m = \begin{bmatrix} r_0 & r_1 & r_2 & \cdots & r_m \\ r_1 & r_0 & r_1 & \cdots & r_{m-1} \\ r_2 & r_1 & r_0 & \cdots & r_{m-2} \\ & & \vdots & & \\ r_m & r_{m-1} & r_{m-2} & \cdots & r_0 \end{bmatrix}$$

and where $\mathbf{f}$ and $\mathbf{g}$ are $(m + 1)$-length column vectors with elements given by the filter coefficients f_τ, $\tau = 0, 1, \ldots, m$, and the crosscorrelation coefficients g_j, $j = 0, 1, \ldots, m$, respectively.

Now it is possible to take advantage of the special form of the Toeplitz matrix $\mathbf{R}_m$ in order to reduce the computational work required for the solution of the set of simultaneous equations (6-13). More specifically, any autocorrelation matrix has the property that all the elements on any given diagonal are the same, and hence is a Toeplitz matrix. That is, the zero-lag autocorrelation coefficient r_0 is the element that appears on the main diagonal; the first-lag autocorrelation coefficient r_1 is the element that appears on the first superdiagonal as well as on the first subdiagonal, and so on. Hence, the entire $(m + 1) \times (m + 1)$ autocorrelation matrix $\mathbf{R}_m$ is one that does not involve $(m + 1)^2$ distinct elements, but only $(m + 1)$ distinct elements, namely $r_0, r_1, r_2, \ldots, r_m$. Let us now see how we can take advantage of such a structure. Specifically, we will show that this Toeplitz structure of the autocorrelation matrix makes possible the following Toeplitz recursive method for the solution of the normal equations (6-13).

The Toeplitz recursive procedure solves the equations in a step-by-step manner. The step $n = 0$ is given as the initial condition, and then the steps $n = 1, n = 2, \ldots, n = m$ are done successively in a recursive manner. The desired filter coefficients are the ones that result on the completion of the final step, namely the step $n = m$. A description of one of these steps will provide the necessary information to enable one to program the method for a digital computer. We will suppose, therefore, that we have completed step n, and we wish to do step $n + 1$.

The completion of step $k = n$ requires that the machine has computed and retained in storage the numerical values of the following quantities:

$$a_{n0}, a_{n1}, \ldots, a_{nn}; \quad \alpha_n, \quad \beta_n$$

and

$$f_{n0}, f_{n1}, \ldots, f_{nn}; \quad \gamma_n$$

By definition, we suppose that these quantities satisfy the matrix equations

$$(a_{n0}, a_{n1}, \ldots, a_{nn}, 0)\mathbf{R}_{n+1} = (\alpha_n, 0, \ldots, 0, \beta_n) \qquad (6\text{-}14)$$

and

$$(f_{n0}, f_{n1}, \ldots, f_{nn}, 0)\mathbf{R}_{n+1} = (g_0, g_1, \ldots, g_n, \gamma_n) \qquad (6\text{-}15)$$

where the parentheses enclose $1 \times (n + 2)$ row vectors, and where $\mathbf{R}_{n+1}$ is the $(n + 2) \times (n + 2)$ autocorrelation matrix.

Because of the Toeplitz structure of the autocorrelation matrix, (6-14) may be manipulated into the equivalent form given by

$$(0, a_{nn}, \ldots, a_{n1}, a_{n0})\mathbf{R}_{n+1} = (\beta_n, 0, \ldots, 0, \alpha_n) \qquad (6\text{-}16)$$

Let us now multiply (6-16) by a constant k_n, as yet undetermined; we then will add the result to (6-14). We obtain

$$(a_{n0}, a_{n1} + k_n a_{nn}, \ldots, a_{nn} + k_n a_{n1}, k_n a_{n0})\mathbf{R}_{n+1}$$
$$= (\alpha_n + k_n \beta_n, 0, \ldots, 0, \beta_n + k_n \alpha_n) \qquad (6\text{-}17)$$

Now we want (6-17) to be identical to

$$(a_{n+1,0}, a_{n+1,1}, \ldots, a_{n+1,n}, a_{n+1,n+1})\mathbf{R}_{n+1} = (\alpha_{n+1}, 0, \ldots, 0, 0) \qquad (6\text{-}18)$$

In order for (6-17) and (6-18) to be identical, we must first make the requirement that

$$\beta_n + k_n \alpha_n = 0$$

This equation allows us to determine the constant k_n; that is, we may compute k_n by the formula

$$k_n = -\frac{\beta_n}{\alpha_n} \qquad (6\text{-}19)$$

The identity of (6-17) and (6-18) together with our knowledge of k_n, then allows us to compute the quantities given by

$$a_{n+1,0} = a_{n0}$$
$$a_{n+1,1} = a_{n1} + k_n a_{nn}$$
$$\cdot$$
$$\cdot$$
$$\cdot$$
$$a_{n+1,n} = a_{nn} + k_n a_{n1}$$
$$a_{n+1,n+1} = k_n a_{n0}$$

$(6\text{-}20)$

as well as

$$\alpha_{n+1} = \alpha_n + k_n \beta_n \tag{6-21}$$

If we define the z transforms $A_n(z)$ and $A_{n+1}(z)$ as

$$A_n(z) = a_{n0} + a_{n1}z + \cdots + a_{nn}z^n$$

and

$$A_{n+1}(z) = a_{n+1,0} + a_{n+1,1}z + \cdots + a_{n+1,n}z^n + a_{n+1,n+1}z^{n+1}$$

we see that equations (6-20) may be encompassed within the confines of one equation, namely

$$A_{n+1}(z) = A_n(z) + k_n z(a_{nn} + \cdots + a_{n1}z^{n-1} + a_{n0}z^n)$$

which is

$$A_{n+1}(z) = A_n(z) + k_n z^{n+1} A_n\left(\frac{1}{z}\right) \tag{6-22}$$

This equation demonstrates the Toeplitz recursion from the known z transform $A_n(z)$ to the unknown z transform $A_{n+1}(z)$.

Next, if we compute

$$\beta_{n+1} = a_{n+1,0}r_{n+2} + a_{n+1,1}r_{n+1} + \cdots + a_{n+1,n+1}r_1 \tag{6-23}$$

we will have obtained all the quantities in

$$(a_{n+1,0}, a_{n+1,1}, \ldots, a_{n+1,n+1}, 0)\mathbf{R}_{n+2} = (\alpha_{n+1}, 0, \ldots, 0, \beta_{n+1}) \tag{6-24}$$

where the parentheses enclose $1 \times (n + 3)$ vectors, and where $\mathbf{R}_{n+2}$ is the $(n + 3) \times (n + 3)$ autocorrelation matrix. Equations (6-14) and (6-24) are counterparts, where (6-14) pertains to step n, whereas (6-24) pertains to step $n + 1$.

Because of the Toeplitz structure of the autocorrelation matrix, (6-18) is the same as

$$(a_{n+1,n+1}, \ldots, a_{n+1,1}, a_{n+1,0})\mathbf{R}_{n+1} = (0, \ldots, 0, \alpha_{n+1}) \qquad (6\text{-}25)$$

Let us now multiply (6-25) by a constant q_n, as yet undetermined, and then add the result to (6-15). We then obtain

$$(f_{n0} + q_n a_{n+1,n+1}, f_{n1} + q_n a_{n+1,n}, \ldots, f_{nn} + q_n a_{n+1,1}, q_n q_{n+1,0})\mathbf{R}_{n+1}$$
$$= (g_0, g_1, \ldots, g_n, \gamma_n + q_n \alpha_{n+1}) \qquad (6\text{-}26)$$

We want (6-26) to be identical with

$$(f_{n+1,0}, f_{n+1,1}, \ldots, f_{n+1,n}, f_{n+1,n+1})\mathbf{R}_{n+1} = (g_0, g_1, \ldots, g_n, g_{n+1}) \quad (6\text{-}27)$$

In order for (6-26) and (6-27) to be identical, first we must take the requirement that

$$\gamma_n + q_n \alpha_{n+1} = g_{n+1}$$

This equation allows us to determine the constant q_n; that is, we may compute q_n by the formula

$$q_n = \frac{g_{n+1} - \gamma_n}{\alpha_{n+1}} \qquad (6\text{-}28)$$

The identify of equations (6-26) and (6-27), together with our knowledge of q_n, allows us to compute the quantities given by

$$f_{n+1,0} = f_{n0} + q_n a_{n+1,n+1}$$
$$f_{n+1,1} = f_{n1} + q_n a_{n+1,n}$$
$$\vdots$$
$$\qquad (6\text{-}29)$$
$$f_{n+1,n} = f_{nn} + q_n a_{n+1,1}$$
$$f_{n+1,n+1} = q_n a_{n+1,0}$$

If we define the z transforms $F_n(z)$ and $F_{n+1}(z)$ as

$$F_n(z) = f_{n0} + f_{n1}z + \cdots + f_{nn}z^n$$

and

$$F_{n+1}(z) = f_{n+1,0} + f_{n+1,1}z + \cdots + f_{n+1,n}z^n + f_{n+1,n+1}z^{n+1}$$

then equations (6-29) may be encompassed within the confines of one equation, namely

$$F_{n+1}(z) = F_n(z) + q_n z^{n+1} A_{n+1}\left(\frac{1}{z}\right) \qquad (6\text{-}30)$$

This equation demonstrates the recursion from the known z transforms $F_n(z)$ and $A_{n+1}(z)$ to the unknown z transform $F_{n+1}(z)$.

Next, if we compute the quantity

$$\gamma_{n+1} = f_{n+1,0} r_{n+2} + f_{n+1,1} r_{n+1} + \cdots + f_{n+1,n+1} r_1 \qquad (6\text{-}31)$$

with the aid of (6-15), we will have computed all the quantities in

$$(f_{n+1,0}, f_{n+1,1}, \ldots, f_{n+1,n+1}, 0)\mathbf{R}_{n+2} = (g_0, g_1, \ldots, g_{n+1}, \gamma_{n+1}) \qquad (6\text{-}32)$$

where the parentheses enclose $1 \times (n + 3)$ row vectors, and where $\mathbf{R}_{n+2}$ is the $(n + 3) \times (n + 3)$ autocorrelation matrix. Equations (6-15) and (6-32) are counterparts, where (6-15) pertains to step n and (6-32) pertains to step $n + 1$. We have thus completed our task; we have obtained step $n + 1$, given that we had step n.

Let us summarize the computations that one would program on a digital computer. We are given the autocorrelation coefficients

$$r_0, r_1, \ldots, r_m$$

and the right-hand-side coefficients

$$g_0, g_1, \ldots, g_m$$

At the completion of step n we have the quantities

$$a_{n0}, a_{n1}, \ldots, a_{nn}; \alpha_n, \quad \beta_n$$

and

$$f_{n0}, f_{n1}, \ldots, f_{nn}; \quad \gamma_n$$

in storage. To perform step $n + 1$, we make the following computations. We first compute k_n by means of (6-19). Then we compute the quantities

$$a_{n+1,0}, a_{n+1,1}, \ldots, a_{n+1,n}, a_{n+1,n+1}; \alpha_{n+1}, \quad \beta_{n+1}$$

by means of (6-20), (6-21), and (6-23). We next compute q_n by means of (6-28). Then we compute the quantities

$$f_{n+1,0}, f_{n+1,1}, \ldots, f_{n+1,n}, f_{n+1,n+1}; \quad \gamma_{n+1}$$

by means of (6-29) and (6-31). This completes step $n + 1$.

To obtain the final result, the solution of the simultaneous equations (6-13), we must start at step $n = 0$; that is, we start with initial quantities which we take as

$$a_{00} = 1 \qquad \alpha_0 = r_0 \qquad \beta_0 = r_1$$

$$f_{00} = \frac{g_0}{r_0} \qquad \gamma_0 = f_{00} r_1$$

Then we do the recursions for $n = 1$ to $n = m$; the final values obtained for the filter coefficients, namely

$$f_{m, 0}, f_{m, 1}, \ldots, f_{m, m}$$

represent the desired solution $f_0, f_1, \ldots, f_m$ of the normal equations (6-13). Instead of fixing the value of m in advance, it is also possible to monitor the recursions according to the value of the error associated with the filter at each step of the recursions. The recursions can be stopped when this error reaches a certain limit.

The machine time required to solve the simultaneous equations (6-13) for a digital filter with m coefficients is proportional to m^2 for the Toeplitz recursive method, as compared to m^3 for conventional methods of solving simultaneous equations. Another advantage of using the Toeplitz recursive method is that it requires computer storage space proportional to m rather than m^2, as in the case of the conventional methods.

The Toeplitz recursive method has been extended to the matrix-valued case (in which case the autocorrelation matrix has the structure of a block Toeplitz matrix) in order to compute the coefficients of multichannel filters. Also, there is the *Simpson sideways recursion* (see Robinson, 1967b, p. 80), which corresponds to shifting the time index of the right-hand side of the simultaneous equations. This sideways recursion is valuable in many applications, for it represents a relatively inexpensive way of determining the filter that incorporates the optimum time lag between the input signal and the desired output signal. Applications of this sideways recursion will be described in Chapter 7. There we shall show that the sideways recursion can be used to determine the optimum spike position for a spiking filter and the optimum positioning of the desired output for a shaping filter.

the design of least-squares digital wavelet filters

Summary

Signal recordings made with standard filters often afford insufficient resolution for overlapping events. In this chapter, we apply the least-squares theory presented in Chapter 6 to the design of high-resolution digital wavelet filters. Under the assumption that an estimate of the wavelet shape is available, we present techniques that allow one to calculate digital filters that transform this wavelet into one which is sufficiently sharp that it can be distinguished against a background of noise. The two design criteria governing filter performance are filter lag (or delay), and filter memory function duration. The performace of the filter is numerically measurable by a quantity called the filter performance parameter P, where $0 \leq P \leq 1$. The quality of the filter output improves as P approaches unity. We thus seek that combination of lag and memory function duration that maximizes P. This goal can be accomplished by the study of a two-dimensional display of P versus lag and memory function duration. The proposed design techniques are illustrated by means of numerical examples.

Introduction

A major difficulty in time-series analysis is the lack of resolution of recorded events. As a result, there is the need to design high-resolution filters that can increase the amount of detail available on a time series. Viewed in the time domain, our objective becomes the design of filters which convert broad, overlapping wavelets into sharp, clearly resolved pulses. Ideally, we would like to compress a wavelet into a delta function; that is, into a *spike*. Except in special cases, this desire can never be satisfied completely in practice. A further serious obstacle is that a filter which converts a broad wavelet into a pure spike must, in most cases of practical interest, have an infinitely long memory function. However, if we are willing to accept filter output wavelet shapes that are approximations to a pure spike, our search is again a meaningful one. In fact, we need not require the output wavelet to be even an approximation to a spike; since our basic objective is increased resolution, any output wavelet shape which is noticeably sharper than the input wavelet will satisfy our needs. Let us call a filter that shapes an input into some desired output a *shaping filter*. Then the spiking filter above becomes a special case of a shaping filter; that is, our desired shape is, in this case, simply a spike.

Our main objective here is to illustrate the power and flexibility of the high-resolution shaping filter by means of a few simple numerical examples. In so doing, we shall examine some of the design criteria needed in order to obtain satisfactory filter performance.

The Elements of Least-Squares Wavelet Filtering

The basic elements of time-domain least-squares filtering are once again summarized in Figure 7-1 (see also Figure 6-1). Given an input wavelet b_t and a desired output wavelet d_t, the problem is to find a filter f_t whose output $c_t = f_t * b_t$ deviates from the desired output d_t in a least-squares sense. The difference between the desired output d_t and the actual output c_t is the error e_t. The least-squares criterion requires that the energy of the error signal

$$J = \sum_t e_t^2$$

be a minimum; as a result, the filter f_t is calculable in a straightforward way. This operator is known as the _optimum least-squares filter._ It is important to realize that the least-squares filter does not guarantee that the error energy be small in any absolute sense, but only that it be as small as possible consistent with the particular situation at hand. The basic computational procedures allowing one to obtain discrete, time-domain least-squares filters

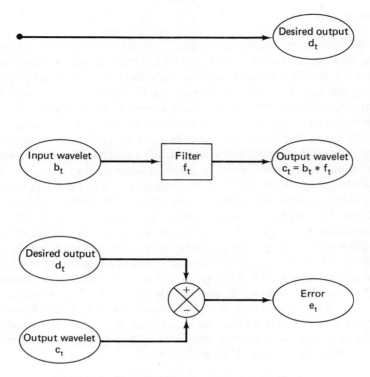

Figure 7-1. Elements of least-squares wavelet filtering.

have already been given in Chapter 6. These methods allow us to calculate time-domain least-squares filters of varying lengths by a fast recursive algorithm, the Toeplitz recursion, which was described in Appendix 6-2.

If the desired output d_t is some arbitrary function of time, then f_t is the memory function of a *shaping* filter. For the special case given by

$$d_t = \begin{cases} 1 & \text{for } t = t_p \\ 0 & \text{for } t \neq t_p \end{cases}$$

where t_p is a particular value of t, then f_t is the memory function of a *spiking filter*.

The Energy Distribution in a Wavelet

Before entering into a discussion of high-resolution-shaping filters, let us consider first how the energy of a wavelet is distributed throughout the time range of its duration. We shall find that the manner in which this energy is distributed provides a basis for understanding how high-resolution filters work. In Figure 7-2, four different wavelets are shown. Each of these wavelets has the same time duration and the same frequency content (see Figure 7-3), so that the amounts of high frequencies, intermediate frequencies, and low frequencies are precisely the same in each wavelet. But, while these wavelets all have a common frequency magnitude spectrum and a common time duration, they differ in their time distribution of energy. The term "magnitude spectrum," as we have used it before, refers to the absolute value of the Fourier transform of the wavelet.

Wavelet (a) has its energy concentrated as closely as possible to its front end; that is, the energy is *delayed* in time the smallest possible amount

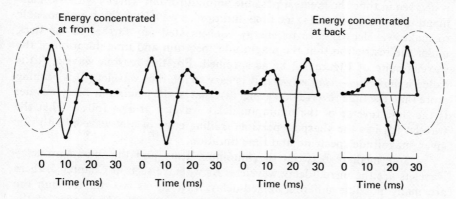

Figure 7-2. Suite of four wavelets having the same magnitude spectrum: (a) minimum delay; (b) and (c) mixed delay; (d) maximum delay.

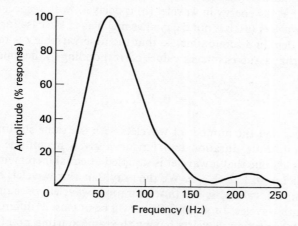

Figure 7-3. Common magnitude spectrum of the wavelet suite of Figure 7-2.

for any wavelet with the magnitued spectrum shown in Figure 7-3. *Under the restriction that the frequency content of the wavelet suite of Figure 7-2 be fixed*, it is not possible for wavelet (a) to have its energy concentrated still closer to its front end. For this reason, the wavelet (a) is called the *minimum-delay* wavelet of the suite. Of course, if we were allowed to add still higher frequencies to the spectrum of Figure 7-3, we could made the wavelet (a) have a still sharper leading edge, but under the restriction that the magnitude spectrum be held constant, the minimum-delay wavelet has a leading edge that is sharper than the leading edge of any other wavelet of the suite.

At the other extreme, we have wavelet (d). This wavelet has its energy concentrated as closely as possible to its back end. In other words, the energy is *delayed* in time the greatest possible amount for *any wavelet* with the same magnitude spectrum and same time duration as wavelet (a). It is not possible to make wavelet (d) have its energy concentrated any farther to the back under the restriction that the magnitude spectrum and time duration of the wavelet suite of Figure 7-2 be as specified. For this reason, wavelet (d) is called a *maximum-delay wavelet*. It is easy to see that wavelet (d) is nothing more than the time reverse of wavelet (a); that is, the maximum-delay wavelet is the time reverse of the minimum-delay wavelet. It also follows that the wavelet (d) has the sharpest possible trailing edge of any wavelet with our given magnitude spectrum and time duration.

Intermediate between the minimum-delay and the maximum-delay wavelets are the mixed-delay wavelets. As we have seen in Chapter 5, there are many possible different mixed-delay wavelets as we run through the transition of energy concentration from the front to the back edge of the wavelet. In Figure 7-2, we show two of the possible mixed-delay wavelets having the same time duration and magnitude spectrum as wavelets (a) and (d). We see that the energy in wavelet (b) is delayed, with respect to the minimum-delay wavelet (a), but not delayed as much as in the case for wavelet (c).

If we deal in discrete time, so that the time variable t is replaced by a time index i that assumes integer values corresponding to the time points that are sampled,

$$i = 0, 1, 2, \ldots,$$

then we know that the number of wavelets with the same starting point and the same finite time duration and having a given magnitude spectrum is finite. Let us assume that a wavelet is sampled at equal increments of time Δt at the points $i = 0, 1, 2, \ldots, n$. We then say that this wavelet, described by $(n + 1)$ sampled values, is an $(n\Delta t)$-duration wavelet, or equivalently an $(n + 1)$-length wavelet. In general, there will be *at most* 2^n different discretely sampled $(n\Delta t)$-duration wavelets having the same starting point and a given magnitude spectrum. There are no other *finite* duration wavelets with the same beginning point and this given magnitude spectrum although there are an infinite number of infinite duration wavelets having the same beginning point and this given magnitude spectrum. In Figure 7-2, the points at which the wavehsapes have been sampled are indicated by heavy black dots. We note that all four of these waveshapes are sampled at 2 ms increments at a total of 16 discrete points, i.e., they are 30 ms duration wavelets. We have here chosen to show only four wavelets of the suite having the magnitude specturm given in Figure 7-3. While all members of a suite have a common magnitude spectrum, their phase spectra are distinct. In fact, the minimum-delay wavelet (a) is the "minimum phase-lag" member of the suite, while the maximum-delay wavelet (d) is the "maximum phase-lag" member of the same suite.

Wavelet Spiking

We wish now to consider the design problem for what we might call a prototype filter, namely, the spiking filter. Let us consider first an isolated wavelet, which is the mixed-delay pulse (b) of Figure 7-2. Our problem is to find a filter that contracts this wavelet to a spike. In theory, this purpose may be achieved *exactly* if we could use a filter with a memory function that is allowed to become infinitely long. For *exact* filter performance, we will also need, in general, to delay the desired spike an infinite amount of time relative to the original pulse. The exception to this rule occurs when the input wavelet is minimum-delay. Then the desired spike does not need to be delayed for *exact* filter performance, although the filter's memory function must still have infinite duration (see Figure 7-11). In practice, we must limit ourselves to digital filters with memory functions of finite duration, and hence, at best, we can achieve our purpose approximately.

Let us suppose that for practical reasons we want first to restrict ourselves to a filter which has a memory function that is of the order of the wavelet's duration. Now we are at liberty to place the desired spike in time wherever we please. For example, in Figure 7-4 we show six possible positions of the spike. In the first case this desired spike is placed at the very beginning of the wavelet we wish to filter; that is, the desired spike's time *lag* with respect to the onset of our wavelet is zero. The five following cases considered are desired spike locations at lags of 8, 16, 24, 32, and 40 ms respectively. The duration of the input wavelet is 30 ms. Spiking filters can be determined for each of the six cases of Figure 7-4. The corresponding filter outputs are hown in the first displays of Figure 7-5; in all cases, the filter duration is 30 ms. We note that the position of the spike is an important factor governing the fidelity with which the ac:ual output resembles the desired spike.

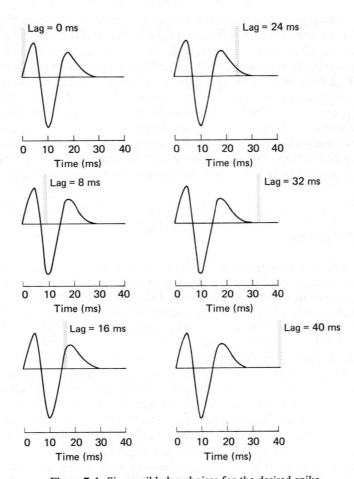

Figure 7-4. Six possible lag choices for the desired spike.

A convenient way to measure the performace of a least-squares fil-ter is to consider the value of its normalized minimum error energy, E. This quantity is the value of the average squared error divided by the zeroth lag of the desired output's autocorrelation function. When the filter performs perfectly, $E = 0$, which means that the desired and actual filter outputs agree for all values of t. On the other hand, the case $E = 1$ corresponds to the worst possible case, that is, there is no agreement at all between desired and actual outputs. Instead of the quantity E, it is desirable for a number of reasons to consider the one's complement of E, namely the filter per-formance parameter, P (see Chapter 6),

$$P = 1 - E$$

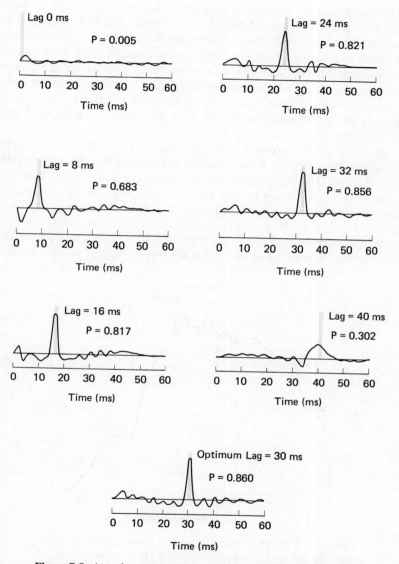

Figure 7-5. Actual outputs of the spiking filters computed for the cases shown in Figure 7-4. Also shown is the actual output of the optimum-lag 30-ms-duration spiking filter of Figure 7-7.

Perfect filter performance then occurs when $P = 1$, while the worst possible situation arises when $P = 0$.

The filter outputs, shown in Figure 7-5, indicate how the value of P varies as the spike is progressively delayed in time. These, as well as all subsequent computations involving a progressive lag or delay of a given desired output are executed with the aid of the Simpson sideways recursion (see Appendix 6-2). We recall that the filters whose outputs are shown here all have 30 ms duration memory functions. *For a constant filter duration*, then, we might suppose that there must exist at least one value of lag at which P is as large as possible. The display of Figure 7-5 shows that the lag of the desired output is of crucial importance in determining filter performance. Our aim must then be to find the *optimum-lag filter* at a given constant (and finite) filter duration.

In Figure 7-6 we show a plot of P versus the lag of the desired output spike for a family of 30 ms duration filters. We observe that this curve exhibits several maxima. The highest point of the curve occurs at a lag of 30 ms, and thus the choice of this lag for the desired spike leads to the optimum-lag 30 ms duration filter. The actual output of this filter is shown in the last picture of Figure 7-5. We note that the performance of the lag 32 ms filter is not much worse than that of the optimum-lag filter; this is so because the P vs. lag curve of Figure 7-6 is nearly flat around its highest point.

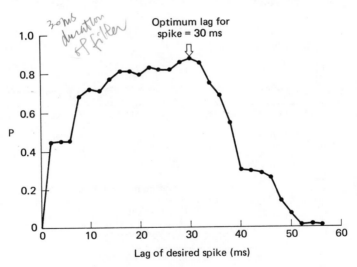

Figure 7-6. Filter performance parameter P as a function of desired spike lag for the 30-ms-duration spiking filters whose outputs are shown in Figure 7-5.

The memory function of the optimum-lag 30 ms duration spiking filter is shown in Figure 7-7(a). Figure 7-7(b) gives the magnitude spectrum of this filter. As we expect, this spectrum is richer in higher frequencies than that of the original pulse, the spectrum of which we also show for comparison. Let us next see what happens as we increase the filter length at con-

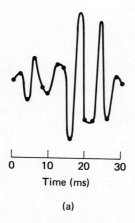

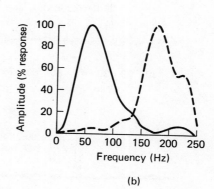

Time (ms)

(a)

Frequency (Hz)

(b)

Figure 7-7. Optimum-lag spiking of the pulse of Figure 7-4. (a) Solid curve—memory function of the optimum-lag 30-ms-duration filter. (b) magnitude spectrum of the pulse of Figure 7-4; dashed curve—magnitude spectrum of the optimum-lag 30-ms-duration filter.

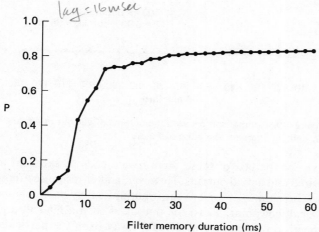

lag = 16 msec

Filter memory duration (ms)

Figure 7-8. Filter performance parameter P as a function of filter duration for a desired spike lag of 16 ms. The pulse to be spiked is the one shown in Figure 7-4.

stant lag. Figure 7-8 shows a plot of P vs. filter length for a desired spike lag of 16 ms. We now observe that this curve *is* monotonic, and that it asymptotically approaches $P = 0.86$ as the filter length becomes larger and larger.

Now, let us examine the performance of some additional optimum-lag filters. In Figure 7-9, we show the actual outputs of the 40, 60, and 100 ms duration optimum-lag spiking filters. In particular, we note that the P

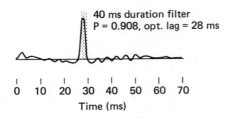

40 ms duration filter
P = 0.908, opt. lag = 28 ms

Time (ms)

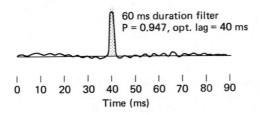

60 ms duration filter
P = 0.947, opt. lag = 40 ms

Time (ms)

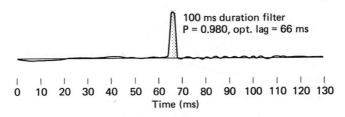

100 ms duration filter
P = 0.980, opt. lag = 66 ms

Time (ms)

Figure 7-9. Comparison between the outputs of various optimum-lag spiking filters for the pulse of Figure 7-4.

value of 0.98 for the last of these three filters leads to excellent agreement between desired and actual outputs. However, a filter response function which is more than three times as long as the input pulse may be highly undesirable from a physical viewpoint. Thus, the output of such a filter at a given time t would depend too much on wavelets far away from the particular wavelet that we wish to spike.

The two important design criteria that we have been discussing here are filter lag and filter length. We can *always* improve performance by increasing the length, but physical considerations prevent us from making this length indefinitely long. On the other hand, we may search for that desired output lag which leads to the highest P value for a given selected filter length. This lag in filter output harms us in no way and, as we have seen, can improve filter performance drastically.

The filter performance parameter P is a function of lag and length. Plots of P vs. lag at constant length (Figure 7-6), or of P vs. length at constant lag (Figure 7-8), are helpful but do not tell us the whole story. Ideally, we would like to inverstigate the dependent of P on lag and length for all physically reasonable values of these variables. One way this can be done is to plot P on a two-dimensional grid, with filter lag as the ordinate and filter length as the abscissa. The array of P values can then be contoured so that we may see at a glance which combinations of lag and length yield optimum filter performance. Such a "contour map" is given by Figure 7-10.

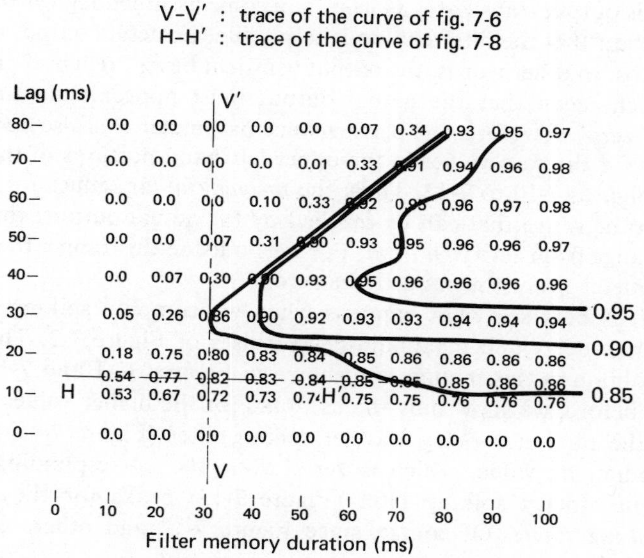

V–V′ : trace of the curve of fig. 7-6
H–H′ : trace of the curve of fig. 7-8

Lag (ms)

			V′							
80—	0.0	0.0	0¦0	0.0	0.0	0.07	0.34	0.93	0.95	0.97
70—	0.0	0.0	0¦0	0.0	0.01	0.33	.91	0.94	0.96	0.98
60—	0.0	0.0	0¦0	0.10	0.33	.92	0.95	0.96	0.97	0.98
50—	0.0	0.0	0¦07	0.31	.90	0.93	.95	0.96	0.96	0.97
40—	0.0	0.07	0¦30	.90	0.93	0.95	0.96	0.96	0.96	0.96
30—	0.05	0.26	0¦86	0.90	0.92	0.93	0.93	0.94	0.94	0.94
20—	0.18	0.75	0¦80	0.83	0.84	0.85	0.86	0.86	0.86	0.86
	0.54	0.77	0¦82	0.83	0.84	0.85	0.85	0.85	0.86	0.86
10— H	0.53	0.67	0¦72	0.73	0.74 H′	0.75	0.75	0.76	0.76	0.76
0—	0.0	0.0	0¦0	0.0	0.0	0.0	0.0	0.0	0.0	0.0

0.95
0.90
0.85

V

0   10   20   30   40   50   60   70   80   90   100
Filter memory duration (ms)

Figure 7-10. P-contour map for spike filtering the mixed delay pulse (b) of Figure 7-2. The gridded numbers correspond to values of the filter performance parameter P.

The input wavelet is again the one shown in Figure 7-4, and our objective is to find a good spiking filter for this wavelet. The section VV' through the contour map then corresponds to the curve of Figure 7-6, while the section HH' corresponds to the curve of Figure 7-8. The map shows only the contours for $P = 0.85$, 0.90, and 0.95. Obviously, we are most interested in the larger P values, for it is there that best filter performance is obtained. This display enables one to select the best combination of filter lag and length by inspection.

If we convolve an input wavelet of length $n + 1$ with a filter memory function of length $m + 1$, the resulting output transient is of length

(Input wavelet length) + (Filter memory length) $- 1 = n + m + 1$

In other words, if we convolve an input wavelet of duration $n\Delta t$ with a filter memory function of duration $m\Delta t$, the resulting output transient is of duration,

(Input wavelet duration) + (Filter memory duration)

$$= (n\Delta t) + (m\Delta t)$$
$$= (n + m)\Delta t.$$

Now, if this output transient is *shorter* than some assumed lag for the desired spike, the best that the filter can do is to produce an actual output consisting only of zeros. In other words, the output transient never "reaches" the desired spike, which mean that the actual output must approximate the desired spike by a zero; therefore, the performance parameter P is also zero. This is the reason for the zero values in the upper left-hand portions of the contour maps of Figures 7-10 to 7-12. Thus, the *meaningful* lag values for a spiking filter lie in the range that can be reached by the actual output; that is, they lie in the range from zero to $n + m$. For lags outside this range, the performance parameter P must necessarily be zero.

Let us next see what happens when we compute spiking filters for the minimum-delay and maximum-delay pulses of Figure 7-2. The contour maps we obtain in this manner are shown in Figures 7-11 and 7-12, respectively. As before, we show only the contours for the higher values of P. The map for the minimum-delay wavelet spiking filter (Figure 7-11) peaks for the minimum lag value, which is zero, while the corresponding map for the maximum-delay spiking filter (Figure 7-12) peaks for the maximum *meaningful* lag values. Of course, since Figure 7-12 and others of its kind

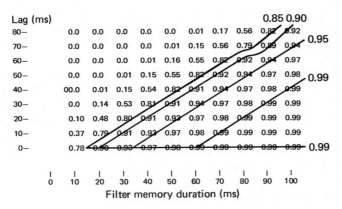

Figure 7-11. *P*-contour map for spike filtering the minimum-delay pulse (a) of Figure 7-2. The gridded numbers correspond to values of the filter performance parameter *P*.

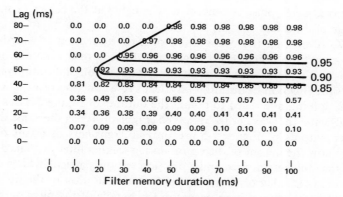

Figure 7-12. *P*-contour map for spike filtering of the maximum-delay pulse (d) of Figure 7-2. The gridded numbers correspond to value of the filter performance parameter *P*.

depict but a portion of the "length–lag" plane, the meaningful lag values for the longer filters are not shown there. Finally, the spiking filter display of Figure 7-10 represents, as we might expect, a case intermediate between the minimum-delay and maximum-delay extremes.

We have previously discussed the problem of energy distribution in a wavelet. Form Figure 7-11 we see that the best filter performance for the minimum-delay case occurs for *zero-lag value*. We are here attempting to spike a minimum-delay wavelet, one whose energy is concentrated near its front end. It is therefore reasonable to expect that a desired spike output at zero-lag value gives better results there than if this spike is more delayed in time. In other words, since the minimum-delay wavelet's energy is concentrated as much as possible near its front end, it makes little sense to lag the desired spike relative to this front end. Similarly, we observe from Figure 7-12 that best performance for the maximum-delay case occurs for the *largest meaningful lag values*. This means that because the energy of the maximum-delay wavelet is concentrated as much as possible near its back end, a spiking filter can do the best job if the desired spike is positioned so that it occurs at the larger values of lag. Strictly speaking, these rules hold only for spiking filters that are sufficiently long in relation to the length of the input wavelet. That is, for sufficiently long spiking filters, the rules are that the best spike position occurs at zero for a minimum-delay input wavelet and at $(n + m)\Delta t$ for a maximum-delay input wavelet. For spiking filters that are short in relation to the length of the input wavelet, there may be deviations from these rules for the best spike position; however, as the length of the spiking filter increases, the best spike position settles down to that position given by the rules. The mixed-delay case of Figure 7-10 then represents an intermediate situation. Inspection of Figure 7-4 tells us that the energy of this mixed-delay pulse is concentrated roughly at the center of the wavelet's range of duration.

We recall that the convolution of a time function of length $m + 1$ with another time function of length $n + 1$ leads to a function of length $m + n + 1$. That is, the convolution of a time function of duration $m\Delta t$, with another time function of duration $n\Delta t$, leads to a function of duration $(m + n)\Delta t$. The spiking filter used to obtain the outputs of Figure 7-5 has a 30 ms duration, while the wavelet to be spiked (Figure 7-4) has the same duration. The filter outputs of Figure 7-5 are thus all transients of 60 ms duration. Consider now the output of the optimum-lag spiking filter of Figure 7-5. We observe that the best desired spike position for the particular filter duration chosen occurs at a lag of 30 ms. Since the total duration of the output is 60 ms, this "best" spike position occurs in the middle of the filter output transient. A visual study of the pulse of Figure 7-4 suggests that its energy is concentrated roughly at the center of its duration range. What happens, then, is that the spiking filter will have the easiest job if the desired spike occupies the same relative position in the filter output function as is taken up by most of the pulse energy in the filter input. This matter is also evident from the contour map of Figure 7-10. Let us consider here vertical lines at constant filter duration. For example, at a filter duration of 50 ms, we expect best performance if we position the desired spike at a lag of roughly $\frac{1}{2}(50 + 30) = 40$ ms. Inspection of the contour map shows that the highest P value at a filter duration of 50 ms indeed occurs at a lag of 40 ms, and is in fact given by $P = 0.93$. If we chose a filter duration of 90 ms, we expect best performance if the desired spike is placed at a lag of $\frac{1}{2}(90 + 30) = 60$ ms. This guess is again confirmed by inspection of the contour map, where we see that the peak P value for a 90 ms duration filter occurs at a lag of 60 ms ($P = 0.97$). This rule of thumb is found to be quite useful in practice. It is evident that contour maps such as the ones shown in Figures 7-10 to 7-12 are useful to display the energy structure of a wavelet. They thus provide a convenient aid for the design of optimum-lag filters.

Wavelet Shaping

Having examined the design problems of the spiking filter in some detail, let us next consider the analogous case of shaping-filter design. The pulse we propose to shape into some desired output waveform is pictured in Figure 7-13(a). It is a symmetric wavelet of 20 ms breadth. The choice of the desired output shape is quite arbitrary, but, since we wish to increase signal resolution, this desired shape should be considerably sharper than the input wavelet. In Figure 7-13(b) and (c) and in Figure 7-14 we show the actual outputs

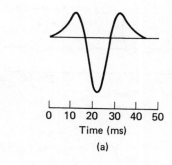

0 10 20 30 40 50
Time (ms)

(a)

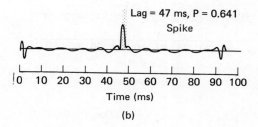

Lag = 47 ms, P = 0.641

Spike

0 10 20 30 40 50 60 70 80 90 100
Time (ms)

(b)

Figure 7-13. (a) Input pulse for shape filtering. (b) Actual output for optimum-lag 50-ms-duration spiking filter. (c) Actual output for optimum-lag 50-ms-duration shaping filter. The desired output shapes are shown.

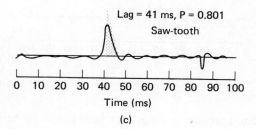

Lag = 41 ms, P = 0.801

Saw-tooth

0 10 20 30 40 50 60 70 80 90 100
Time (ms)

(c)

of a series of 50 ms duration *optimum-lag* shaping filters that were computed for the input pulse of Figure 7-13(a). The desired output shapes are also shown in these figures so that these may be compared directly with the actual outputs. In every case we indicate the value of the filter performance parameter P and of the optimum lag at which this peak P value is obtained. The 50 ms duration filters we have computed here are slightly longer than the input pulse, which is a 44 ms duration wavelet. The resulting filter outputs are therefore 94 ms duration transients.

The desired shape for the output shown in Figure 7-13(b) is a spike. We observe that, although the actual output does peak sharply at the optimum lag of 47 ms, there are sharp events of lesser amplitude at the beginning and end of the output. These spurious signals may often be undesirable, as we shall show below. The comparatively low value of P for this case sug-

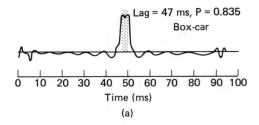

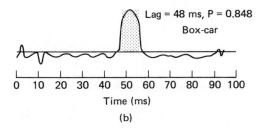

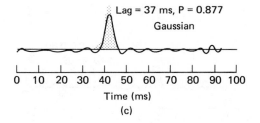

Figure 7-14. Actual outputs for three different 50-ms-duration shaping filters for the pulse of Figure 7-13(a). The desired output shapes are shown.

gests that the spiking filter here could do better if the filter duration were increased. This is indeed confirmed by a study of the appropriate *P*-contour map shown in Figure 7-15.

Figure 7-13(c) gives the result for the choice of a desired sawtooth output. We note that the value of *P* has increased, and that only one spurious signal occurs near the end of the output transient. The breadth of this spurious signal is, however, very different from the actually obtained approximation to the desired shape. We therefore expect that this shaping filter will be a better one to use than the spiking filter of Figure 7-13(a).

Figure 7-14 shows three additional cases. Of the set of five desired shapes treated here and in Figure 7-13, we see that the Gaussian desired waveshape assumed in Figure 7-14(c) yields the highest *P* value for our chosen filter duration of 50 ms. In Figure 7-16 we give the *P*-contour map for the sawtooth waveshape of Figure 7-13(c), while in Figure 7-17, we show the *P*-contour map corresponding to the desired Gaussian waveshape of Figure 7-14(c). A comparison of the contour maps of Figures 7-15, 7-16, and 7-17

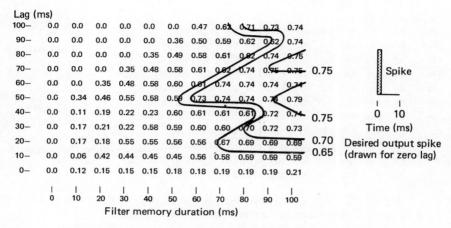

Figure 7-15. *P*-contour map for spiking the pulse of Figure 7-13(a). The gridded numbers correspond to values of the filter performance parameter *P*.

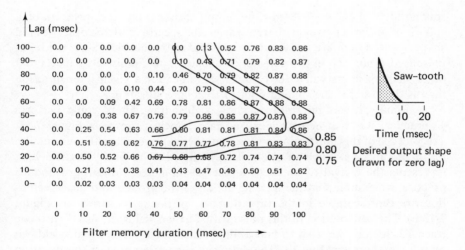

Figure 7-16. *P*-contour map for shaping the pulse of Figure 7-13(a) into a desired output as shown. The gridded numbers correspond to values of the filter performance parameter *P*.

shows that these differ significantly among themselves. This is, of course, to be expected, since each contour map corresponds to a different desired output waveshape. The convenience of using such maps in optimum filter design again becomes evident.

It might be asked why the shaping filters with actual outputs shown in Figure 7-13(c) and in Figure 7-14(a), (b), and (c) perform better than the

Lag (ms)

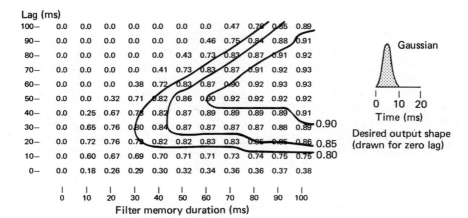

100–	0.0	0.0	0.0	0.0	0.0	0.0	0.0	0.47	0.79	0.85	0.89
90–	0.0	0.0	0.0	0.0	0.0	0.0	0.46	0.75	0.84	0.88	0.91
80–	0.0	0.0	0.0	0.0	0.0	0.43	0.73	0.83	0.87	0.91	0.92
70–	0.0	0.0	0.0	0.0	0.41	0.73	0.83	0.87	0.91	0.92	0.93
60–	0.0	0.0	0.0	0.38	0.72	0.83	0.87	0.90	0.92	0.93	0.93
50–	0.0	0.0	0.32	0.71	0.82	0.86	0.90	0.92	0.92	0.92	0.92
40–	0.0	0.25	0.67	0.78	0.82	0.87	0.89	0.89	0.89	0.89	0.91
30–	0.0	0.65	0.76	0.80	0.84	0.87	0.87	0.87	0.87	0.88	0.89
20–	0.0	0.72	0.76	0.78	0.82	0.82	0.83	0.83	0.85	0.85	0.86
10–	0.0	0.60	0.67	0.69	0.70	0.71	0.71	0.73	0.74	0.75	0.75
0–	0.0	0.18	0.26	0.29	0.30	0.32	0.34	0.36	0.36	0.37	0.38
	0	10	20	30	40	50	60	70	80	90	100

Filter memory duration (ms)

0.90

0.85

0.80

Gaussian

0   10   20
Time (ms)

Desired output shape
(drawn for zero lag)

Figure 7-17. *P*-contour map for shaping the pulse of Figure 7-13(a) into a desired output as shown. The gridded numbers correspond to values of the filter performance parameter *P*.

spiking filter of Figure 7-13(b). One reason is that, while the spectrum of a spike is of uniform power at all frequencies, the spectra of all the other desired shapes treated here are not as rich in the higher frequencies. Since the input pulse is also not rich in these higher frequencies, one intuitively expects that other shaping filters will lead to higher *P* values for a given filter duration than is the case for the spiking filter of that same duration.

Finally, it is of interest to see how such shaping filters would perform in practice. We treat here an idealized case, which is shown in Figure 7-18. Trace 1 (T1) of this illustration gives a hypothetical layered-earth unit impulse response function. Each of the spikes, depicted by a heavy vertical line, represents the arrival of an event at the seismometer. Following standard practice, we assume that the filtering action of the ground and instrumentation are representable in the time domain by the pulse shown in Figure 7-13(a). The convolution of this pulse with T1 then yields the usual synthetic trace T2. Ideally, we wish to find a filter which, with T2 as input, yields an output closely resembling T1. The output of the optimum-lag 50 ms duration spiking filter is given by T4. We observe that many of the spikes of T1 have been recovered, but that the noise level is very high. This behavior is to be expected because the spiking filter used here generates strong spurious signals [see Figure 7-13(b)]. The output of the optimum-lag 50 ms duration shaping filter with a saw-tooth waveform as desired output is shown in T3, while T5 gives the output for such a shaping filter with a Gaussian waveform as desired output (see also Figures 7-13 and 7-14). These letter two traces are not as noisy as T4. Moreover, the choice of desired output shapes other than simple spikes leads to results in which the signal has more diagnostic features that

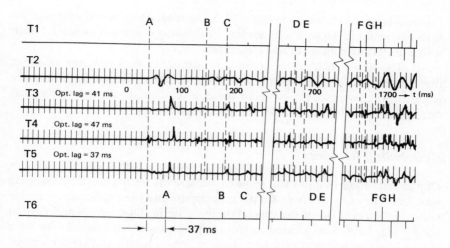

T1: Layered earth unit impulse response.
T2: (T1)∗(pulse of Fig. 7-13(a)).
T3: Output from shaping filter; desired output
    signal: saw-tooth waveform of Fig. 7-13(c).

T4: Output from spiking filter.
T5: Output from shaping filter; desired output
    signal: Gaussian waveform of Fig. 7-14(c).
T6: T1 lagged by 37 ms.

Figure 7-18. A hypothetical application of shaping filters.

can aid the interpreter in its recognition in a noise background. In order to illustrate this point, let us consider the events labelled "A" through "H" in Figure 7-18. In order to locate events on T2 through T5 that correspond to spikes on T1, it is necessary to add the optimum lag to the arrival time of a given spike on T1. The respective optimum lags are indicated at the beginning of each trace. Event A, since it occurs some 150 ms from its nearest neighbor, yields output on all traces uncorrupted by adjacent signals, and the shapes here obtained are those also shown in Figures 7-13(b) and (c) and 7-14(c). Events B and C are closer together, but are still distinguishable as separate signals on the synthetic trace T2. The corresponding shaped signals on T3, T4, and T5 are indicated by appropriate shading. We observe that good resolution has been obtained. On the other hand, events D and E are sufficiently close together so that they are no longer distinguishable as separate signals on the synthetic trace T2. However, our shaping filters do a good job in resolving this pair of pulses on traces 3 and 5; the spiking filter (T4) does not perform well in this case, as one might expect from what has been said before. A similar situation occurs for events F, G, and H, which are so close together that they are not distinguishable as separate pulses on T2. These events become clearly resolved on T3 and T5; the corresponding result on the spiking filter output of T4 is again of much poorer quality. Many other such examples are observable on the traces, but these have not been explicitly indicated for the sake of brevity. In order to help the reader judge the quality of the results obtained on T5, the ideal unit impulse response T1 is shown again in T6 with a 37 ms lag. In this way dominant events on both traces are brought into vertical alignment.

The above example is admittedly idealized; we know, for example, that the choice of a single seismic waveshape for the entire length of the trace is unrealistic. Even so, the synthetic traces given here do show roughly what one might expect to see in actual seismic applications.

Concluding Remarks

Digital least-squares filters can serve to increase the resolution of overlapping wavelets. They are relatively simple to compute and are applicable in a wide variety of ways. Two of the crucial design criteria, filter length and desired output lag, have been treated in this chapter considerable detail. We have shown how one can select the best combination of length and lag for a given problem by a study of an appropriate P-contour map. We have stated that the choice of the desired output shape is arbitrary, although, of course, we wish to see signals in the actual output which are sharper than their counterparts in the input. Nevertheless, given the input seismic pulse, it might perhaps be possible to arrive by analytical means at desired output pulse shapes that are themselves optimum. In other words, we wish to find that desired signal shape, given an input pulse, which is not only "best" distinguishable in a noise background, but which is achievable also by a short filter with a high value of the performance parameter P. Such an investigation is perhaps logically the next step to take in this field.

the error in least-squares filtering

Summary

Consideration of many factors is necessary in order to optimize the performance of a digital filter. Optimization depends upon the purpose of the filter, the nature of the input data, the criterion used for evaluating performance, and allowable tolerances in accuracy. In this chapter, we consider the problems involved in preserving, enhancing, or inverting the waveshape of a signal and in predicting its future values. All such filters fall under the general heading of wave-shaping filters or briefly, shaping filters. Although Chapters 6 and 7 have already dealt with many aspects of the shaping filter, our emphasis here is with an explicit error analysis of the filter design problem.

The Wave-Shaping Filter

For digital computing applications it is important to develop the theory of shaping filters in terms of finite-length operators. Given a finite-length input signal, one wishes to find a finite-length operator that shapes the input signal into the form of a given desired finite-length output signal. Except in special circumstances this shaping operation cannot be perfect, so there will be an error between the actual output and the desired output. The desideratum is to choose that operator which makes this error as small as possible in a least-squares sense. In this chapter, we study the error in such least-squares filtering operations.

Let $b_0, b_1, \ldots, b_n$ represent the given real finite-length *input signal*, and let $d_0, d_1, \ldots, d_{m+n}$ represent the real finite-length *desired output signal*. Here m and n are each nonnegative integers. The problem is to determine the real finite-length *operator* or *filter* $f_0, f_1, \ldots, f_m$ such that the *actual output signal* $c_0, c_1, \ldots, c_{m+n}$ is the least-squares approximation to the desired output. The actual output is equal to the convolution of the input with the operator. This convolution can be written as

$$c_t = \sum_{s=0}^{m} f_s b_{t-s} \qquad \text{for } t = 0, 1, 2, \ldots, m+n$$

The error between desired and actual output is

$$d_t - c_t \qquad \text{for } t = 0, 1, 2, \ldots, m+n$$

and the sum of squared errors, J, is

$$J = \sum_{t=0}^{m+n} (d_t - c_t)^2 = \sum_{t=0}^{m+n} \left(d_t - \sum_{s=0}^{m} f_s b_{t-s} \right)^2 \qquad (8\text{-}1)$$

From the theory of least squares we know that the sum of squared errors is a minimum if and only if

$$\frac{\partial J}{\partial f_i} = 0 \qquad \text{for } i = 0, 1, 2, \ldots, m$$

If we carry out the differentiation, we obtain the *normal equations*

$$-2 \sum_{t=0}^{m+n} \left(d_t - \sum_{s=0}^{m} f_s b_{t-s} \right) b_{t-i} = 0 \qquad (8\text{-}2)$$

or

$$\sum_{t=0}^{m+n} d_t b_{t-i} - \sum_{s=0}^{m} f_s \left(\sum_{t=0}^{m+n} b_{t-s} b_{t-i} \right) = 0$$

Let r_s denote the autocorrelation of the input signal, that is,

$$r_s = \sum_{t=0}^{n} b_{t+s} b_t = b_s b_0 + b_{1+s} b_1 + \cdots + b_n b_{n-s}$$

and let g_s denote the crosscorrelation of the desired output with the input signal; that is,

$$g_s = \sum_{t=0}^{n} d_{t+s} b_t = d_s b_0 + d_{1+s} b_1 + \cdots + d_{n+s} b_n$$

where $s = 0, 1, 2, \ldots, m$. The autocorrelation is symmetrical; that is,

$$r_s = r_{-s}$$

Then the normal equations in terms of this autocorrelation and crosscorrelation become

$$\sum_{s=0}^{m} f_s r_{i-s} = g_i \qquad \text{for } i = 0, 1, 2, \ldots, m$$

Let us now find an expression for the minimum value attained by the sum of squared errors. We have from (8-1) that

$$J = \sum_{t=0}^{m+n} d_t \left(d_t - \sum_{s=0}^{m} f_s b_{t-s} \right) - \sum_{j=0}^{m} f_j \left[\sum_{t=0}^{m+n} (d_t - \sum_{s=0}^{m} f_s b_{t-s}) b_{t-j} \right]$$

But, because of equation (8-2), the expression in the square brackets is zero, so the (minimum) sum of squared errors is

$$J = \sum_{t=0}^{m+n} d_t^2 - \sum_{s=0}^{m} f_s \sum_{t=0}^{m+n} d_t b_{t-s}$$

or

$$J = \sum_{t=0}^{m+n} d_t^2 - \sum_{s=0}^{m} f_s g_s$$

This equation states that the sum of squared errors for the shaping filter f is equal to the sum of squares of the desired output minus the dot product of the filter with the crosscorrelation.

Matrix Notation

The foregoing derivation of the wave-shaping filter can be reformulated in terms of matrix theory. Let

$$\mathbf{b} = (b_0, b_1, \ldots, b_n)$$

be the input wavelet, let

$$\mathbf{f} = (f_0, f_1, \ldots, f_m)$$

be the filter, and let

$$\mathbf{c} = (c_0, c_1, \ldots, c_{m+n})$$

be the actual output. We define the $(m + 1) \times (m + n + 1)$ matrix $\mathbf{B}$ as the matrix whose rows are successively delayed replicas of the input wavelet; that is,

$$\mathbf{B} = \begin{bmatrix} b_0 & b_1 & b_2 & \cdots & b_n & & \\ & b_0 & b_1 & b_2 & \cdots & b_n & \text{zeros} \\ & & & \ddots & & & \\ \text{zeros} & & & & & \\ & & & b_0 & b_1 & b_2 & \cdots & b_n \end{bmatrix}$$

Then convolution, or filtering, can be defined by the matrix multiplication

$$\mathbf{c} = \mathbf{f}\mathbf{B}$$

Let the desired output be

$$\mathbf{d} = (d_0, d_1, \ldots, d_{m+n})$$

Then the error between desired and actual outputs is $d - c$, and the sum of squared errors is

$$J = (\mathbf{d} - \mathbf{c})(\mathbf{d} - \mathbf{c})^T$$

In order to find the required filter $\mathbf{f}$, it is instructive to think of $\mathbf{f}$ as a set of regression coefficients. Here the term "regression" is used in the same sense as in the classical theory of statistics. We can write the regression equation as

$$\mathbf{d} = \mathbf{f}\mathbf{B} + (\mathbf{d} - \mathbf{c})$$

where the matrix $\mathbf{B}$ is the regressor and the error $(\mathbf{d} - \mathbf{c})$ is the regression residual. From the theory of regression analysis we know that the sum of squared errors is a minimum if and only if the regressor $\mathbf{B}$ is normal (i.e., orthogonal) to the residual, that is, if and only if

$$(\mathbf{d} - \mathbf{c})\mathbf{B}^T = 0$$

The resulting equations, namely

$$\mathbf{c}\mathbf{B}^T = \mathbf{d}\mathbf{B}^T$$

or

$$\mathbf{f}\mathbf{B}\mathbf{B}^T = \mathbf{d}\mathbf{B}^T$$

are called the normal equations. The solution of the normal equations gives the required optimum wave-shaping filter $\mathbf{f}$. In the normal equations, we recognize that $\mathbf{BB}^T$ is the $(m+1) \times (m+1)$ symmetric autocorrelation matrix $\mathbf{R}$ of the input wavelet; that is,

$$\mathbf{BB}^T = \mathbf{R} = \begin{bmatrix} r_0 & r_1 & r_2 & \cdots & r_m \\ r_1 & r_0 & r_1 & \cdots & r_{m-1} \\ r_2 & r_1 & r_0 & \cdots & r_{m-2} \\ & & \vdots & & \\ r_m & r_{m-1} & r_{m-2} & \cdots & r_0 \end{bmatrix}$$

Also, we recognize that $\mathbf{dB}^T$ is the row vector of the crosscorrelation of the desired output with the input wavelet; that is,

$$\mathbf{dB}^T = \mathbf{g} = (g_0, g_1, \ldots, g_m)$$

Thus, the normal equations can be written in matrix form as

$$\mathbf{f\,R} = \mathbf{g}$$

The minimum value of the sum of squared errors is

$$J = (\mathbf{d} - \mathbf{c})(\mathbf{d} - \mathbf{c})^T$$
$$= (\mathbf{d} - \mathbf{c})\mathbf{d}^T - (\mathbf{d} - \mathbf{c})\mathbf{c}^T$$
$$= (\mathbf{d} - \mathbf{f\,B})\mathbf{d}^T - [(\mathbf{d} - \mathbf{c})\mathbf{B}^T]\mathbf{f}^T$$

Because the expression in the square brackets is zero, we have

$$J = \mathbf{dd}^T - \mathbf{fg}^T$$

which is the same result we obtained before without matrix notation.

The Set of All Possible Spiking Filters

A special case of the shaping filter is the *spiking filter*. A spiking filter is a shaping filter for which the desired output is a unit spike. An alternative term for "spiking filter" is "least-squares inverse filter." If one looks at the desired output,

$$d_0, d_1, \ldots, d_{m+n}$$

one sees that any one of the $m + n + 1$ values may represent the spike, while the remaining values are set equal to zero. Hence, there are $m + n + 1$

different spike filters possible in our given model, one spike filter for each of the $m + n + 1$ different spike positions. Let

$$\mathbf{a}_0 = (a_{00}, a_{01}, \ldots, a_{0m})$$

be the spiking operator for the zero-delay desired output spike

$$\mathbf{d}_0 = (1, 0, 0, \ldots, 0)$$

Similarly, let

$$\mathbf{a}_1 = (a_{10}, a_{11}, \ldots, a_{1m})$$

be the spiking operator for the one-delay desired output spike

$$\mathbf{d}_1 = (0, 1, 0, \ldots, 0)$$

and so on, until we let

$$\mathbf{a}_{m+n} = (a_{m+n,0}, a_{m+n,1}, \ldots, a_{m+n,m})$$

wavelet $(n+1)$

be the spiking operator for the $(m + n)$-delay desired output spike

operator is $(m+1)$ long

$$\mathbf{d}_{m+n} = (0, 0, 0, \ldots, 0, 1)$$

We see that these successively delayed spikes represent the rows of an $(m + n + 1) \times (m + n + 1)$ identity matrix $\mathbf{I}$. Let A be the $(m + n + 1) \times (m + 1)$ matrix whose rows are the spike operators from zero delay to $(m + n)$ delay; the *spiking operator matrix* is

$m+1$

$m+n+1$

$$\mathbf{A} = \begin{bmatrix} a_{00} & a_{01} & \cdots & a_{0m} \\ a_{10} & a_{11} & \cdots & a_{1m} \\ & & \cdot & \\ & & \cdot & \\ & & \cdot & \\ a_{m+n,0} & a_{m+n,1} & \cdots & a_{m+n,m} \end{bmatrix}$$

If we look at the normal equations $\mathbf{a}_j\mathbf{R} = \mathbf{g}$ for any of these spiking operators, $j = 0, 1, 2, \ldots, m + n$, we see that the matrix $\mathbf{R}$ does not depend upon the spike position. However, the right-hand side *does* depend upon the spike position, in the following way.

For spike position 0, we have

$$\mathbf{g} = (b_0, 0, 0, 0, \ldots, 0)$$

For spike position 1, we have

$$\mathbf{g} = (b_1, b_0, 0, 0, \ldots, 0)$$

For spike position 2, we have

$$\mathbf{g} = (b_2, b_1, b_0, 0, \ldots, 0)$$

We continue in this way, until finally for spike position $m + n$ we have $\mathbf{g} = (b_{m+n}, b_{m+n-1}, \ldots, b_n)$ which (because $b_t = 0$ for $t > n$) reduces to

$$\mathbf{g} = (0, 0, 0, \ldots, 0, b_n)$$

The right-hand vectors $\mathbf{g}$ for all the spike positions can be incorporated into a $(m + n + 1) \times (m + 1)$ matrix, which we recognize as the matrix $\mathbf{B}^T$.

Thus, the normal equations for each of the spiking operators can be encompassed in one equation,

$$\mathbf{AR} = \mathbf{B}^T$$

where the superscript T indicates matrix transpose. Let

$$c_{00}, c_{01}, \ldots, c_{0,m+n}$$

be the actual output of the zero-delay spiking operator. Let

$$c_{10}, c_{11}, \ldots, c_{1,m+n}$$

be the actual output of the one-delay spiking filter $\mathbf{a}_1$, and so on. Then the $(m + n + 1) \times (m + n + 1)$ square matrix $\mathbf{C}$ defined by

$$\mathbf{C} = \begin{bmatrix} c_{00} & c_{01} & \cdots & c_{0,m+n} \\ c_{10} & c_{11} & \cdots & c_{1,m+n} \\ & & \cdot & \\ & & \cdot & \\ & & \cdot & \\ c_{m+n,0} & c_{m+n,1} & \cdots & c_{m+n,m+n} \end{bmatrix}$$

is called the *spiking output matrix*, and satisfies the equation $\mathbf{AB} = \mathbf{C}$. The normal equation $\mathbf{AR} = \mathbf{B}^T$ may be written as

$$\mathbf{ABB}^T = \mathbf{B}^T$$

which by postmultiplying by $\mathbf{A}^T$ becomes

$$\mathbf{ABB}^T\mathbf{A}^T = \mathbf{B}^T\mathbf{A}^T$$

which is

$$\mathbf{CC}^T = \mathbf{C}^T$$

Let J_i be the sum of squared errors for the spiking filter of delay i.

The grand sum of squared errors (i.e., the sum of the sum of squared errors for each spiking operator) is

$$V = J_0 + J_1 + \cdots + J_{m+n}$$

It can be shown (Robinson and Treitel, 1978, Chap. 7) that the grand sum of squared errors is equal to

$$V = (m + n + 1) - (m + 1) = n$$

Hence, we come to the conclusion that the grand sum of squared errors for the spiking filters for all possible delays is equal to n, where $n + 1$ is the length of the input wavelet $b_0, b_1, \ldots, b_n$. Moreover, we see that the grand sum of squared errors V is independent of the filter length, $m + 1$. If the sum of squared errors J_i were the same for all possible delays, it would be $J_i = n/(m + n + 1)$ for each spiking filter $\mathbf{a}_i$, where $i = 0, 1, 2, \ldots, m + n$. Generally, the J_i will not be equal for all the spiking filters. The largest possible J_i is unity, because the zero filter $\mathbf{a}_i = [0, 0, \ldots, 0]$ would have $J_i = 1$, and any filter computed by least squares could not exceed this sum of squared errors. This maximum error would be obtained, for example, by the zero-delay spiking filter for an input signal whose leading term is zero: $\mathbf{b} = (0, b_1, b_2, \ldots, b_n)$; for in this case the right-hand side of the normal equations is zero, and hence the filter is zero and produces maximum error. Consider next the case where all the terms of the input signal are zero except the last: $\mathbf{b} = (0, 0, 0, \ldots, 0, b_n)$. In this case the first n spiking filters produce maximum $J_i = 1$ $(i = 0, 1, \ldots, n - 1)$. The sum of squared errors for these first n spiking filters is therefore n, and since this quantity is identically equal to the grand sum of squared errors, it follows that the last $m + 1$ spike filters must produce minimum $J_i = 0$ $(i = n, n + 1, \ldots, n + m)$.

In any case there is some delay i for which the sum of squared errors J_i is a minimum. This minimum need not be unique. The value of i that produces the minimum J_i is called the *optimum delay*, or *optimum spike position*, and the corresponding spiking filter $\mathbf{a}_i$ is called the optimum spiking filter for the given input signal $\mathbf{b}$. Numerical examples of this behavior were given in Chapter 7.

For very short spiking filters it appears that no general rules can be established. However, for sufficiently long spiking filters we can state the following rules:

1. The optimum delay for a minimum-delay input signal is the smallest possible delay, namely 0.
2. The optimum delay for a maximum-delay input signal is the largest possible delay, namely $m + n$.
3. The optimum delay for a mixed-delay input signal is intermediate (i.e., between the smallest and largest possible delays).

In fact, these rules can actually be used to define the concepts of minimum delay, maximum delay, and mixed delay.

The Error for the Shaping Filter

Let us now return to the case of a shaping filter $\mathbf{f}$ for an arbitrary desired output $\mathbf{d}$. The matrix normal equation is

$$\mathbf{f}\mathbf{R} = \mathbf{d}\mathbf{B}^T$$

But $\mathbf{A}\mathbf{R} = \mathbf{B}^T$, so

$$\mathbf{f}\mathbf{R} = \mathbf{d}\mathbf{A}\mathbf{R}$$

Hence, the shaping filter $\mathbf{f}$ can be expressed in terms of the desired output $\mathbf{d}$ and the spiking operator matrix $\mathbf{A}$ as

$$\mathbf{f} = \mathbf{d}\mathbf{A}$$

or

$$\mathbf{f} = d_0\mathbf{a}_0 + d_1\mathbf{a}_1 + \cdots + d_{m+n}\mathbf{a}_{m+n}$$

That is, the shaping filter $\mathbf{f}$ is a weighted sum of the spiking filters for all possible delays, the weighting factors being the values of the desired output at times corresponding to the respective delays.

We recall that the sum of squared errors for the shaping filter $\mathbf{f}$ is equal to the sum of squared errors of the desired output minus the dot product of the filter with the crosscorrelation. Therefore, the sum of squared shaping errors is the quadratic form

$$J = \mathbf{d}\mathbf{d}^T - (\mathbf{d}\mathbf{A})(\mathbf{d}\mathbf{B}^T)$$
$$= \mathbf{d}\mathbf{d}^T - \mathbf{d}\mathbf{A}\mathbf{B}\mathbf{d}^T$$
$$= \mathbf{d}\mathbf{d}^T - \mathbf{d}\mathbf{C}\mathbf{d}^T$$
$$= \mathbf{d}(\mathbf{I} - \mathbf{C})\mathbf{d}^T$$

where the matrix of the quadratic form is $\mathbf{I} - \mathbf{C}$, the difference between the desired and actual outputs for all possible spiking filters.

Least-Squares Inverse Filtering

A spiking filter may be described as the least-squares inverse filter for the input wavelet. Least-squares inverse filtering always involves consideration of the error. Under certain conditions the error will go to zero as the length of the filter tends to infinity. In this section we will show that the error will go to zero if either (1) the waveform being inverted is minimum delay, or (2) if the output is chosen to come after a sufficiently long time delay. If the waveform being inverted is not minimum-delay and if, in addition, the output is not chosen to be delayed, the error will be finite and may be large.

The equations for least-squares inverse filtering are given in the preceding sections. In this section, the following conditions will be shown: (1) If the filter is a zero-delay inverse (i.e., a zero-delay spiking filter), the error will tend to zero as the filter length becomes infinite if and only if the wavelet being inverted is minimum-delay; (2) If the spike position is sufficiently delayed, the energy in the error will tend to zero as $1/m$, where m is the length index of the filter.

First, we see that if the waveform to be inverted is minimum-delay (its z transform has no zeros inside the unit circle), the error from a zero-delay least-squares inverse will tend to zero as the filter length tends to infinity. This follows from the results of Chapter 4, where we showed that if the wavelet to be inverted is minimum-delay, an exact inverse can be found by the method of polynomial division. Since this exact inverse has zero error, a least-squares error method must also give zero error as the filter length tends to infinity.

Next, if the wavelet is not minimum-delay, the error for the zero-delay inverse will not go to zero. Let us now consider the normal equations $\mathbf{aR} = \mathbf{g}$. The matrix $\mathbf{R}$ has rows and columns that are the autocorrelation of the wavelet $\mathbf{b}$, the zero lag of the autocorrelation being on the main diagonal. The matrix $\mathbf{R}$ has no information about the phase spectrum of $\mathbf{b}$. If the desired output is the spike $(1, 0, 0, \ldots)$, the crosscorrelation column vector $\mathbf{g}$ is $(b_0, 0, 0, \ldots)$. A scale factor, b_0, in the inhomogeneous part (right side) of the normal equations can only affect the solution filter within a scale factor. Thus, there is no information about the phase spectrum of $\mathbf{b}$ in these normal equations. Any member of a class of wavelets with a given correlation will produce the same normal equations as any other member, except for a scale factor. Hence, the solution filters are all the same except for a scale factor. It can be shown (Robinson, 1962) that the wavelet $\mathbf{b}$ can be represented in the form

$$\mathbf{b} = \mathbf{p} * \mathbf{w}$$

where $\mathbf{p}$ is an all-pass filter and $\mathbf{w}$ is a minimum-phase wavelet with the same autocorrelation function as $\mathbf{b}$. An all-pass filter does not attenuate any frequency component, but may delay its phase (see Chapter 5). Now suppose that we convolve $\mathbf{b}$ with the zero-delay spiking filter $(a_0, a_1, \ldots, a_m)$. As m tends to infinity, we have noted that $\mathbf{a}$ tends to be proportional to the exact zero-delay inverse of $\mathbf{w}$. Thus, as $m \to \infty$,

$$\mathbf{b} * \mathbf{a} = \mathbf{p} * \mathbf{w} * \mathbf{a} \to (\text{constant}) \, \mathbf{p}$$

This can equal $(1, 0, 0, \ldots)$ only if $\mathbf{p}$ is the trivial phase-shift filter (no phase shift). Hence, we have the conclusion that the error cannot tend to zero unless **b** is minimum delay.

Finally, if the desired output of the filter **a** is a delayed spike $\mathbf{d} = (0, \ldots, 0, 0, 1, 0, 0, \ldots, 0)$ with the sufficient delay, then the sum-of-squared errors will tend to zero regardless of the phase characteristic of **b**. We consider the class of least-squares inverse filters with output at all possible different lags.

The grand sum of squared errors for the filters with all the possible spike positions is

$$V = n$$

as we have seen earlier in this chapter. In turn, this means that the grand sum of squared errors for inverse filters at all possible lags is independent of the filter length m. In fact, it is just equal to n, the length index of the wavelet to be inverted.

As the length, m, of the filter goes to infinity, the total error $V = n$ is spread out over a larger and larger interval of size $n + m + 1$. Hence, given n, and letting $v(m)$ be the smallest error for some delay for a given m, we must have

$$v(m) \leq \frac{n}{m + n + 1}$$

Thus, as $m \longrightarrow \infty$, $v \longrightarrow 0$.

If the sum of squared errors were the same for all possible delays, it would be $n/(m + n + 1)$ for any particular inverse filter. This need not be the case, however. It may be considerably greater for small and/or large delays. For example, when one tries to invert a nonminimum-delay wavelet with a zero-delay spike as desired output, one obtains a finite and possibly large error (see Figure 7-5).

The largest possible sum of squared errors for any inverse filter is unity. This is because the zero filter, namely $\mathbf{a} = (0, 0, \ldots, 0)$, would have a sum of squared errors of unity, and any least-squares inverse would produce no more than this error. The maximum error would actually be obtained if one were to try to invert without delay in the spike position a wavelet that had undergone a pure delay, say $\mathbf{b} = (0, b_1, b_2, \ldots, b_n)$. In this case the right side of the normal equations is zero, and hence the filter is zero and no good at all.

Delay Properties of Spiking Filters

Let us consider the shaping filter shown in Figure 8-1. The filter a_0, a_1 is determined so that the error energy

$$J = \sum_{t=0}^{2} (d_t - c_t)^2 = (d_0 - c_0)^2 + (d_1 - c_1)^2 + (d_2 - c_2)^2$$

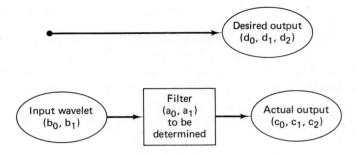

Figure 8-1. Flow diagram for the shaping filter.

is a minimum. Now the actual output is

$$c_0 = a_0 b_0$$
$$c_1 = a_0 b_1 + a_1 b_0$$
$$c_2 = a_1 b_1$$

where the filter coefficients are given by the solution of the normal equations

$$a_0 r_0 + a_1 r_1 = d_0 b_0 + d_1 b_1$$
$$a_0 r_1 + a_1 r_0 = d_1 b_0 + d_2 b_1$$

where $r_0 = b_0^2 + b_1^2$ and $r_1 = b_0 b_1$.

    We obtain three spiking filters as follows. The desired outputs for the three filters are:

$$(\text{Filter for spike at } 0) = S_0: \quad (d_0, d_1, d_2) = (1, 0, 0)$$
$$(\text{Filter for spike at } 1) = S_1: \quad (d_0, d_1, d_2) = (0, 1, 0)$$
$$(\text{Filter for spike at } 2) = S_2: \quad (d_0, d_1, d_2) = (0, 0, 1)$$

The right-hand side of the normal equations is

$$\begin{pmatrix} d_0 b_0 + d_1 b_1 \\ d_1 b_0 + d_2 b_1 \end{pmatrix}$$

For S_0, the right-hand side is

$$\begin{pmatrix} b_0 \\ 0 \end{pmatrix}$$

For S_1, the right-hand side is

$$\begin{pmatrix} b_1 \\ b_0 \end{pmatrix}$$

For S_2, the right-hand side is

$$\begin{pmatrix} 0 \\ b_1 \end{pmatrix}$$

Thus, the normal equations are:

For S_0: $a_{00}r_0 + a_{01}r_1 = b_0$
$\quad\quad\quad a_{00}r_1 + a_{01}r_0 = 0$

For S_1: $a_{10}r_0 + a_{11}r_1 = b_1$
$\quad\quad\quad a_{10}r_1 + a_{11}r_0 = b_0$

For S_2: $a_{20}r_0 + a_{21}r_1 = 0$
$\quad\quad\quad a_{20}r_0 + a_{21}r_0 = b_1$

where the first subscript on the filter coefficient denotes the position of the appropriate desired output spike. The solutions of the normal equations are (note that $\Delta = r_0^2 - r_1^2$):

For S_0:

$$a_{00} = \frac{\begin{vmatrix} b_0 & r_1 \\ 0 & r_0 \end{vmatrix}}{\Delta} = \frac{b_0 r_0}{\Delta}$$

$$a_{01} = \frac{\begin{vmatrix} r_0 & b_0 \\ r_1 & 0 \end{vmatrix}}{\Delta} = -\frac{b_0 r_1}{\Delta}$$

For S_1:

$$a_{10} = \frac{\begin{vmatrix} b_1 & r_1 \\ b_0 & r_0 \end{vmatrix}}{\Delta} = \frac{r_0 b_1 - r_1 b_0}{\Delta}$$

$$a_{11} = \frac{\begin{vmatrix} r_0 & b_1 \\ r_1 & b_0 \end{vmatrix}}{\Delta} = \frac{r_0 b_0 - r_1 b_1}{\Delta}$$

For S_2:

$$a_{20} = \frac{\begin{vmatrix} 0 & r_1 \\ b_1 & r_0 \end{vmatrix}}{\Delta} = \frac{-b_1 r_1}{\Delta}$$

$$a_{21} = \frac{\begin{vmatrix} r_0 & 0 \\ r_1 & b_1 \end{vmatrix}}{\Delta} = \frac{b_1 r_0}{\Delta}$$

The delay properties of the wavelets are determined by the following conditions:

(b_0, b_1) is minimum-delay if $|b_0| > |b_1|$

(b_0, b_1) is maximum-delay if $|b_0| < |b_1|$

(a_0, a_1) is minimum-delay if $|a_0| > |a_1|$

(a_0, a_1) is maximum-delay if $|a_0| < |a_1|$

(c_0, c_1, c_2) is minimum-delay if both (b_0, b_1) and (a_0, a_1) are minimum-delay

(c_0, c_1, c_2) is maximum-delay if both (b_0, b_1) and (a_0, a_1) are maximum-delay

Otherwise, (c_0, c_1, c_2) is mixed-delay

Let us construct a delay table in which we let $\mathbf{a} = (a_0, a_1)$, $\mathbf{b} = (b_0, b_1)$, $\mathbf{c} = (c_0, c_1, c_2)$, and minD = minimum-delay, mixD = mixed-delay, and maxD = maximum-delay. The zero for the z transform of each of the spiking filters is:

For S_0:

$$a_{00} + a_{01}z = 0$$

so the zero is

$$z = \frac{-b_0 r_0}{-b_0 r_1} = \frac{r_0}{r_1} = \frac{b_0^2 + b_1^2}{b_0 b_1}$$

For S_1:

$$a_{10} + a_{11}z = 0$$

so the zero is

$$z = \frac{r_1 b_0 - r_0 b_1}{r_0 b_0 - r_1 b_1} = \frac{b_1 b_0^2 - (b_0^2 + b_1^2)b_1}{(b_0^2 + b_1^2)b_0 - b_0 b_1 b_1} = \frac{b_1^3}{b_0^3} = \left(\frac{b_1}{b_0}\right)^3$$

For S_2:

$$a_{20} + a_{21}z = 0$$

so the zero is

$$z = \frac{b_1 r_1}{b_1 r_0} = \frac{r_1}{r_0} = \frac{b_0 b_1}{b_0^2 + b_1^2}$$

Accordingly, we have the following delay table:
For spike at 0:

If $\mathbf{b}$ is minD, then $\mathbf{a}$ is minD and $\mathbf{c}$ is minD

If $\mathbf{b}$ is maxD, then $\mathbf{a}$ is minD and $\mathbf{c}$ is mixD

For spike at 1:

> If **b** is minD, then **a** is maxD and **c** is mixD
> If **b** is maxD, then **a** is minD and **c** is mixD

For spike at 2:

> If **b** is minD, then **a** is maxD and **c** is mixD
> If **b** is maxD, then **a** is maxD and **c** is maxD

Let us now introduce matrix notation. Let

$$\mathbf{B} = \begin{pmatrix} b_0 & b_1 & 0 \\ 0 & b_0 & b_1 \end{pmatrix}$$

$$\mathbf{d}_0 = (1, 0, 0)$$
$$\mathbf{d}_1 = (0, 1, 0)$$
$$\mathbf{d}_2 = (0, 0, 1)$$

$$\mathbf{D} = \begin{pmatrix} 1 & 0 & 0 \\ 0 & 1 & 0 \\ 0 & 0 & 1 \end{pmatrix} = \text{identity matrix} = \mathbf{I}$$

and

$$\mathbf{A} = \begin{pmatrix} a_{00} & a_{01} \\ a_{10} & a_{11} \\ a_{20} & a_{21} \end{pmatrix}$$

The normal equations for all three spiking filters are

$$\mathbf{ABB}^T = \mathbf{DB}^T$$

which is

$$\mathbf{ABB}^T = \mathbf{B}^T \qquad (\text{since } \mathbf{D} = \mathbf{I})$$

or

$$\mathbf{A} = \mathbf{B}^T(\mathbf{BB}^T)^{-1}$$

The actual output for all three spiking filters is

$$\mathbf{C} = \begin{pmatrix} c_{00} & c_{01} & c_{02} \\ c_{10} & c_{11} & c_{12} \\ c_{20} & c_{21} & c_{22} \end{pmatrix} = \mathbf{AB}$$

The error for the spike at 0 is

$$\mathbf{e}_0 = (e_{00}, e_{01}, e_{02}) = (d_{00}, d_{01}, d_{02}) - (c_{00}, c_{01}, c_{02})$$
$$= (1, 0, 0) - (c_{00}, c_{01}, c_{02})$$

with sum of squares

$$J_0 = \mathbf{e}_0\mathbf{e}_0^T = (1 - c_{00})^2 + c_{01}^2 + c_{02}^2$$
$$= 1 - 2c_{00} + c_{00}^2 + c_{01}^2 + c_{02}^2$$

The actual output of the zero-delay spiking filter is

$$\mathbf{c}_0 = (c_{00}, c_{01}, c_{02}) = (a_{00}, a_{01})\mathbf{B} = \mathbf{a}_0\mathbf{B}$$

Because $\mathbf{e}_0$ is orthogonal to $\mathbf{a}_0\mathbf{B}$ as seen in Figure 8-2, we have

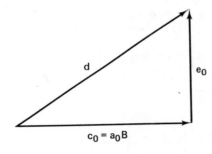

Figure 8-2. Orthogonality of the error vector $\mathbf{e}_0$ to the actual output $\mathbf{c}_0 = \mathbf{a}_0\mathbf{B}$ of the zero-delay spiking filter.

$$\mathbf{d}\mathbf{d}^T = \mathbf{e}_0\mathbf{e}_0^T + (\mathbf{a}_0\mathbf{B})(\mathbf{a}_0\mathbf{B})^T$$

or

$$\mathbf{e}_0\mathbf{e}_0^T = \mathbf{d}\mathbf{d}^T - \mathbf{a}_0\mathbf{B}\mathbf{B}^T\mathbf{a}_0^T = 1 - (c_{00}^2 + c_{01}^2 + c_{02}^2)$$

Now

$$\mathbf{a}_0 = (a_{00}, a_{01}) = (1, 0, 0)\mathbf{B}^T(\mathbf{B}\mathbf{B}^T)^{-1}$$

so

$$\mathbf{e}_0\mathbf{e}_0^T = \mathbf{d}\mathbf{d}^T - \mathbf{a}_0\mathbf{B}\mathbf{B}^T\mathbf{a}_0^T$$
$$= (1, 0, 0)\begin{pmatrix} 1 \\ 0 \\ 0 \end{pmatrix} - (1, 0, 0)\mathbf{B}^T(\mathbf{B}\mathbf{B}^T)^{-1}\mathbf{B}\mathbf{B}^T(\mathbf{B}\mathbf{B}^T)^{-1}\mathbf{B}\begin{pmatrix} 1 \\ 0 \\ 0 \end{pmatrix}$$

Let us now denote $J_0 = \mathbf{e}_0\mathbf{e}_0^T$ simply by $(\mathbf{e}\mathbf{e}^T)_0$, and similarly for J_1 and J_2.

That is, for the spike at 0, we have

$$(ee^T)_0 = (1, 0, 0)\begin{pmatrix} 1 \\ 0 \\ 0 \end{pmatrix} - (1, 0, 0)\mathbf{B}^T(\mathbf{BB}^T)^{-1}\mathbf{B}\begin{pmatrix} 1 \\ 0 \\ 0 \end{pmatrix}$$

Similarly, for the spike at 1, we have

$$(ee^T)_1 = (0, 1, 0)\begin{pmatrix} 0 \\ 1 \\ 0 \end{pmatrix} - (0, 1, 0)\mathbf{B}^T(\mathbf{BB}^T)^{-1}\mathbf{B}\begin{pmatrix} 0 \\ 1 \\ 0 \end{pmatrix}$$

Likewise, for the spike at 2,

$$(ee^T)_2 = (0, 0, 1)\begin{pmatrix} 0 \\ 0 \\ 1 \end{pmatrix} - (0, 0, 1)\mathbf{B}^T(\mathbf{BB}^T)^{-1}\mathbf{B}\begin{pmatrix} 0 \\ 0 \\ 1 \end{pmatrix}$$

The *total* of these three is the grand sum of squared errors

$$V = \text{Total} = (ee^T)_0 + (ee^T)_1 + (ee^T)_2 = \text{trace}\left[\begin{pmatrix} 1 & 0 & 0 \\ 0 & 1 & 0 \\ 0 & 0 & 1 \end{pmatrix} - \mathbf{B}^T(\mathbf{BB}^T)^{-1}\mathbf{B}\right]$$

where the term "trace" denotes the sum of the elements on the main diagonal of the matrix.

Let the inverse of the matrix

$$\mathbf{BB}^T = \begin{bmatrix} r_0 & r_1 \\ r_1 & r_0 \end{bmatrix} = \mathbf{R}$$

be

$$\boldsymbol{\phi} = (\mathbf{BB}^T)^{-1} = \begin{bmatrix} \phi_{00} & \phi_{01} \\ \phi_{10} & \phi_{11} \end{bmatrix} = \mathbf{R}^{-1}$$

Then the main diagonal of $\mathbf{B}^T(\mathbf{BB}^T)^{-1}\mathbf{B}$ is

$$\begin{bmatrix} b_0^2\phi_{00} & * & * \\ * & b_1^2\phi_{00} + b_0 b_1\phi_{10} + b_0 b_1\phi_{01} + b_0^2\phi_{11} & * \\ * & * & b_1^2\phi_{11} \end{bmatrix}$$

where the asterisk * indicates unspecified entries. Hence, the *total* is

$$\text{Total} = 3 - b_0^2\phi_{00} - (b_1^2\phi_{00} + b_0b_1\phi_{10} + b_0b_1\phi_{01} + b_0^2\phi_{11}) - b_1^2\phi_{11}$$

Recalling that $b_0^2 + b_1^2 = r_0$ and $b_0b_1 = r_1$, we have

$$\text{Total} = 3 - r_0\phi_{00} - r_1\phi_{10} - r_1\phi_{01} - r_0\phi_{11} = 1$$

which shows that the *total* is equal to $n = 1$, as previously stated in this chapter (Robinson and Treitel, 1978, Chap. 7). Nevertheless, the sum of squared errors for each of the three different spikes *does* depend in general upon the delay properties (i.e., the shape) of b_0, b_1; that is,

Spike at 0:

$$(ee^T)_0 = 1 - b_0^2\phi_{00} \qquad \text{(depends upon } b_0)$$

Spike at 1:

$$(ee^T)_1 = 1 - b_1^2\phi_{00} - b_0b_1\phi_{10} - b_0b_1\phi_{01} - b_0^2\phi_{11}$$

[In general, $(ee^T)_1$ depends on b_0 and b_1, although in the special case of a two-term filter $\mathbf{a} = (a_0, a_1)$, the value of $(ee^T)_1$ depends only upon the autocorrelation of (b_0, b_1). For we see that $\phi_{00} = \phi_{11}$ and $\phi_{01} = \phi_{10}$ because

$$\phi = (\mathbf{BB}^T)^{-1} = \begin{pmatrix} r_0 & r_1 \\ r_1 & r_0 \end{pmatrix}^{-1} = \frac{1}{\Delta}\begin{pmatrix} r_0 & -r_1 \\ -r_1 & r_0 \end{pmatrix} = \begin{pmatrix} \phi_{00} & \phi_{01} \\ \phi_{10} & \phi_{11} \end{pmatrix}$$

and hence

$$(ee^T)_1 = 1 - (b_0^2 + b_1^2)\phi_{00} - 2b_0b_1\phi_{01}$$
$$= 1 - r_0\phi_{00} - 2r_1\phi_{01}$$

so that $(ee^T)_1$ does not depend upon the shape of (b_0, b_1).]

Spike at 2:

$$(ee^T)_2 = 1 - b_1^2\phi_{11} \qquad \text{(depends on } b_1)$$

Let us finally note that

$$e = d - aB$$
$$ee^T = (d - aB)(d - aB)^T$$
$$= (d - aB)d^T \qquad \text{(since } e \text{ is orthogonal to } aB)$$
$$= dd^T - aB^Td^T$$
$$= dd^T - dB^T(BB^T)^{-1}Bd^T$$
$$= d[I - B^T(BB^T)^{-1}B]d^T$$
$$\rightarrow 0 \qquad \text{for any } d \text{ as } m \rightarrow \infty \text{ if } b_0, b_1, \ldots, b_n \text{ is minimum delay.}$$

In summary, the spiking filter may be described as follows. For the finite case the desired output is

$$\mathbf{d}_\alpha = (0, 0, 0, \ldots, 1, 0, \ldots, 0)$$
$$\uparrow \uparrow$$
$$t = 0 t = \alpha$$

where α is the lag of the spike. As we have seen, the spiking filter for this lag is

$$\mathbf{a} = \underbrace{(b_\alpha, \ldots, b_1, b_0, 0, 0, \ldots, 0)}_{(m+1)\text{ terms}} \begin{pmatrix} r_0 & r_1 & \ldots & r_m \\ r_1 & r_0 & \ldots & \\ r_m & & \ldots & r_0 \end{pmatrix}^{-1}$$

which depends on the initial portion $b_0, b_1, \ldots, b_\alpha$ of the wavelet as well as on the autocorrelation $r_0, r_1, \ldots, r_m$. The minimum sum of squared errors also depends on the shape of the wavelet, except in special cases.

Wavelet Prediction

An important special case of the shaping filter is the wavelet prediction filter. Here the desired output is equal to the input signal advanced by a certain time distance α, called the *prediction distance*. The advanced signal consists of two parts: the irreducible part,

$$b_0, b_1, \ldots, b_{\alpha-1}$$

which occurs before time instant zero, and thus is outside the range of the filter, and the reducible part,

$$b_\alpha, b_{\alpha+1}, \ldots, b_n$$

which is within the range of the filter. Hence, the reducible part is the part that represents the desired output of the filter; that is, the desired output is the $1 \times (m + n + 1)$ row vector given by

$$\mathbf{d}_\alpha = (b_\alpha, b_{\alpha+1}, \ldots, b_n, 0, \ldots, 0)$$

The prediction filter for prediction distance α is

$$\mathbf{f}_\alpha = \mathbf{d}_\alpha \mathbf{A} = b_\alpha \mathbf{a}_0 + b_{\alpha+1} \mathbf{a}_1 + \cdots + b_n \mathbf{a}_{n-\alpha}$$

The sum of squared errors for shaping the input signal into the reducible part is

$$v_\alpha = \mathbf{d}_\alpha (\mathbf{I} - \mathbf{C}) \mathbf{d}_\alpha^T$$
$$= \sum_{i=\alpha}^{n} b_i^2 - \sum_{i=0}^{n-\alpha} \sum_{j=0}^{n-\alpha} b_{i+\alpha} c_{ij} b_{j+\alpha}$$

The prediction error is the sum of the irreducible part, which is all error, and the error between the reducible part and the actual output of the filter. If we let w_α denote the sum of squared prediction errors, we have

$$w_\alpha = \sum_{i=0}^{\alpha-1} b_i^2 + v_\alpha$$

where the first term on the right is the contribution of the irreducible part and the second the reducible part.

For a given input signal **b** and a given operator length $m + 1$, let us consider the sum of squared prediction errors as a function of the prediction distance; that is, let us consider w_α as a function of α. For sufficiently long operators, w_α is a monotone-increasing function of α (where $\alpha = 1, 2, 3,$ ...); for short operators this is not always so. The irreducible component of w_α is the partial energy of the input signal up to time $\alpha - 1$; thus, the irreducible component is a monotonically nondecreasing function of α. No general statement can be made about the reducible component v_α, except that it is zero for α greater than n. As a result, the curve w_α will have a minimum value for one or more values of α; such a value of α is called the *optimum prediction distance*.

It is interesting to note that whereas the optimum delay (i.e., optimum spike position) of a signal is fundamentally related to the delay properties (i.e., minimum delay, mixed delay, or maximum delay) of the signal, the optimum prediction distance in no way depends on the delay properties of the signal. We can establish this result by considering the normal equation for the determination of the prediction operator. The matrix normal equation is

$$\mathbf{f}_\alpha \mathbf{R} = \mathbf{d}_\alpha \mathbf{B}^T$$

The right-hand side is the crosscorrelation of the reducible part of the signal with the signal; that is, the right-hand side is that portion of the autocorrelation given by

$$\mathbf{d}_\alpha \mathbf{B}^T = (r_\alpha, r_{\alpha+1}, \ldots, r_{\alpha+m})$$

where it is understood that $r_s = 0$ for $s > n$. Hence, the normal equation for the prediction operator $\mathbf{f}_\alpha$ involves only the autocorrelation of the input wavelet **b**, and since the autocorrelation does not depend on the delay properties of the signal, neither does the prediction operator. Thus, the optimum prediction distance is independent of the delay properties of the input signal **b**.

The counterparts of the prediction filters are the hindsight filters. Here the desired output is equal to the input signal delayed by a certain time distance, called the *hindsight distance*. The delay distance may be represented as $-\alpha$, where α is an intrinsically negative number. That is, delay distance may be thought of as a negative prediction distance. The reverse of the hind-sight operator for hindsight distance $-\alpha = m + 1$ is the same as the prediction operator for prediction distance $\alpha = 1$. More generally, the reverse of the hindsight operator for hindsight distance $-\alpha = m + k$ is the same as the prediction operator for prediction distance $\alpha = k$.

Wavelet Prediction with Infinitely Long Filters

Let us now consider the rationale of predicting the future values of a wavelet from its past values by means of a causal digital filter. Let the wavelet $(b_0, b_1, b_2, \ldots)$ be the input to a causal digital filter $(f_0, f_1, f_2, \ldots)$. This filter may have an infinite number of coefficients. As output, we desire the predicted values of the input wavelet. If we denote the prediction distance by the integer α, the desired output is a replica of the input wavelet advanced in time by α units. The sum of squared errors for prediction distance α is given by

$$w_\alpha = b_0^2 + b_1^2 + \cdots + b_{\alpha-1}^2 + (b_\alpha - c_0)^2 + (b_{\alpha+1} - c_1)^2 + (b_{\alpha+2} - c_2)^2 + \cdots$$

The contribution

$$b_0^2 + b_1^2 + \cdots + b_{\alpha-1}^2$$

is due to the fact that the filter is causal; hence, it can produce no output before the input, which starts at time zero. The contribution

$$(b_\alpha - c_0)^2 + (b_{\alpha+1} - c_1)^2 + (b_{\alpha+2} - c_2)^2 + \cdots$$

is due to the error between actual and predicted values. A causal time-invariant linear filter can only form the sum of delayed replicas of the input weighted by the filter coefficients; the actual output

$$c_t = \sum_{s=0}^{\infty} f_s b_{t-s} \qquad (\text{where } t = 0, 1, 2, 3, \ldots)$$

is the weighted sum of the delayed replicas b_{t-s} for $s = 0, 1, 2, \ldots$, where the coefficients f_s are the weights.

A particular set of wavelets is called *complete* if any wavelet whatsoever may be expressed as a linear combination of members of this particular set. It can be shown that the set of delayed replicas b_{t-s} for $s = 0, 1, 2, \ldots$ is a complete set if the input wavelet b_t is minimum-delay, whereas this set is not complete if the input wavelet is not minimum-delay. Let us consider the case where the input wavelet is minimum-delay, which we represent as $(b_{0,0}, b_{0,1}, b_{0,2}, \ldots)$. The desired output before time zero is unpredictable. The desired output from time zero on is the wavelet $(b_{0,\alpha}, b_{0,\alpha+1}, b_{0,\alpha+2}, \ldots)$, which can be expressed as a linear combination of the delayed replicas of the minimum-delay input wavelet. Thus, if the input wavelet is minimum-delay, we can find filter coefficients $(f_0, f_1, f_2, \ldots)$ such that the actual output wavelet $(c_0, c_1, c_2, \ldots)$ is exactly equal to the wavelet $(b_{0,\alpha}, b_{0,\alpha+1}, b_{0,\alpha+2}, \ldots)$. In this case the contribution

$$(b_{0,\alpha} - c_0)^2 + (b_{0,\alpha+1} - c_1)^2 + (b_{0,\alpha+2} - c_2)^2 + \cdots$$

is zero, and hence the sum of squared errors becomes

$$w_\alpha = b_{0,0}^2 + b_{0,1}^2 + \cdots + b_{0,\alpha-1}^2 \qquad \text{(minimum-delay case)}$$

which is a minimum. The operator $(f_0, f_1, f_2, \ldots)$ which achieves this minimum is the optimum prediction operator. We recognize this sum of squared errors as the energy buildup curve of the minimum-delay input wavelet up to time $\alpha - 1$.

Let us next consider the case when the input wavelet is not minimum-delay. In this case there is a discrepancy between the actual output $(c_0, c_1, c_2, \ldots)$ and the wavelet $(b_\alpha, b_{\alpha+1}, b_{\alpha+2}, \ldots)$. However, the same optimum filter coefficients as in the minimum-delay case yield the minimum value of the sum of squared errors. This minimum value is numerically the same as the minimum value as in the minimum-delay case, except that now both contributions to the sum of squared errors are present. The contribution due to the energy buildup is less than in the minimum-delay case (as the energy buildup curve of a nonminimum-delay wavelet falls below the energy buildup curve of the minimum-delay wavelet), whereas the contribution due to the discrepancy between the wavelets $(b_\alpha, b_{\alpha+1}, b_{\alpha+2}, \ldots)$ and $(c_0, c_1, c_2, \ldots)$ exactly makes up for the decrease in the first contribution.

Concluding Remarks

A detailed analysis of the error associated with the least squares shaping and spiking filters sheds much light on the design problem. Not surprisingly, the error behavior is governed by the delay properties of the input signal to be filtered. Additional insight is achieved by consideration of the overall error associated with all possible spiking filters for a given input signal, and for the expression of the shaping filter error in terms of this set of all possible spiking filters.

stationary time series

Summary

Largely through the impetus of Norbert Wiener, statistical communication theory has emphasized the generalized harmonic analysis, or spectral representation, of a time series which results in its representation in terms of its harmonic, or sinusoidal, components. This chapter, on the other hand, develops in an expository manner the generalized regression analysis, or innovational representation, of a time series. This representation gives the time series at any moment as the sum of two components. The first component is the output of a minimum-delay filter subject to a white-noise input, which constitutes the innovations. The second component is a purely deterministic time series. For a purely nondeterministic time series, this second component is absent.

Introduction

There are two basic approaches to treating data observed in nature, and in particular the data represented on a seismogram. One is the deterministic approach and the other is the statistical approach. Many people think of these

213

two approaches as conflicting, but actually this is not the case. Investigations and experiments in many different sciences indicate that each approach is fundamentally equivalent to the other.

The approach in classical seismology has been almost exclusively deterministic. In this approach deterministic methods are used to investigate laws connecting seismological phenomena. These laws are considered to be precise in action even though the observations on the quantities involved may be inaccurate and are certain to be incomplete.

On the other hand, the statistical approach utilizes quantities in the form in which they are observed. Distributions and statistical functions of these quantities are examined in such combinations as one chooses. Of course, one has considerable freedom in the selection of the quantities that are to form the subject of a statistical investigation. Actually, in an ideally complete survey, one should investigate all possible statistical parameters and combinations of parameters, not merely a selection from among them. Unfortunately, such an undertaking would be impossible because of its sheer magnitude.

Therefore, in a statistical investigation one should look for groups of parameters which are connected with each other by rigid dynamic laws and with the nature of the desired information. For such a group, some of the parameters would be determined by a knowledge of the remaining ones, and the dynamics would be expressed as a statistical fact. If the dynamics are not so expressible, one can conclude either that a sufficient number of the significant statistical parameters have not been considered to give a true picture of the situation, or that these parameters have been observed so inaccurately that they cannot give the true picture.

It is unlikely for a significant dynamic relationship not to be brought out by a proper statistical examination of the relevant quantities. In fact, if certain simplifying assumptions have to be made in the derivation of dynamic laws by a deterministic approach, it is frequently the case that the statistical approach actually yields more information.

The Basic Problems

Exploration seismology can be broken down into logical steps, which are presented graphically in Figure 9-1. We see that there are two mathematical approaches for the treatment of data, the deterministic approach and the statistical approach. The *deterministic approach* consists of utilizing physical theories of wave propagation involving the solutions of integral and differential equations satisfying boundary and initial conditions. The *statistical approach* consists of utilizing statistical theories of time series leading to the expression of the dynamics as a statistical fact. The basic problems of seismic research will never be all solved, but nevertheless significant steps have been taken in that direction.

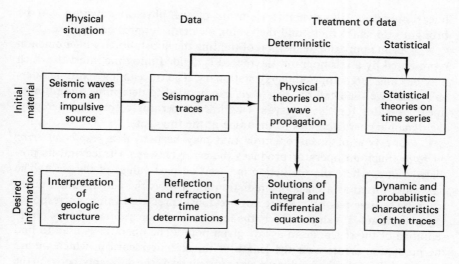

Figure 9-1. Deterministic and statistical approaches to seismology.

Briefly, the basic problems deal with the determination of (1) statistical methods adequate to separate desired information from the total information present on the seismogram, (2) the relationship of desired information in statistical form with significant seismologic variables and the geologic structure, and (3) the interrelation of the deterministic and statistical approaches.

The Introduction of Statistical Methods to Seismology

This section deals with the first basic problem, the determination of statistical methods adequate to separate desired information from the total information present on the seismogram. We consider statistical methods because a seismogram as recorded is a statistical time series, and all the traces on a seismic record constitute a set of multiple time series. This set of multiple time series is tied down to a specific origin in time, the time of source initiation. Time series with such a time origin dependence are called nonstationary, as opposed to stationary time series, which are not linked to a specific origin in time.

Statistical research in seismology is concerned with the evaluation of valid statistical methods in order to obtain the goal set forth in the first basic problem. One of the most useful statistical methods is the application of linear operators to seismic records. The method of utilizing linear operators is mathematically equivalent to the solution of a system of differential equations in space and time, but is more powerful in that it can cope effectively with disturbing influences of a random or quasi-random nature. By the use of linear operators, the dynamic elements of the physical situation can be brought into sharp focus and disturbing elements suppressed.

As an approximate method of treating the nonstationary phenomenon represented by a seismogram, the record is divided into time intervals which may be considered approximately stationary. The trace in a given time interval, which we shall call the *time gate*, can be used to determine an optimum linear operator for this interval. This linear operator contains, inherently, the dynamic characteristics of the trace in the time gate.

One type of linear operator that may be used is a *prediction error operator*. Such an operator produces the error between a trace and its predicted value, where the prediction is based on past values of the trace. The past values used as a basis of prediction are always the actual values taken from the seismic record, and are not previously predicted values corresponding to these actual values. Hence, the input into the operator is always representative of the seismogram at any given point. The operator is used to find the prediction errors over the length of the seismic record in which we are interested. Because the prediction error is defined as the difference between the predicted value and the actual value of the trace, this error gives a measure of the innovational structure of the seismogram. We use the word "innovation" in the sense that low error, or good predictability, indicates small innovation, and large error indicates considerable innovation, at that instant. Each prediction error operator generates a sequence of errors that forms a time series, which we may call the *prediction error time series*. An investigation of these error time series, as we shall see, exposes the innovations as a statistical fact. Such information about the innovational characteristics constitutes desired information.

The Relationship of Statistical Information to Geologic Information

Once the innovational characteristics of a seismic record are rendered by statistical methods into a usable form, the next problem is to translate this statistical information into meaningful geologic information. This makes up the second basic problem of seismic research. By computing linear operators over different sections of the record, it is found that the error of prediction gives a measure of the reflection coefficients of the subsurface geologic interfaces. Therefore, by examining either error time series or statistics computed from these error time series, one is able to pick off the arrival of reflected energy at places of high error. Such a procedure of analysis has been found to have two advantages: (1) the qualitative characteristics of reflections are better defined, and (2) quantitatively, more reflections may be distinguished than by visual inspection of the raw seismogram. Many illustrations of these concepts are given in the next three chapters.

The problems to be explored in this direction include a more complete analysis of the reason why the dynamics of the seismograms change as they do. Various types of operators might be explored to determine which types are more sensitive to the location and discrimination of reflections. The variables that come under consideration may be subdivided into geologic variables, instrumental variables, and mathematical variables. The geologic variables include such factors as underground structure, physical constants of the earth, and source effects. Instrumental variables include geophone layouts and response characteristics of the instruments used. Mathematical variables include the parameters of the linear operator and the statistics used to characterize the time series under consideration. It is necessary to investigate the effects of all these variables and to optimize those under our control.

The Interrelation of the Deterministic and the Statistical Approaches

The third basic problem, the interrelation of the statistical and deterministic approaches, is fundamental and also the most difficult. One method of approach is the following. We are given a physical situation that can be treated by exact physical theory, such as the layered model of the earth under certain simplifying assumptions. From such a model a theoretical time-series analysis can be carried out. In particular, the form of such statistical functions as the autocorrelation and the crosscorrelation can be derived from the physical situation.

Another method of approach is to determine what makes up the predictable component of the seismogram traces in terms of finite-difference approximations to the wave equation. The degree of predictability is intimately related to the stability of the wave-equation operators. The stability of these functions depends upon the type of nonstationary time series generated by the geologic situation. The nonstationary character of these time series means that any differential equation which is set up to explain this phenomenon has variable coefficients which depend either upon time or upon the phenomenon itself. For optimum results the solution of the proper wave-equation formulation might be incorporated into the prediction mechanism.

Time-Series Analysis

We shall now present some concepts from the theory of stationary time series. We are concerned chiefly with stating definitions and important theorems, and we shall by no means present a complete summary of the theory of time-series analysis. Those readers who are familiar with these concepts we present here may proceed directly to Chapter 10 without losing continuity.

The concepts given in this section depend upon the assumption that the time series is stationary. A time series is said to be stationary if the probabilities involved are not tied down to a specific origin in time, and if the series is conceived to run from minus infinity to plus infinity in time.

The particular feature that makes the application of traditional statistical methods to time series difficult is the absence of independence between successive observations. This lack of independence between successive observations is the fundamental characteristic of a time series. In the time series observed in nature, such interdependence is caused by neither completely random nor completely deterministic factors, but instead the motivating factors lie somewhere between these two extremes.

As a result, the most direct procedure in the analysis of time series is one that exploits the lack of independence between successive observations. The use of an operator that depends upon the interdependence of observations in the time series is one such procedure. Operators may be linear or nonlinear. Since the concept of linearity that is used is of the most general type and includes a wide range of time series, we deal with linear operators almost exclusively in this book, although conceivably the use of nonlinear operators may some day prove of greater applicability to seismogram analysis.

As an introduction to the concept of a linear operator, we shall consider the pure sine series u_t of angular frequency ω_0,

$$u_t = A \sin (\omega_0 t + \theta) \qquad (9\text{-}1)$$

where A is amplitude and θ is phase.

Such a series is completely deterministic, for it contains no random element. As a result, there exists an identity connecting three consecutive observations, which is given by

$$u_{t+2\Delta t} = a u_{t+\Delta t} + b u_t \qquad (9\text{-}2)$$

The constant a is equal to $2 \cos \omega_0 \Delta t$ and the constant b is equal to -1. The constants a and b constitute a linear operator whereby, from two consecutive values of the time series, all future values may be obtained. It should be noted that the constants a and b of the linear operator are independent of the phase θ of the sine series. Hence, the linear operator reveals the stationary character of the sine series in that the operator is not tied down to any particular origin in time. Also, the operator is independent of the amplitude A, which means that it is independent of the measurement used in the observations. Therefore we see that the linear operator represents an intrinsic property of the sine series, and it is not linked to the time origin or scale of the individual observations. These properties of a linear operator carry over for a large group of functions other than sines and cosines.

The Relationship of the Linear Operator and the Autocorrelation

Such an operator, which is linear, invariant with respect to the origin in time, and dependent only on the past history of the time series, is the type that Wiener considered in his theory of prediction for stationary time series. He approximated the future values $x_{t+\alpha}$ of the times series by applying a linear operator k_τ to the past values $x_{t-\tau}$ by means of the expression

$$\hat{x}_{t+\alpha} = \sum_{\tau=0}^{\infty} k_\tau x_{t-\tau} \tag{9-3}$$

The linear operator k_τ is determined by minimizing the mean square error, which is given by

$$\lim_{T \to \infty} \frac{1}{2T+1} \sum_{t=-T}^{T} \left(x_{t+\alpha} - \sum_{\tau=0}^{\infty} k_\tau x_{t-\tau} \right)^2 \tag{9-4}$$

Utilizing the calculus of variations for this minimization process, he obtained the normal equations

$$\sum_{s=0}^{\infty} k_s \phi_{\tau-s} = \phi_{\tau+\alpha} \qquad \text{for } \tau \geq 0 \tag{9-5}$$

Here ϕ_τ is defined by

$$\phi_\tau = \lim_{T \to \infty} \frac{1}{2T+1} \sum_{t=-T}^{T} x_{t+\tau} x_t \tag{9-6}$$

and is called the *autocorrelation function*. Hence, equation (9-5) tells us that the prediction operator k_τ is determined from the autocorrelation function ϕ_τ. The autocorrelation function represents the intrinsic dynamic properties of the time series.

The Spectrum

The fundamental theorem of generalized harmonic analysis, due to Wiener, relates the autocorrelation function ϕ_τ with a monotone-nondecreasing function $\Lambda(\omega)$. More precisely, if the autocorrelation function ϕ_τ exists, there exists a monotone-nondecreasing function $\Lambda(\omega)$ which is given by the Fourier transform

$$\phi_\tau = \frac{1}{2\pi} \int_{-\pi}^{+\pi} e^{i\omega\tau} \, d\Lambda(\omega) \tag{9-7}$$

We say that the function $\Lambda(\omega)$ is a monotone-nondecreasing function, instead of a monotone increasing function, because there may be plateaus.

The function $\Lambda(\omega)$ is called the *integrated spectrum* of x_t, and represents the total power in the spectrum of x_t between the frequency $\omega = -\pi$ and the frequency ω. The integrated spectrum $\Lambda(\omega)$ may have a series of jumps or other singularities if there are exact frequencies or spectral lines in the time series x_t. Otherwise, $\Lambda(\omega)$ will be absolutely continuous, and this is the usual case met in applications.

If $\Lambda(\omega)$ is absolutely continuous, its derivative, $\Lambda'(\omega) = \Phi(\omega)$, is called the *power spectrum* of the time series x_t. We may then write the Fourier transform

$$\phi_\tau = \frac{1}{2\pi} \int_{-\pi}^{+\pi} e^{i\omega\tau} \Phi(\omega) \, d\omega \tag{9-8}$$

and, in the case of simple functions, the inverse transform

$$\Phi(\omega) = \sum_{\tau=-\infty}^{\infty} \phi_\tau e^{-i\omega\tau} \tag{9-9}$$

Since in this case, both ϕ_τ and $\Phi(\omega)$ are even functions, we may rewrite equations (9-8) and (9-9) as

$$\phi_\tau = \frac{1}{\pi} \int_0^{\pi} \Phi(\omega) \cos \omega\tau \, d\omega \tag{9-10}$$

and

$$\Phi(\omega) = \phi_0 + 2 \sum_{\tau=1}^{\infty} \phi_\tau \cos \omega\tau \tag{9-11}$$

Thus, the autocorrelation function gives information about x_t, which is equivalent to the information given by the spectrum. More precisely, information about the amplitudes of the frequencies of x_t is preserved, and information about the phases of the individual frequencies is lost, both in the autocorrelation function and in the spectrum.

By setting $\tau = 0$, equations (9-6) and (9-8) reduce to

$$\phi_0 = \lim_{T \to \infty} \frac{1}{2T+1} \sum_{t=-T}^{T} x_t^2 = \frac{1}{2\pi} \int_{-\pi}^{\pi} \Phi(\omega) \, d\omega \tag{9-12}$$

Hence, we see that the total power in the spectrum is given by ϕ_0. The customary statistical practice is to normalize the autocorrelation function and spectrum by normalizing x_t so that it has zero mean and unit variance in the time-average sense. Then $\phi_0 = 1$ and $|\phi_\tau| \leq 1$, and the total power in the spectrum is equal to 1.

The Relationship between the Autocorrelation and the Spectrum

We shall now give four examples illustrating the relationship as given in equations (9-10) and (9-11) between the autocorrelation function and the spectrum.

The first example is that of the *pure sine series* given by equation (9-1). This series is completely deterministic, for it contains no random elements. The normalized autocorrelation of this series, computed from equation (9-6), is given by

$$\phi_\tau = \cos \omega_0 \tau \qquad (9\text{-}13)$$

which is an undamped cosine wave. Classical Fourier series methods show that the spectrum is a line spectrum in which all the power is concentrated at the frequency ω_0.

This example allows us to give a heuristic interpretation to the relationship between the autocorrelation function and the spectrum given in equation (9-10). Consider the spectrum $\Phi(\omega)$ of an arbitrary time series x_t. Each small band of frequencies between ω and $\omega + d\omega$ acts with the differential power $\Phi(\omega) \, d\omega$. In view of equation (9-13), the differential transform of the small band of frequencies is given by $\Phi(\omega) \cos \omega\tau \, d\omega$. This differential is the contribution of the small band of frequencies between ω and $\omega + d\omega$ to the autocorrelation function. Summing these differential transforms from $\omega = -\pi$ to $\omega = \pi$, we obtain the integral for the autocorrelation ϕ_τ given in equation (9-10).

The second example is that of a *random series*. A random series is conceived to have a white-light spectrum; that is, the spectrum is given by a rectangular distribution. Let the spectrum be

$$\Phi(\omega) = 1 \qquad (9\text{-}14)$$

for the range given by $-\pi \leq \omega \leq \pi$, and let $\Phi(\omega) = 0$ for values of ω outside this range. Then the autocorrelation is given by

$$\phi_\tau = \begin{cases} 1 & \text{if } \tau = 0 \\ 0 & \text{if } \tau \neq 0 \end{cases} \quad \text{or} \quad \phi_\tau = \delta_\tau \qquad (9\text{-}15)$$

where δ_τ is the Kronecker delta function. Such a random series is called an *uncorrelated series*, or *white noise*.

The third example is the case of *pure persistence* in a time series, and is taken from Yaglom (1962, p. 57). In this case the autocorrelation function is given by the exponential

$$\phi_\tau = ca^{|\tau|}, \qquad c > 0, \qquad |a| < 1 \tag{9-16}$$

and the spectrum by the curve

$$\Phi(\omega) = c\,\frac{1 - a^2}{|1 - ae^{-i\omega}|^2} \tag{9-17}$$

Hence, we see that all frequencies exist in the range $-\pi \le \omega \le \pi$.

The last example also comes from Yaglom (1962, pp. 58–59), and is the case in which the autocorrelation is the weighted sum of the autocorrelations (9-15) and (9-16) given by

$$\phi_\tau = \frac{(a - b)(1 - ab)}{a(1 - a^2)}a^{|\tau|} + \frac{b}{a}\delta_\tau \tag{9-18}$$

where $|a| < 1$ and $|b| < 1$. Then the spectrum is given by the curve

$$\Phi(\omega) = \frac{|1 - be^{-i\omega}|^2}{|1 - ae^{-i\omega}|^2} \tag{9-19}$$

The Relationship between the Linear Operator and the Crosscorrelation

The discussion to this point has concerned itself with the statistical properties of a single stationary time series x_t. We now wish to extend these concepts to the case where we have multiple stationary time series.

A linear operator for this case is defined in a way analogous to the case of single time series. It predicts the future of one time series from its past values and the past values of the other time series. The minimization of the mean square error for the general case was carried out by Wiener (1949). It was shown there that the linear operator depends only on the autocorrelation and crosscorrelations of the time series considered. The crosscorrelation function is a property of two time series, x_{1t} and x_{2t}, and is defined in a way similar to the autocorrelation function by

$$\phi_{12}(\tau) = \lim_{T \to \infty} \frac{1}{2T + 1} \sum_{t=-T}^{T} x_{1,t+\tau}x_{2,t} \tag{9-20}$$

The crosscorrelation between x_{2t} and x_{1t} is defined as

$$\phi_{21}(\tau) = \lim_{T \to \infty} \frac{1}{2T + 1} \sum_{t=-T}^{T} x_{2,t+\tau}x_{1t} \tag{9-21}$$

From the definitions (9-20) and (9-21) it follows that

$$\phi_{12}(\tau) = \phi_{21}(-\tau) \qquad (9\text{-}22)$$

In statistical practice the crosscorrelation function is usually normalized by letting both x_{1t} and x_{2t} have zero mean and unit variance in the time-average sense. Then we have $|\phi_{11}(\tau)| \leq 1$ and $|\phi_{22}(\tau)| \leq 1$. Using the Schwarz inequality, we have the desired normalization of the crosscorrelation, which is $|\phi_{12}(\tau)| \leq 1$.

The Cross-Spectrum

The crosscorrelation function $\phi_{12}(\tau)$ of x_{1t} and x_{2t} may be expressed as the Fourier transform

$$\phi_{12}(\tau) = \frac{1}{2\pi} \int_{-\pi}^{\pi} e^{i\omega\tau}\Phi_{12}(\omega)\, d\omega \qquad (9\text{-}23)$$

Here $\Phi_{12}(\omega)$ is defined to be the cross-spectrum of x_{1t} and x_{2t}. In the case of simple functions, the inverse transform may be written as

$$\Phi_{12}(\omega) = \sum_{\tau=-\infty}^{\infty} e^{-i\omega\tau}\phi_{12}(\tau) \qquad (9\text{-}24)$$

In general, the crosscorrelation function $\phi_{12}(\tau)$ is not an even function of τ, and hence equation (9-24) tells us that the cross-spectrum $\Phi_{12}(\omega)$ has real and imaginary parts. Equations analogous to (9-23) and (9-24) hold for the crosscorrelation $\phi_{21}(\tau)$ and the cross-spectrum $\Phi_{21}(\omega)$ between the time series x_{2t} and x_{1t}. From these relations we find that

$$\Phi_{12}(\omega) = \Phi_{21}^{*}(\omega) \qquad (9\text{-}25)$$

where the asterisk indicates the complex conjugate.

Since the cross-spectrum $\Phi_{12}(\omega)$ is a complex-valued function of the real variable ω, we may write

$$\Phi_{12}(\omega) = \mathrm{Re}\,[\Phi_{12}(\omega)] + i\,\mathrm{Im}\,[\Phi_{12}(\omega)] \qquad (9\text{-}26)$$

where $\mathrm{Re}\,[\Phi_{12}(\omega)]$ designates the real part, and $\mathrm{Im}\,[\Phi_{12}(\omega)]$ designates the imaginary part, of the cross-spectrum. We may also express the cross-spectrum by

$$\Phi_{12}(\omega) = |\Phi_{12}(\omega)|\, e^{i\theta(\omega)} \qquad (9\text{-}27)$$

Here $|\Phi_{12}(\omega)|$ designates the absolute value of the cross-spectrum, and is given by

$$|\Phi_{12}(\omega)| = \sqrt{[\text{Re } \Phi_{12}(\omega)]^2 + [\text{Im } (\Phi_{12})]^2} \qquad (9\text{-}28)$$

The argument $\theta(\omega)$ of the cross-spectrum is a function of the frequency ω, and is given by

$$\theta(\omega) = \tan^{-1}\left[\frac{\text{Im } \Phi_{12}(\omega)}{\text{Re } \Phi_{12}(\omega)}\right] \qquad (9\text{-}29)$$

Let the spectrum of x_{1t} be $\Phi_{11}(\omega)$ and the spectrum of x_{2t} be $\Phi_{22}(\omega)$. It can be shown that the absolute value of the crosssectrum is less than or equal to the geometric mean of the individual spectra; that is,

$$|\Phi_{12}(\omega)| \leq \sqrt{\Phi_{11}(\omega)}\sqrt{\Phi_{22}(\omega)} \qquad (9\text{-}30)$$

Hence, we see that the cross-spectrum preserves at most only the common frequencies of x_{1t} and x_{2t}.

The matrix

$$\begin{bmatrix} \Phi_{11}(\omega) & \Phi_{12}(\omega) \\ \Phi_{21}(\omega) & \Phi_{22}(\omega) \end{bmatrix}$$

is called the *coherency matrix*. In order for x_{1t} and x_{2t} not to be completely dependent upon each other, the determinant of the coherency matrix must be positive for all frequencies. That is, the coefficient of coherency $X_{12}(\omega)$ defined by

$$\Phi_{12}(\omega) = X_{12}(\omega)\sqrt{\Phi_{11}(\omega)\Phi_{22}(\omega)}$$

must have absolute value $|X_{12}(\omega)|$ less than 1 for all frequencies. In other words, for two nondeterministic stationary time series, the Fourier series in exp $(-i\omega)$ with Fourier coefficients that are the crosscorrelation of their autocorrelations must exceed almost everywhere the Fourier series in exp $(-i\omega)$ with Fourier coefficients that are the autocorrleation of their crosscorrelation.

Let us consider two observational finite time series x_{jt}, where $j = 1, 2$ and $t = 0, 1, \ldots, n$. If one assumes that x_{jt} vanishes outside this time range and estimates the spectral densities by

$$\Phi_{jk}(\omega) = \sum_{t=0}^{n} x_{jt}e^{i\omega t} \sum_{s=0}^{n} x_{ks}e^{-i\omega s} = \sum_{\tau=-n}^{n}\left(\sum_{t=0}^{n} x_{jt}x_{k,t+\tau}\right)e^{-i\omega\tau}$$

then the determinant of the coherency matrix will vanish identically in ω. In other words, the estimation formula above forces the two time series to be completely coherent. This unfortunate situation is frequently called

Simpson's paradox. The resolution of the paradox is that under the hypothesis that two observational time series are not completely coherent, one should utilize formulas that provide estimates of the theoretical coherency which actually exists between the two time series.

The concept of *coherency* is an important one in the study of seismic records. Computations indicate that seismic traces are more coherent on the average in an interval containing a major reflection than in an adjacent nonreflection interval. This coherency property of reflections assists the visual detection of reflections on a seismogram, and hence may be exploited in the detection of weak reflections by statistical methods.

In closing this section, which deals with concepts from the theory of stationary time series, we mention the following interesting example. Consider the purely random series $u_1, u_2, u_3, \ldots$ and the purely random series $v_1, v_2, v_3, \ldots$, in which the v_t series is defined by the relationship $v_t = u_{t-j}$. Then it is seen that the crosscorrelation of the two series is zero everywhere except at the jth lag, where the crosscorrelation is equal to 1. Such an example illustrates the value of the crosscorrelation function to determine phase relationships.

Ensemble Averages and Time Averages

The techniques that we present in this chapter have proven to be of great value in many applications. The linear systems for prediction and filtering described here are basic to filter theory and must be known to those who wish to design more complicated systems. In treating the discrete time case, we retain the fundamental ideas involved in the continuous time case, and yet are able to keep the mathematical argument at a relatively elementary level.

Any observational time series $x_t (-\infty < t < \infty)$ may be considered to be a realization of a *random process*, or *stochastic process*, which is a mathematical abstraction defined with respect to a probability field. For any stochastic process, one may form averages with respect to the statistical population or *ensemble* of realizations x_t for a fixed value of time t. Such averages are called *ensemble averages*, and we shall denote such an averaging process by the expectation symbol E. If the mean value $m = E(x_t)$ and the (unnormalized) autocorrelation coefficients

$$\phi_\tau = E(x_{t+\tau} x_t)$$

are finite and independent of t, the process is called stationary in the wide sense, or second-order stationary, or covariance stationary. Without loss of generality we assume $E(x_t)$ to be zero. There is another type of average, known as the *time average*, in which the averaging process is carried out with respect to all values of time t for a fixed realization x_t ($-\infty < t < \infty$) of the stochastic process. For a large class of stationary processes, called *ergodic processes*, ensemble averages and the corresponding time averages are equal with probability 1. Consequently, the autocorrelation of an ergodic process may be expressed as the time average

$$\phi_\tau = \lim_{T \to \infty} \frac{1}{2T+1} \sum_{t=-T}^{T} x_{t+\tau} x_t \tag{9-31}$$

taken over a single realization of the time series x_t. The autocorrelation function is a nonnegative definite function; that is,

$$\sum_{s=0}^{n} \sum_{t=0}^{n} a_s a_t \phi_{t-s} \geq 0 \tag{9-32}$$

for any n and any sequence of (real) constants $a_0, a_1, a_2, \ldots, a_n$. The nonnegative definiteness property of the autocorrelation is mathematically equivalent to the monotonic-nondecreasing property of the integrated spectrum $\Lambda(\omega)$.

Linear Prediction

Let us define the random variable $\hat{x}_{t+\alpha}$ to be the linear least-squares prediction of $x_{t+\alpha}$ in terms of the complete past $\ldots, x_{t-2}, x_{t-1}, x_t$ of the time series up to time t. That is, $\hat{x}_{t+\alpha}$ is given by equation (9-3), which is

$$\hat{x}_{t+\alpha} = \sum_{\tau=0}^{\infty} k_\tau x_{t-\tau} \tag{9-3}$$

where $k_0, k_1, k_2, \ldots$ is the prediction operator, with Fourier transform

$$K(\omega) = \sum_{\tau=0}^{\infty} k_\tau e^{-i\omega\tau} \tag{9-33}$$

The prediction error at time $t + \alpha$ is defined as $x_{t+\alpha} - \hat{x}_{t+\alpha}$.

The prediction operator is determined by requiring the mean square prediction error, given by expression (9-4), to be a minimum. In terms of the ensemble average, expression (9-4) becomes

$$\sigma_\alpha^2 = E[(x_{t+\alpha} - \hat{x}_{t+\alpha})^2] = \frac{1}{2\pi} \int_{-\pi}^{\pi} |e^{i\omega\alpha} - K(\omega)|^2 \, d\Lambda(\omega) \tag{9-34}$$

The x_t process is called deterministic if $\sigma_1^2 = 0$, in which case the future $x_{t+\alpha}$ is completely determined from the remote past $x_\tau, x_{\tau-1}, \ldots$, where τ

is allowed to approach minus infinity. The process is called nondeterministic if $\sigma_1^2 > 0$, in which case the future cannot be completely determined by a linear operation on the past.

White Noise

Two real random variables x and y with finite variances are said to be uncorrelated if $E(xy) = E(x)E(y)$, orthogonal if $E(xy) = 0$, and orthonormal if $E(xy) = 0$, $E(x^2) = 1$, and $E(y^2) = 1$. A mutually uncorrelated process is a stationary process for which the observations ϵ_t are uncorrelated in pairs; that is, $E(\epsilon_t \epsilon_s) = E(\epsilon_t)E(\epsilon_s)$ for t not equal to s. In what follows we shall consider uncorrelated variables to be normalized such that $E(\epsilon_t) = 0$ and $E(\epsilon_t^2) = 1$, in which case the ϵ_t forms an orthonormal sequence of random variables. As we have seen, the autocorrelation of the ϵ_t process as given by equation (9-15) vanishes except for zero lag, and the power spectrum as given by equation (9-14) is constant for the interval $(-\pi, \pi)$. These processes therefore have flat spectra, and they are called white noise. If the ϵ_t are independent random variables with the same Gaussian probability density function, then ϵ_t is (discrete-time parameter) Brownian motion.

Given a white-noise process ϵ_t, the corresponding process of moving summation is defined as

$$x_t = \sum_{s=-\infty}^{\infty} c_s \epsilon_{t-s} \qquad (-\infty < t < \infty) \qquad (9\text{-}35)$$

It is supposed that the operator c_s is stable, that is, $\sum_{s=-\infty}^{\infty} c_s^2 < \infty$. The mean of the x_t process is zero and the autocorrelation coefficients are given by

$$\phi_\tau = E(x_{t+\tau} x_t) = \sum_{t=-\infty}^{\infty} c_{t+\tau} c_t$$

Let $C(\omega)$ be the Fourier transform of the filter coefficients c_s. Then it may be shown that the power spectrum $\Phi(\omega)$ of x_t is given by

$$\Phi(\omega) = |C(\omega)|^2$$

Let us interpret these results. Equation (9-35) is in the form of a stable two-sided filter, where the white noise ϵ_t is the input and the time series x_t is the output. The transfer function of this linear system is $C(\omega)$. The square of the magnitude spectrum of this filter, that is, $|C(\omega)|^2$, is the power spectrum $\Phi(\omega)$ of the time series x_t. Since, in general, the filter c_s is two-sided, this filter is not necessarily causal. We now wish to investigate the conditions under which the filter coefficients c_s may be replaced in equation (9-35) by a unique minimum-delay filter b_s. That is, we want to find a filter b_s which is stable, causal, and has minimum phase lag.

The Problem of Spectral Factorization

From a realization of a stationary time series x_t we may compute the autocorrelation function ϕ_τ as a time average by means of equation (9-6) and then the power spectrum $\Phi(\omega)$ as the Fourier transform of the auto-correlation, as given by equation (9-9). As we have seen, the power spectrum is equal to the squared magnitude spectrum of a linear filter into which white noise ϵ_t is passed in order to obtain the time series x_t as output. Thus, we know that the magnitude spectrum of this filter is equal to $\sqrt{\Phi(\omega)}$. Now any filter with this magnitude spectrum, and with arbitrary phase spectrum, would be an admissible system to describe the time series x_t. However, let us specify that the particular filter which we desire is one that is causal and stable, with minimum phase lag. In other words, we desire the minimum-delay filter, with magnitude spectrum $\sqrt{\Phi(\omega)}$ and phase spectrum $\theta(\omega)$. The phase spectrum $\theta(\omega)$ must be determined in such a way that $-\theta(\omega)$ is a minimum in the class of all causal filters with the given magnitude spectrum $\sqrt{\Phi(\omega)}$. Thus, the transfer function $B(\omega)$ of the desired filter may be expressed as

$$B(\omega) = \sqrt{\Phi(\omega)}e^{i\theta(\omega)} = \sum_{s=0}^{\infty} b_s e^{-i\omega s} \qquad (9\text{-}36)$$

where $\theta(\omega)$ is the desired phase spectrum and the set $b_0, b_1, b_2, \ldots$ are the desired minimum-delay filter coefficients. The problem of factoring the power spectrum $\Phi(\omega)$ is the problem of expressing the power spectrum as

$$\Phi(\omega) = |B(\omega)|^2$$

where $B(\omega)$ is the transfer function of the desired minimum-delay filter.

Let us first consider a special case, namely a *moving-average* (MA) *process* x_t. An MA process is a stationary process for which the autocorrelation ϕ_τ vanishes for τ greater than m. Thus, its power spectrum is

$$\Phi(\omega) = \sum_{\tau=-m}^{m} \phi_\tau e^{-i\omega\tau}$$

If we let $z = e^{-i\omega}$, we see that

$$\Phi(z) = \sum_{\tau=-m}^{m} \phi_\tau z^\tau$$

so

$$z^m \Phi(z) = \sum_{s=0}^{2m} \phi_{s-m} z^s$$

is a polynomial of degree $2m$. Since we are dealing with real-valued processes, it follows that $\Phi(\omega)$ is a real function of ω, so that if z_k is a root of this polynomial, then z_k^{*-1} is also a root. Moreover, since $\Phi(\omega)$ is an even function of ω, it follows that if z_k is a root of this polynomial, then z_k^* is also a root; see Robinson (1954, Chap. 2) for proofs of these statements. Since $\Phi(\omega)$ is a nonnegative function of ω, it follows that any root of modulus 1 must appear an even number of times. Let γ_k and γ_k^* denote the complex roots of the polynomial $z^m \Phi(z)$ with modulus greater than 1, and also half of those complex roots with modulus equal to 1. Similarly, let ρ_j denote the real roots of this polynomial with modulus greater than 1, and also half of those real roots with modulus equal to 1. Thus, this polynomial may be factored into

$$
z^m \Phi(z) = \left[b_m \prod_{k=1}^{h} (z - \gamma_k)(z - \gamma_k^*) \prod_{j=1}^{l} (z - \rho_j) \right]
$$
$$
\cdot \left[b_0 \prod_{k=1}^{h} (z - \gamma_k^{-1})(z - \gamma_k^{*-1}) \prod_{j=1}^{l} (z - \rho_j^{-1}) \right]
$$

(9-37)

where any root of order p is repeated p times and where $2h + l = m$. Let us denote the first factor in brackets in equation (9-37) by $B(z)$; then the second factor in brackets is seen to be $z^m B(z^{-1})$. We see that $B(z)$ is a polynomial in z with real coefficients, and so we may represent $B(z)$ by $b_0 + b_1 z + \cdots + b_m z^m$. Moreover, we see that $B(z)$ has no zeros within the unit circle. In those cases in which there are no γ_k and ρ_j of modulus one, the polynomial $B(z)$ has no zeros within or on the unit circle and hence is the z transform of a strictly minimum-delay filter. The coefficients $b_0, b_1, \ldots, b_m$ of this minimum-delay filter are the required filter coefficients. Suppose, on the other hand, that we did not choose the roots of the polynomial $z^m \Phi(z)$ in the foregoing fashion. Because there are at most 2^m different ways of choosing the roots, then $B(z)$ would have roots, some of which have modulus greater than 1, and some of which have modulus less than 1. Consequently, the filter $B(\omega)$ would not be minimum-delay.

Let us now consider another special case, the case of an *autoregressive* (AR) *process*. The power spectrum of an AR process may be described as follows. The reciprocal of an AR power spectrum has the same mathematical form as an MA power spectrum with no roots on the unit circle. That is, the reciprocal of an AR power spectrum has no roots γ_k and ρ_k of modulus 1. Accordingly, the reciprocal of the AR power spectrum may be factored in the same way into the form $|A(\omega)|^2$, where the polynomial $A(z)$ has no zeros within and on the unit circle. We note that the reason that $A(z)$ can have no zeros on the unit circle is that $\Phi(\omega)$ is integrable on the interval $(-\pi, \pi)$. Thus, the power spectrum $\Phi(\omega)$ of the autoregressive process may be factored as $\Phi(\omega) = |B(\omega)|^2$, where the factor $B(\omega)$ is the reciprocal of $A(\omega)$, and, like $A(\omega)$, is minimum-delay. The factor $B(\omega)$ is the transfer function of the desired minimum-delay filter.

More generally, any stationary process whose power spectrum is a rational function in z is a hybrid between an AR process and a MA process and so is called an *ARMA process*. Accordingly, the numerator and denominator each may be factored in the foregoing way to give

$$\Phi(\omega) = \frac{|G(\omega)|^2}{|H(\omega)|^2} = \frac{\left| \sum_{s=0}^{M} g_s e^{-i\omega s} \right|^2}{\left| \sum_{s=0}^{N} h_s e^{-i\omega s} \right|^2} \tag{9-38}$$

where, letting $z = \exp(-i\omega)$, the polynomials $G(z)$ and $H(z)$ have no common factors, the roots of $H(z)$ have modulus greater than 1, and the roots of $G(z)$ have modulus greater than or equal to 1. Thus, the factor $B(\omega) = G(\omega)/H(\omega)$ is the transfer function of the desired minimum-delay system.

General Solution of the Spectral Factorization Problem

Let us now take up the general solution of the factorization problem for a discrete stationary process with an arbitrary power spectrum $\Phi(\omega)$. Let us first turn our attention to the properties of the desired minimum-delay filter with transfer function $B(\omega) = |B(\omega)| \exp i\theta(\omega)$. Here $|B(\omega)|$ is the magnitude spectrum and $\theta(\omega)$ is the phase spectrum, under the restriction that the phase lag $-\theta(\omega)$ is a minimum in the class of all causal filters with the same magnitude spectrum. Since $B(z)$ has no singularities or zeros within the unit circle, $\log B(z)$ is analytic within the unit circle. Consequently, $\log B(z)$ has a power-series representation within the unit circle which, as $|z|$ approaches 1, converges to

$$\log B(\omega) = \beta_0 + 2 \sum_{t=1}^{\infty} \beta_t e^{-i\omega t} = \beta_0 + 2 \sum_{t=1}^{\infty} \beta_t \cos \omega t - 2i \sum_{t=1}^{\infty} \beta_t \sin \omega t \tag{9-39}$$

where we have let $z = \exp(-i\omega)$. Let us now turn our attention to the power spectrum $\Phi(\omega)$. The following conditions on the power spectrum must be satisfied: (1) $\Phi(\omega)$ must be nonzero almost everywhere on the interval $(-\pi, \pi)$, (2) the integral of $\Phi(\omega)$ over the interval $(-\pi, \pi)$ must be finite, and (3) the integral of $\log \Phi(\omega)$ over the interval $(-\pi, \pi)$ must be finite. Under these conditions, $\log \sqrt{\Phi(\omega)}$, which is an even real function of ω, may be expanded in a real, symmetric Fourier cosine series,

$$\log \sqrt{\Phi(\omega)} = \tfrac{1}{2} \log \Phi(\omega) = \sum_{t=-\infty}^{\infty} \delta_t \cos \omega t = \delta_0 + 2 \sum_{t=1}^{\infty} \delta_t \cos \omega t \tag{9-40}$$

where the Fourier coefficients δ_t are given by

$$\delta_t = \delta_{-t} = \frac{1}{2\pi} \int_0^\pi \cos \omega t \, \log \Phi(\omega) \, d\omega \qquad (9\text{-}41)$$

By taking the logarithm of each side of $B(\omega) = \sqrt{\Phi(\omega)} \exp i\theta(\omega)$ and utilizing equation (9-40), we have

$$\log B(\omega) = \log \sqrt{\Phi(\omega)} + i\theta(\omega) = \delta_0 + 2 \sum_{t=1}^\infty \delta_t \cos \omega t + i\theta(\omega) \qquad (9\text{-}42)$$

Now equation (9-39) gives an expression for $\log B(\omega)$ which was derived from the minimum-delay condition, namely, that $\log B(z)$ be analytic within the unit circle. On the other hand, equation (9-42) gives an expression for $\log B(\omega)$ derived from the knowledge that the magnitude spectrum $|B(\omega)|$ be equal to $\sqrt{\Phi(\omega)}$. Setting these two equations equal to each other, we find that $\delta_t = \beta_t$. Thus, the required phase spectrum is given in terms of the power spectrum by

$$\theta(\omega) = -2 \sum_{t=1}^\infty \delta_t \sin \omega t = -\frac{1}{\pi} \sum_{t=1}^\infty \sin \omega t \int_0^\pi \cos ut \, \log \Phi(u) \, du \qquad (9\text{-}43)$$

That is, $\theta(\omega)$ is the discrete Hilbert transform of $\log \sqrt{\Phi(\omega)}$. Since $B(z) = \exp[\log B(z)]$, we have, by letting $z = \exp(-i\omega)$ in equation (9-39), that

$$B(z) = \sum_{s=0}^\infty b_s z^s = \exp\left(\delta_0 + 2 \sum_{t=1}^\infty \delta_t z^t\right) \qquad (|z| < 1) \qquad (9\text{-}44)$$

By means of this equation we may solve for the desired linear operator b_s in terms of the δ_t given by equation (9-41). Letting $z = \exp(-i\omega)$ in $B(z)$, we obtain $B(\omega)$ for which $\Phi(\omega) = |B(\omega)|^2$. Thus, the power spectrum has been factored, where the factor $B(\omega)$ has magnitude spectrum $\sqrt{\Phi(\omega)}$ and phase spectrum $\theta(\omega)$ such that the phase-lag spectrum $-\theta(\omega)$ is minimum in the class of all causal filters with the same magnitude spectrum. Thus, $B(\omega)$ is the transfer function and $b_0, b_1, b_2, \ldots$ is the memory function of the required minimum-delay filter.

Innovational Representation of a Stationary Process

For a nondeterministic stationary process x_t we define the prediction error ϵ_t as $\sigma_1 \epsilon_t = x_t - \hat{x}_t$, where $\hat{x}_t$, given by equation (9-3) for $\alpha = 1$, is the least-squares prediction of x_t from the past values $x_{t-1}, x_{t-2}, \ldots$, and σ_1 is the positive square root of σ_1^2, the minimum mean square error (9-34).

Equivalently, we may express this prediction error ϵ_t in terms of the prediction error operator

$$\epsilon_t = a_0 x_t + a_1 x_{t-1} + a_2 x_{t-2} + \cdots \qquad (9\text{-}45)$$

where $a_0 = \sigma_1^{-1}$, $a_1 = -\sigma_1^{-1} k_0$, $a_2 = -\sigma_1^{-1} k_1$, and so forth. Because the prediction error represents the innovation at time t, it is uncorrelated with the past of the time series; that is, $E(x_{t-s} \epsilon_t) = 0$ for $s > 0$. Consequently, $E(\epsilon_t \epsilon_s) = 0$ for $t \neq s$, so ϵ_t forms an orthonormal set (i.e., a white-noise process). We may now regress x_t on $\epsilon_t, \epsilon_{t-1}, \ldots$. We therefore obtain the *innovational representation of a nondeterministic stationary time series* x_t as the sum of a Fourier series u_t plus a residual v_t; that is,

$$x_t = u_t + v_t \qquad \text{where } u_t = \sum_{s=0}^{\infty} b_s \epsilon_{t-s}, \quad \sum_{s=0}^{\infty} b_s^2 < \infty, \quad b_s = E(x_t \epsilon_{t-s})$$

$$(9\text{-}46)$$

Because of the correlation properties of regression residuals, we have $E(\epsilon_t v_s) = E(u_t v_s) = 0$. Now substitute (9-46) into (9-45) to obtain

$$\epsilon_t = \sum_{r=0}^{\infty} a_r x_{t-r} = \sum_{r=0}^{\infty} a_r (u_{t-r} + v_{t-r}) = \sum_{r=0}^{\infty} a_r \sum_{s=0}^{\infty} b_s \epsilon_{t-r-s} + \sum_{r=0}^{\infty} a_r v_{t-r}$$

If we define $\tau = r + s$, this equation becomes

$$\epsilon_t = \sum_{\tau=0}^{\infty} \left(\sum_{r=0}^{\infty} a_r b_{\tau-r} \right) \epsilon_{t-\tau} + \sum_{r=0}^{\infty} a_r v_{t-r} \qquad (9\text{-}47)$$

In order for this equation to hold identically, we must make two requirements:

$$\sum_{r=0}^{\infty} a_r b_{\tau-r} = \begin{cases} 1 & \text{when } \tau = 0 \\ 0 & \text{when } \tau \neq 0 \end{cases} \qquad (9\text{-}48)$$

and

$$\sum_{r=0}^{\infty} a_r v_{t-r} = 0 \qquad (9\text{-}49)$$

The first of these two requirements states that the filters a_s and b_s are inverse to each other, and the second states that v_t is perfectly predictable from its past, so the time series v_t is deterministic. The innovational representation tells us that every nondeterministic stationary process may be decomposed into a purely nondeterministic component u_t plus a deterministic component v_t. By a purely nondeterministic component we mean a nondeterministic process with no deterministic component. The spectral distributions of the

u_t and v_t processes are, respectively, the absolutely continuous and the singular components of the spectral distribution $\Lambda(\omega)$ of the x_t process.

Let us now consider a purely nondeterministic process. We may replace the operator coefficients c_t in the moving summation (9-35) by the minimum-delay filter b_t found by the factorization of the power spectrum. We thus obtain the *innovational representation of a purely nondeterministic stationary time series* x_t as

$$x_t = \sum_{s=0}^{\infty} b_s \epsilon_{t-s} = \sum_{s=-\infty}^{t} \epsilon_s b_{t-s} = b_0 \epsilon_t + b_1 \epsilon_{t-1} + b_2 \epsilon_{t-2} + \cdots \qquad (9\text{-}50)$$

This equation renders the time series x_t as the output of a minimum-delay filter b_t with input given by the innovations ϵ_t. Since the filter is causal, the value of x_t is expressed in terms of the present value ϵ_t and past values $\epsilon_{t-1}, \epsilon_{t-2}, \ldots$, but no future values $\epsilon_{t+1}, \epsilon_{t+2}, \ldots$ of the innovations. This representation corresponds to the purely nondeterministic component u_t in the innovational representation (9-46).

Let us now consider the innovational representation (9-50) in the language of the engineer. The dynamic structure of a stationary process may be represented by a minimum-delay filter $B(\omega)$. This filter has a minimum phase-lag spectrum (i.e., minimum negative phase) and magnitude spectrum $\sqrt{\Phi(\omega)}$. The coefficients b_t represent the impulsive response of the filter. The random elements of the stationary process are represented by the innovations ϵ_t $(-\infty < t < \infty)$, which is a mutually uncorrelated sequence (i.e., white noise). The time series x_t $(-\infty < t < \infty)$ is the output of the filter in response to the white noise input ϵ_t $(-\infty < s \leq t)$. That is, ϵ_s may be regarded as an impulse of strength ϵ_s, which will produce a response $\epsilon_s b_{t-s}$ at the subsequent time t. By adding the contributions of all the impulses ϵ_s $(-\infty < s \leq t)$, we obtain the total response, which is the time series x_t given by the representation (9-50). Since the input ϵ_t is an orthonormal process, let us note that its power spectrum is equal to 1 and its autocorrelation vanishes except for lag zero. Then the spectral factorization, $|B(\omega)|^2 = \Phi(\omega)$, states that the power spectrum of the input ϵ_t multiplied by the power transfer function $|B(\omega)|^2$ of the filter yields the power spectrum $\Phi(\omega)$ of the output x_t. The power transfer function $|B(\omega)|^2$ may then be called the energy spectrum of the transient b_t. Thus the spectral factorization states that the energy spectrum of the minimum-delay transient b_t is equal to the power spectrum of the time series x_t. In the time domain, the spectral factorization becomes

$$\sum_{t=0}^{\infty} b_t b_{t+\tau} = E(x_t x_{t+\tau}) = \phi_\tau \qquad (9\text{-}51)$$

which states that the autocorrelation of the transient b_t is equal to the autocorrelation of the time series x_t.

Let us now examine the innovational representation of an ARMA time series x_t with a rational power spectrum (9-38). The output x_t is obtained by passing white noise ϵ_t through the filter with transfer function $B(\omega) = G(\omega)/H(\omega)$. This linear operation is equivalent to first passing the white noise ϵ_t through the filter $G(\omega)$ and then passing the output of the $G(\omega)$ filter through the filter $1/H(\omega)$. Let $f_0, f_1, f_2, \ldots$ be the impulsive response of the $1/H(\omega)$ filter; that is, $f_0, f_1, f_2, \ldots$ is the minimum-delay inverse operator to the minimum-delay operator $h_0, h_1, \ldots, h_N$. Then the innovational representation (9-50) becomes

$$x_t = \sum_{s=0}^{\infty} b_s \epsilon_{t-s} = \sum_{s=0}^{\infty} f_s \sum_{r=0}^{\infty} g_r \epsilon_{t-r-s} = \sum_{s=0}^{\infty} \left(\sum_{r=0}^{M} g_r f_{s-r} \right) \epsilon_{t-s} \qquad (9\text{-}52)$$

which may be written

$$\sum_{s=0}^{N} h_s x_{t-s} = \sum_{s=0}^{M} g_s \epsilon_{t-s} \qquad (9\text{-}53)$$

where ϵ_t is the orthonormal input and the time series x_t with spectrum (9-38) is the output. By setting $h_0 = 1$ and the other h's equal to zero, equation (9-53) represents an MA process. On the other hand, by setting $g_0 = 1$ and the other g's equal to zero, equation (9-53) represents an AR process.

Explicit Prediction Formula

Let us now find an explicit prediction formula for purely nondeterministic stationary time series. The value of $x_{t+\alpha}$ is given by the innovational representation

$$x_{t+\alpha} = (b_0 \epsilon_{t+\alpha} + b_1 \epsilon_{t+\alpha-1} + \cdots + b_{\alpha-1} \epsilon_{t+1}) + (b_\alpha \epsilon_t + b_{\alpha+1} \epsilon_{t-1} + \cdots) \qquad (9\text{-}54)$$

If time t is the present time, the present and past values $x_t, x_{t-1}, x_{t-2}, \ldots$ are known. Consequently, the values $\epsilon_t, \epsilon_{t-1}, \epsilon_{t-2}, \ldots$ at and prior to time t may be obtained by use of the inverse a_t of the minimum-delay operator b_t. Thus, the component $(b_\alpha \epsilon_t + b_{\alpha+1} \epsilon_{t-1} + \cdots)$ of equation (9-54) can be computed at time t, and this component is the optimum least-squares prediction, $\hat{x}_{t+\alpha}$, of equation (9-3). Explicitly, by making use of equation (9-45) we may write equation (9-3) as

$$\hat{x}_{t+\alpha} = \sum_{s=0}^{\infty} b_{\alpha+s} \epsilon_{t-s} = \sum_{s=0}^{\infty} b_{\alpha+s} \sum_{n=0}^{\infty} a_n x_{t-n-s} = \sum_{r=0}^{\infty} \left(\sum_{s=0}^{\infty} b_{\alpha+s} a_{r-s} \right) x_{t-r} \qquad (9\text{-}55)$$

Comparing this with equation (9-3), we see that the prediction coefficient k_r is given by

$$k_r = \sum_{s=0}^{\infty} b_{\alpha+s} a_{r-s} \qquad (9\text{-}56)$$

This expression gives the prediction coefficients k_r to be used in equation (9-3) to yield the optimum least-squares prediction for a purely nondeterministic stationary time series. Computing the transfer function $K(\omega)$ of the operator k_r of equation (9-56), we obtain Wiener's well-known formula

$$K(\omega) = \sum_{r=0}^{\infty} k_r e^{-i\omega r} = \frac{\sum_{s=0}^{\infty} b_{\alpha+s} e^{-i\omega s}}{\sum_{s=0}^{\infty} b_s e^{-i\omega s}} = \frac{1}{2\pi B(\omega)} \sum_{s=0}^{\infty} e^{-i\omega s} \int_{-\pi}^{\pi} B(u) e^{iu(\alpha+s)} \, du$$

$$(9\text{-}57)$$

for the transfer function of the optimum predictor. The other component of the decomposition (9-54), namely, the component $(b_0 \epsilon_{t+\alpha} + \cdots + b_{\alpha-1} \epsilon_{t+1})$, involves future values of the innovations and hence cannot be computed at time t. This component is the prediction error for the prediction distance α, and its mean square value, $(b_0^2 + b_1^2 + \cdots + b_{\alpha-1}^2)$, is the minimum mean square value σ_α^2 of equation (9-34). Although the orthonormal variable ϵ_t is often called the prediction error, we see that the prediction error for unit prediction distance is actually $b_0 \epsilon_t$.

Let us now consider the problem of separating the message from a time series made up of the message plus noise. The innovational representation of the time series x_t consisting of message m_t plus noise n_t is

$$x_t = m_t + n_t = \sum_{s=0}^{\infty} b_s \epsilon_{t-s} \qquad (9\text{-}58)$$

Since we assume that the message is purely nondeterministic (i.e., the message has an absolutely continuous spectral distribution), it may be represented by the process of moving summation given by

$$m_t = \sum_{s=-\infty}^{\infty} q_s \epsilon_{t-s} + \sum_{s=-\infty}^{\infty} r_s \gamma_{t-s} \qquad (9\text{-}59)$$

where ϵ_t and γ_t each represent an orthonormal sequence of random variables, and $E(\epsilon_t \gamma_s) = 0$. Let $\Phi(\omega)$ be the power spectrum of x_t, $\Phi_{11}(\omega)$ be the power spectrum of the message, $\Phi_{12}(\omega)$ be the cross-spectrum of message and noise, and $\Phi_{22}(\omega)$ be the power spectrum of the noise. Using equation (9-58), we have

$$\Phi(\omega) = \Phi_{11}(\omega) + \Phi_{12}(\omega) + \Phi_{21}(\omega) + \Phi_{22}(\omega) = |B(\omega)|^2 \qquad (9\text{-}60)$$

where $B(\omega)$ is the minimum-delay factor of $\Phi(\omega)$. Letting $Q(\omega)$ and $R(\omega)$ be the transfer functions of the smoothing operators q_s $(-\infty < s < \infty)$ and r_s $(-\infty < s < \infty)$, respectively, it follows from equations (9-58) and (9-59) that $\Phi_{11}(\omega) = QQ^* + RR^*$ and $\Phi_{12}(\omega) = BQ^* - QQ^* - RR^*$. Consequently, we have $BQ^* = \Phi_{12}(\omega) + \Phi_{11}(\omega)$, so that

$$q_t = \frac{1}{2\pi} \int_{-\pi}^{\pi} \frac{\phi_{11}(\omega) + \phi_{21}(\omega)}{B(\omega)^*} e^{i\omega t} \, d\omega \qquad (9\text{-}61)$$

Following our usual notation, b_t is the memory function of the minimum-delay filter $B(\omega)$, and a_t is the inverse of b_t; that is, a_t is the memory function of the minimum-delay filter $A(\omega) = 1/B(\omega)$. The operator a_t is obtained from b_t by means of equation (9-48). Since the present and past values, $x_t, x_{t-1}, x_{t-2}, \ldots$, of the time series are known at time t, we may obtain the present and past values $\epsilon_t, \epsilon_{t-1}, \epsilon_{t-2}, \ldots$, of the white noise by means of the a_s filter. By considering the moving summation (9-59) at the time $t + \alpha$, we see that the predictable part of the message $m_{t+\alpha}$, where α is the prediction distance or lead, is

$$\hat{m}_{t+\alpha} = \sum_{s=0}^{\infty} q_{s+\alpha}\epsilon_{t-s} = \sum_{s=0}^{\infty} q_{s+\alpha} \sum_{n=0}^{\infty} a_n x_{t-n-s} = \sum_{r=0}^{\infty} \left(\sum_{s=0}^{\infty} q_{s+\alpha} a_{r-s} \right) x_{t-r} \qquad (9\text{-}62)$$

The nonpredictable part, or filtering error, is $m_{t+\alpha} - \hat{m}_{t+\alpha}$, which has mean square value

$$E[(m_{t+\alpha} - \hat{m}_{t+\alpha})^2] = \sum_{s=-\infty}^{\alpha} q_s^2 + \sum_{s=-\infty}^{\infty} r_s^2 \qquad (9\text{-}63)$$

which is a minimum. In equation (9-63) we see that the first term on the right depends on the lag $-\alpha$, whereas the second term does not. Note that the lag is defined as minus the lead. Thus, the optimum linear operator in the sense of the principle of least squares to be used in separating message and noise from a nondeterministic time series x_t has coefficients h_r given by the expression in parentheses on the right-hand side of equation (9-62); that is,

$$h_r = \sum_{s=0}^{\infty} q_{s+\alpha} a_{r-s} \qquad (9\text{-}64)$$

The transfer function is then

$$H(\omega) = \sum_{r=0}^{\infty} h_r e^{-i\omega r} = \frac{\sum_{s=0}^{\infty} q_{s+\alpha} e^{-i\omega s}}{\sum_{s=0}^{\infty} b_s e^{-i\omega s}} \qquad (9\text{-}65)$$

$$= \frac{1}{2\pi B(\omega)} \sum_{s=0}^{\infty} e^{-i\omega s} \int_{-\pi}^{+\pi} \frac{\Phi_{11}(u) + \Phi_{21}(u)}{B^*(u)} e^{iu(s+\alpha)} \, du$$

This equation is the transfer function of what electrical engineers know as a "wave filter."

Concluding Remarks

This chapter has focused on some of the more fundamental aspects of stationary time series. The analysis can be carried out either in the time domain, where the dominant concepts are the autocorrelation and the crosscorrelation functions, or in the frequency domain, where the corresponding dominant concepts are the spectrum and the crossspectrum. A stationary time series can in general be represented as the sum of an innovative component and a deterministic component. If the deterministic component is absent, the resulting time series is purely nondeterministic. The present formulation will be used in subsequent chapters as the theoretical framework for the method of predictive deconvolution.

predictive deconvolution
of seismic traces

Summary

This chapter develops a model of a seismic trace additively composed of many overlapping seismic wavelets that arrive as time progresses. Each wavelet has the same stable, causal, minimum-delay shape. The arrival times and strengths of these wavelets are represented by an innovation sequence of uncorrelated random variables. We show how the wavelet shape may be extracted from the trace, leaving as a residual the strengths of the wavelets at their respective arrival times. This process is called *deconvolution*.

Introduction

A large part of basic seismic research has traditionally been directed toward a better understanding of the physical processes involved in the seismic method. Such an approach is fundamentally sound. From this point of view, the seismic trace is the response of the system consisting of the earth and recording apparatus to the impulsive source, the explosion. This system, although usually very complicated, is susceptible to a deterministic (non-

238

statistical) approach toward its analysis. To this end, controlled experiments may be carried out, and mathematical and physical models may be set up from the resulting data. Careful replication of the experiment and high precision of measurement can render such data very accurate. On the other hand, large numbers of seismic records are needed to carry out an exploration program over a geographic area. This quantity of data necessarily requires the consideration of each record as a member of a larger group, or ensemble of records. Thus, the reliability of a single record is considerably less than the reliability of the ensemble of records in connection with the description of the geologic conditions existing in that area. Also, from an economic standpoint, the amount of control in such an exploration program must be kept at the bare minimum consistent with worthwhile results.

As a rule, the controlled experiment aspect of exploration seismology, although possible, falls short of the needs of a research scientist who wishes to set up a mathematical or physical model. As a result, in these cases the working geophysicist must proceed to fit his empirical information into the larger overall framework without the aid of elaborate mathematical or physical models. Since the geologic structure is physically fixed and constant in nature, and has no intrinsic random characteristics, any statistical approach to this problem immediately encounters difficulties which are commonly associated in the statistical literature with Bayes' theorem. Nevertheless, modern statistical theory admits the bypassing of these difficulties, and hence the working geophysicist may be considered to be faced with a situation that is essentially statistical. For example, a reflection that may be followed from trace to trace and record to record usually has more value to the seismic interpreter and hence is statistically more significant than a reflection that appears on only a few traces. Such a procedure in picking reflections does not imply that the reflection which appears only on a few traces is necessarily spurious information, but only that economic limitations preclude further examination and experimentation which may render it in a more useful form. In the final analysis, the potential usefulness of the statistical approach depends upon the coordination of statistical methods with knowledge of practical and theoretical seismology.

Wavelet Theory

From a physical point of view, the seismic trace is the response of the system consisting of the earth and recording apparatus to the impulsive source, the explosion. This system, although usually very complicated, is susceptible to a deterministic approach toward its analysis. Nevertheless, the complicated nature of seismograms recorded in seismic exploration many times precludes the study of the overall response of the earth and recording system as a whole.

Also, in the final analysis one is interested in the various components of this total response; for example, one wishes to separate components of reflected energy from those of nonreflected energy. One can model a seismogram as an elaborate wavelet complex, and the analysis of a seismogram consists in breaking the record down into its components. According to the theory of the propagation of normal-incidence plane waves in a horizontally layered medium, a wave due to a sharp impulse such as an explosion is propagated with change in form due to the multiple reflections and is received at a distance as time-varying wavelets with arrival times and strengths governed by the reflecting interfaces. Consequently, in layered media a seismogram should consist of a succession of time-varying wavelets, each wavelet associated with a reflecting interface. If there were no multiple reflections, the interfaces would yield primary reflections only, and the net effect would be a seismic trace made up of a sharp and clear-cut series of impulses at the times of the primary reflections. In such an idealized case, many of the difficulties in seismic prospecting would disappear. As we know, however, only in ideal situations is such a simple seismogram received in the propagation of seismic waves through the earth. (We will discuss this ideal situation at great length in Chapter 13.) Usually, we obtain the more complicated seismograms which are familiar to every geophysicist.

Thus, the seismogram may be visualized as the totality of time-varying responses to sharp impulses, each sharp impulse being associated with a reflecting horizon. These time-varying responses are the seismic wavelets. The analysis of a seismogram consists in breaking down this elaborate wavelet complex into its component wavelets. In particular, we desire the time sequence of theoretical sharp impulses that produce these wavelets. This time sequence is made up of the innovations that represent the reflecting horizons. There are two basic approaches which one may use toward the solution of this problem: the deterministic approach and the statistical approach. In the deterministic approach one utilizes basic physical laws, for example, to determine the shape of the wavelet. At all stages in such an investigation one may compare mathematical results with direct and indirect observation of the physical phenomenon. In this chapter we are concerned with the statistical approach. Such an approach in no way conflicts with the deterministic approach, although each approach has certain advantages and disadvantages, which do not necessarily coincide. In practice the two approaches may be utilized in such a manner as to complement each other.

The Statistical Approach

Let us apply the statistical approach to a specific problem. We assume that a given section of seismic trace is additively composed of wavelets, where each wavelet has the same shape or form. In other words, we assume that the time-varying character of the wavelet can be neglected over a certain time interval on the seismic record, so that within this interval the wavelet has essentially constant shape. We shall assume that this constant wavelet shape is given by a stable, causal, minimum-delay time function. Hence, each wavelet is a one-sided transient that damps with a certain degree of rapidity. Further, we assume that from knowledge of the arrival time of one wavelet we cannot predict the arrival time of another wavelet; and we assume that from knowledge of the strength of one wavelet we cannot predict the strength of another wavelet. Finally, let us assume that over the time interval in question the seismic trace has a constant standard deviation (or variance) with time. The specific problem that we wish to consider is the following: given the seismic trace described in this paragraph, determine the arrival times and strengths of the wavelets and determine the basic wavelet shape. We shall discuss a theoretical solution of this problem, and shall also discuss a practical solution that involves statistical estimation.

Let us translate our assumptions about the seismic trace into mathematical notation for discrete time t. First, we let the discrete-time function b_t represent the ordinates of the fundamental constant wavelet at discrete, equally spaced, integer-valued times t. Our assumption about the nature of this wavelet is that the b_t are the coefficients of a stable, causal, minimum-delay linear operator. In other words, we assume that $b_t = 0$ for t less than zero so that the wavelet is one-sided, and that

$$\sum_{s=0}^{\infty} b_s^2 < \infty; \quad \sum_{s=0}^{\infty} b_s z^s \neq 0 \quad (|z| < 1) \quad (10\text{-}1)$$

hold so that the wavelet is a damped, minimum-phase-lag time function. Let the strength, or weighting factor, of the wavelet that arrives at time t be given by ϵ_t. Thus, at time t, this wavelet has ordinate $b_0\epsilon_t$; at time $t + 1$, it has ordinate $\epsilon_t b_1$; at time $t + 2$, it has ordinate $\epsilon_t b_2$; and so forth. The variable ϵ_t is the theoretical knife-sharp innovation for which the particular wavelet (i.e., the one that arrives at time t) is the response. For example, if no wavelet arrives at a particular time t, then $\epsilon_t = 0$. The innovation ϵ_t gives the strength, or reflection coefficient, of the reflecting interface with (primary) reflection time t.

In our discussion of the nature of the seismic trace, we shall call the knife-sharp impulses ϵ_t "random variables." Our use of the term "random variable ϵ_t," does not imply that the variable ϵ_t is one whose value is uncertain and can be determined by a "chance" experiment. That is, the variable ϵ_t is not random in the sense of the frequency interpretation of probability, but is fixed by the geologic structure. Frechet describes this type of variable as "nombre certain" and "fonction certaine," and Neyman translates these terms "sure number" and "sure function." Another example of a sure number is the ten-thousandth digit of the expansion $\pi = 3.1415926 \ldots$, which, although unknown, is a definite fixed number. Since the knowledge of the working geophysicist about the entire deterministic setting is far from complete, we shall treat this incomplete knowledge from a statistical point of view. We thus call ϵ_t a random variable, although we keep in mind that it is a sure number.

Our assumption about the unpredictability of the arrival times and strengths of wavelets means mathematically that the knife-sharp impulses ϵ_t are mutually uncorrelated random variables; that is, $E(\epsilon_t \epsilon_s) = E(\epsilon_t)E(\epsilon_s)$ for t not equal to s. Our assumption that the knife-sharp impulses ϵ_t are mutually uncorrelated with each other is an orthogonality assumption, and is a weaker assumption than the assumption that the ϵ_t are statistically independent, which we need not make. Returning again for the moment to our discussion about the "sure" nature of the knife-sharp impulses ϵ_t, we see that the assumption that they are mutually uncorrelated does not hold in a completely deterministic system. Nevertheless, such an assumption is a reasonable one for the working geophysicist, whose knowledge of the entire deterministic setting is far from complete and who is faced with essentially a statistical problem. In other words, we assume that knowledge of the arrival time and strength of one wavelet does not allow us to predict the arrival time and strength of any other wavelets. In particular, we assume that an arrival time and magnitude of a reflection from a certain reflecting horizon does not allow us to predict the arrival time and magnitude of a reflection from another reflecting horizon.

Predictive Deconvolution

We have assumed that our seismogram trace is additively composed of wavelets, all with the same shape b_s, but weighted by their respective strengths ϵ_t. That is, at the time t, the wavelet that arrives then gives the contribution $\epsilon_t b_0$; the wavelet that arrived at $t - 1$ gives the contribution $\epsilon_{t-1}b_1$; the wavelet that arrived at $t - 2$ gives the contribution $\epsilon_{t-2}b_2$; and so forth. The seismic trace x_t at time t is then the summation of all these contributions, and hence we may write this wavelet complex mathematically as

$$x_t = b_0\epsilon_t + b_1\epsilon_{t-1} + b_2\epsilon_{t-2} + \cdots = \sum_{s=0}^{\infty} b_s\epsilon_{t-s} \qquad (10\text{-}2)$$

for the time interval (t_1, t_2), called the time interval or time gate, which comprises our basic section of seismic trace. This equation includes tails of wavelets with shape b_t, these wavelets being due to knife-sharp impulses $\epsilon_{t_1-1}, \epsilon_{t_1-2}, \ldots$ which occur before time t_1. Without loss of generality, we may center the knife-sharp impulses ϵ_t so that their mean $E(\epsilon_t)$ is equal to zero.

Our assumption that the seismic trace have constant variance, that is, that $E(x_t^2)$ be constant, means that the strengths ϵ_t of the impulses must have constant variance; that is, $E(\epsilon_t^2)$ must be constant, which without loss of generality we shall take to be unity. Thus, the ϵ_t represent an orthonormal sequence of random variables. For the purposes of our theoretical discussion, let us assume that our assumptions about the time series x_t, equation (10-2), now hold for all time. Thus, equation (10-2) becomes the innovational representation of a purely nondeterministic stationary time series, as given by equation (9-50).

In equation (10-2) the wavelet shape b_t represents the "dynamics" of the time series, whereas the innovations ϵ_t represent the "random" nature of the time series. Our problem then consists of the extraction of the wavelet shape b_s from the trace x_t so as to yield the innovations ϵ_t, which represent the wavelet strengths at their respective arrival times t. The theoretical procedure for infinite time series is as follows.

First, we wish to average the random components ϵ_t out of the time series x_t so as to yield the wavelet shape b_t. To do so, we may use the following procedure. From the realization of the time series $x_t(-\infty < t < \infty)$, compute the autocorrelation function ϕ_τ. Then take the Fourier transform of the autocorrelation to yield the power spectrum $\Phi(\omega)$. Next factor the power spectrum into $\Phi(\omega) = |B(\omega)|^2$, where $B(\omega)$ is the minimum-delay factor. The factor $B(\omega)$ is the spectrum of the desired wavelet b_t, so we find b_t from $B(\omega)$ by the Fourier transform

$$b_t = \frac{1}{2\pi} \int_{-\pi}^{\pi} B(\omega) e^{i\omega t} \, d\omega \qquad (10\text{-}3)$$

Let us now examine this procedure in more detail. The computation of the autocorrelation function averages out the random, uncorrelated elements ϵ_t and preserves the dynamic wavelet shape b_t in the form of the autocorrelation of the wavelet shape. In other words, the autocorrelation ϕ_τ of the time series x_t is the same function as the autocorrelation of the wavelet b_t. In the computation of the power spectrum $\Phi(\omega)$ we are, in effect, computing the energy spectrum $|B(\omega)|^2$ of the wavelet, since the power spectrum $\Phi(\omega)$ of the time series is the same function as the energy spectrum of the wavelet b_t. Since the magnitude spectrum $|B(\omega)|$ is the real positive square root of the energy spectrum $|B(\omega)|^2$, we may immediately find $|B(\omega)|$, which is equal to $\sqrt{\Phi(\omega)}$. Now there are many different wavelet shapes that have this same magnitude spectrum $|B(\omega)|$, but have different phase spectra. Nevertheless, only one of these wavelet shapes is a stable, causal minimum-delay transient time function; the spectrum of this unique wavelet shape is the minimum-delay factor $B(\omega)$. It follows that the factor $B(\omega)$ must be the spectrum of our required wavelet. The factor $B(\omega)$ is equal to $\sqrt{\Phi(\omega)} \exp i\theta(\omega)$, where

$-\theta(\omega)$ is the minimum negative phase characteristic given by equation (9-43), namely

$$-\theta(\omega) = \frac{1}{\pi} \sum_{t=1}^{\infty} \sin \omega t \int_0^{\pi} \cos ut \log \Phi(u) \, du \qquad (10\text{-}4)$$

Hence, the wavelet b_t may be found from $B(\omega)$ by means of equation (10-3).

Next, using the wavelet shape b_t thus found, we wish to remove this wavelet shape from x_t, thereby leaving, as a residual, the innovations ϵ_t. To do so, we use the following procedure. From the wavelet shape b_t, we find the inverse wavelet shape a_t by means of

$$a_0 b_0 = 1, \qquad \sum_{s=0}^{n} a_s b_{n-s} = 0 \qquad (n = 1, 2, 3, \ldots) \qquad (10\text{-}5)$$

We now use the inverse wavelet shape a_t to remove the wavelets, all of which are of shape b_t, from the time series x_t by means of the linear operation

$$\sum_{s=0}^{\infty} a_s x_{t-s} = \sum_{s=0}^{\infty} a_s \sum_{r=0}^{\infty} b_r \epsilon_{t-r-s} = \epsilon_t \qquad (10\text{-}6)$$

In other words, the operator a_s acting on the time series x_t yields the innovations ϵ_t, which represent the primary reflections. The inverse wavelet shape a_t is the optimum least-squares prediction error operator for unit-prediction distance and the $b_0 \epsilon_t$ are the prediction errors. Thus, by these theoretical steps we may separate the dynamic component, represented by the transient response function or wavelet shape b_t, from the random component, represented by the innovations ϵ_t, which are the strengths of the wavelets at their respective arrival times.

Statistical Estimation

The practical solution of the problem of separating the "dynamic" and the "random" components of a finite section of a seismic trace involves statistical estimation. One method consists of estimating the prediction-error operator, or inverse wavelet shape, directly from the finite section of seismic trace. For this purpose one may use the method of least squares. The unit-step prediction operator is given by

$$\hat{x}_{i+1} = c + \sum_{s=0}^{M} k_s x_{i-s} \qquad (10\text{-}7)$$

We note that a constant c appears in equation (10-7) to take account of the mean value of the time series, since now we do not require the mean to be zero. The time interval or time gate is chosen to be the time interval of the section of trace, which we assume to be a section of a stationary time series.

In this expression, $\hat{x}_{i+1}$ is the approximated value of the future value x_{i+1}, where the prediction distance is 1. The prediction is based on the present value x_i and the M past values $x_{i-1}, x_{i-2}, \ldots, x_{i-M}$. The $M + 2$ constants given by $c, k_0, k_1, \ldots, k_M$ are to be determined in an optimum sense. As is customary, these constants are determined by the method of least squares. Hence, we wish to find the values of $c, k_0, k_1, \ldots, k_M$ that minimize the sum of squared errors between the actual value x_{i+1} and the predicted value $\hat{x}_{i+1}$. The gate is defined as that time interval of length n which includes the values x_{i+1} from $i + 1 = N + 1$ through $i + 1 = N + n$. We wish to minimize the sum of squared prediction errors within this gate, so the summation index $i + 1$ should run from $i + 1 = N + 1$ through $i + 1 = N + n$, or equivalently the summation should run from $i = N$ through $i = N - 1 + n$. Hence, we wish to minimize the sum of squared prediction errors given by

$$J = \sum_{i=N}^{N+n-1} (x_{i+1} - \hat{x}_{i+1})^2 \tag{10-8}$$

with respect to c, and with respect to k_s $(s = 0, 1, \ldots, M)$. Substituting equation (10-7) into equation (10-8), we have

$$J = (\sum_i x_{i+1} - c - \sum_s k_s x_{i-s})^2 \tag{10-9}$$

All summations on the index i in equation (10-9) as well as in what follows are for $i = N$ to $i = N + n - 1$, and all summations on the index s are for $s = 0$ to $s = M$.

To carry out the minimization, we set the partial derivatives of J with respect to the coefficients c and k_s $(s = 0, 1, \ldots, M)$ equal to zero. We obtain $M + 2$ linear algebraic equations in the $M + 2$ unknowns given by the coefficients c and k_s (where $s = 0, 1, \ldots, M$). This set of simultaneous normal equations is

$$cn + \sum_s (k_s \sum_i x_{i-s}) = \sum_i x_{i+1}$$

$$c \sum_i x_{i-r} + \sum_s (k_s \sum_i x_{i-r} x_{i-s}) = \sum_i x_{i-r} x_{i+1} \quad \text{for } r = 0, 1, \ldots, M$$

$$\tag{10-10}$$

Expanding equation (10-9) and using equations (10-10), we see that the minimum value of J is given by

$$J = \sum_i x_{i+1}^2 - c \sum_i x_{i+1} - \sum_s (k_s \sum_i x_{i-s} x_{i+1}) \tag{10-11}$$

This minimum value of J is one measure of how well the operator reproduces the x series in the operator time interval.

We define the sample variance J_0 to be

$$J_0 = \sum_i (x_{i+1} - \bar{x})^2 \qquad (10\text{-}12)$$

where $\bar{x} = (1/n) \sum x_{i+1}$ is the sample mean of the x series in the operator time interval. Then the percent reduction of the sample variance about the sample mean is defined to be

$$\%R = 100 \left(1 - \frac{J}{J_0}\right) \qquad (10\text{-}13)$$

The percent reduction is another measure of how well the operator reproduces the x trace in the operator time interval, where 100% reduction is perfect reproduction.

The solution of the normal equations (10-10) then yields the operator coefficients $c, k_0, k_1, \ldots, k_M$. Since the prediction error for unit prediction distance is given by $b_0 \epsilon_t = x_t - \hat{x}_t$, we see that by comparing the operators (10-6) and (10-7) their operator coefficients are related by $a_0 = b_0^{-1}$, $a_1 = -a_0 k_0$, $a_2 = -a_0 k_1, \ldots, a_m = -a_0 k_M$, and $a_s = 0$ for s greater than m, where $m = M + 1$. This relation allows us to determine immediately the inverse wavelet shape a_t from the solution k_t of the normal equations since for convenience we may let $a_0 = b_0 = 1$, at the expense of the ϵ_t no longer having unit variance. The constant c of equation (10-7), which adjusts for the mean value of the empirical trace, is not used in determining the shape a_t of the wavelet. Provided that a_t is a minimum-phase operator, the shape b_t of the wavelet may be readily computed from a_t by means of equation (10-5). In terms of the prediction operator k_s of equation (10-7) we have $b_t = 0$ for t less than zero, $b_0 = a_0^{-1}$, and

$$b_{t+1} = -a_0^{-1} \sum_{s=1}^{m} a_s b_{t+1-s} = \sum_{s=0}^{M} k_s b_{t-s}, \qquad t = 0, 1, 2, \ldots . \quad (10\text{-}14)$$

That is, the dynamic wavelet shape b_t for $t > 0$ is determined by successive step-by-step predictions from its past values, where we let the initial values be $b_t = 0$ for $t < 0$ and $b_0 = a_0^{-1}$.

As we have seen, the method of least squares described yields an empirical estimate of the theoretical prediction error operator or inverse wavelet shape. This empirical estimate has certain optimum statistical properties under general conditions. Provided that trace in the operator time interval is approximately stationary, a good estimate of the prediction operator should yield prediction errors within the operator time interval which are not significantly autocorrelated. In other words, the prediction errors ϵ_t within the operator time interval should be mutually uncorrelated at some preassigned level of significance. In Figure 10-1, in the left-hand diagram, we show

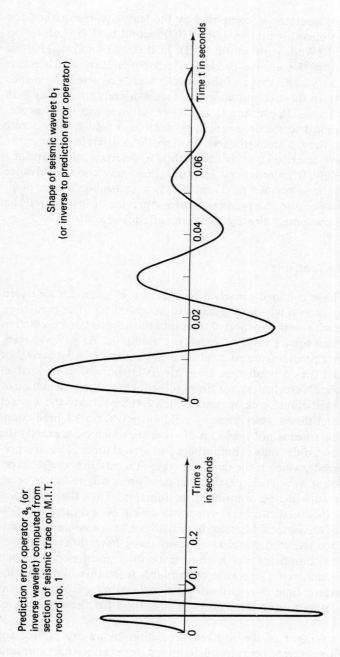

Figure 10-1. Computation of wavelet from section of trace on M.I.T. record No. 1.

Shape of seismic wavelet b_1
(or inverse to prediction error operator)

Time t in seconds

Prediction error operator a_s (or inverse wavelet) computed from section of seismic trace on M.I.T. record no. 1

Time s in seconds

the prediction error operator a_s computed by the least-squares method for trace N650 for the operator time interval 0.350 second to 0.475 second on MIT Record No. I (Wadsworth et al., 1953). In the right-hand diagram of Figure 10-1 we show the inverse prediction error operator, which is the shape of the wavelet b_t. The shape of the wavelet was "predicted" by means of equation (10-14). In these computations, we used discrete time series with the spacing $\Delta t = 2.5$ ms. In plotting a_s and b_t we followed our usual procedure which is to plot discrete time functions, such as a_s and b_t, as discrete points and then to draw a smooth curve through these discrete points.

Here we have described a statistical method to determine the shape of a wavelet. Alternatively, from other considerations, one may know in advance the shape of the seismic wavelet b_t. Then, if b_t is a realizable, stable, minimum-phase linear operator, the prediction error operator or inverse wavelet shape, a_t, may be computed directly by means of equation (10-5).

Detection of Reflections

So far we have confined ourselves to a section of seismic trace called the operator time interval which we assume to be stationary. The prediction error operator transforms this section of trace into the uncorrelated prediction errors $b_0\epsilon_t$, the mean square value of which is a minimum. As we have seen, the operator cannot predict from past values of the trace the initial arrival of a new wavelet, and thus a prediction error $b_0\epsilon_t$ is introduced at the arrival time of each wavelet. Nevertheless, for times subsequent to the arrival time of the wavelet, the prediction error operator which is the inverse to the wavelet can perfectly predict this wavelet, thereby yielding zero error of prediction. However, a seismic trace is not made up of wavelets which have exactly the same form and which differ only in amplitudes and arrival times. Thus if a prediction error operator, which is the unique inverse of a certain wavelet shape, encounters a different wavelet shape, the prediction error will no longer be an impulse, but instead will be a transient time function. Thus the prediction errors yielded by this prediction error operator acting on a time series additively composed of wavelets of different shapes will not have a minimum mean-square value. Since reflected wavelets in many cases have different shapes than the wavelets comprising the seismic trace in a given non-reflection interval, a prediction error operator determined from this non-reflection interval (the operator time interval) will yield high errors of prediction at such reflections. Such a procedure provides a method for the detection of reflections.

Running averages of the squared prediction errors are plotted for least-square prediction error operators determined from various operator time intervals. The peaks on these prediction error curves indicate reflections on the seismogram. When two-trace operators are used, the empirical coherency existing between the two traces is utilized in the determination of these prediction errors. Since only the information existing in the operator time interval is utilized in the determination of linear operators by this method, one may expect greater resolving power if more information on the seismogram is utilized in the determination of various other types of operators.

The Measure of Error of Prediction

Once numerical values of the coefficients c and k_s $(s = 0, 1, \ldots, M)$ of the filter are determined, they are used to predict the trace x by means of equation (10-7). In using equation (10-7), it must be remembered that the past values of x_{i-s} (where $s = 0, 1, \ldots, M$) are the actual values taken from the seismogram trace, not the previously predicted values.

In order to determine how well trace x is being predicted by the operator, the error between the actual value x_{i+1} and the predicted value $\hat{x}_{i+1}$ should be measured. These errors, or residuals, may be studied individually or statistically in order to measure the effectiveness of the operator in reconstructing the x trace.

One such statistical quantity that may be used is the running average E_i of square errors given by

$$E_i = \frac{1}{2p + 1} \sum_{j=i-p}^{i+p} (x_j - \hat{x}_j)^2 \qquad (10\text{-}15)$$

where $2p + 1$ is the number of elements in each average. If we let

$$V_i = \frac{1}{2p + 1} \sum_{j=i-p}^{i+p} (x_j - \bar{x})^2 \qquad (10\text{-}16)$$

denote the running variance of trace x, then another measure of the goodness of the prediction of the x trace is the running reduction R_i given by

$$R_i = 1 - \frac{E_i}{V_i}$$

The averages given in equations (10-15) and (10-16) represent time averages of single realizations of time series. If we have many operators operating on many traces, another type of averaging is possible on the ensemble of time series generated by these operators. An ensemble average, or an average across these series rather than along an individual one, is another measure that can be used.

Let us suppose that we have taken a series of operators on a record that consists of traces from equally spaced seismometers. Suppose that there are T traces, and on the lth trace ($l = 1, 2, \ldots T$) we have chosen N_l operators. For the kth operator on this trace ($k = 1, \ldots N_l$) there is an associated error time series which we define as $e_i(kl)$. Then, for example, we may construct a single sequence ϵ_i^l to be associated with the lth trace by the expression

$$\epsilon_i^l = \sum_{k=1}^{Nl} [e_i^{(kl)}]^2 \qquad (10\text{-}17)$$

We may then average these sequences over the various traces. But since we are interested not only in finding reflection times but also the associated moveouts, we construct the sequence $\delta_i^{(\alpha)}$ with an arbitrary lag or lead α:

$$\delta_i^{(\alpha)} = \sum_{l=1}^{T} \epsilon_{i-\alpha l}(l) \qquad (\alpha = 0, \pm 1, \pm 2, \ldots) \qquad (10\text{-}18)$$

with the expectation that a peak on this sequence, corresponding to a certain reflection, should be highest and narrowest for that value of α most closely corresponding to the true moveout of the given reflection.

Concluding Remarks

We have assumed that the seismic trace is composed of a superposition of wavelets of constant shape, but weighted by random strengths. The constant wavelet shape represents the dynamic part of the trace, while the random strengths represent the innovative, or random part of this trace. The method of predictive deconvolution enables us to obtain estimates of both the constant wavelet shape as well as of the innovative, or random strengths. In the next two chapters, we shall see how these concepts can be given geological meaning.

deconvolution for the elimination of ghost reflections and reverberations

Summary

In the standard deterministic model of water reverberation generation, the reverberation pulse train resulting from a deep reflection is minimum-delay. Even in the more complex physical situations encountered in the field, there is evidence that in many cases the reverberation pulse-train waveforms are minimum-delay, or at least approximately so. The reason for this minimum-delay property is that a pulse-train waveform results from multiple reflections and transmissions within the layered earth. Because reflection coefficients are less than unity in magnitude, the concentration of energy in a pulse train must appear at its beginning rather than its end. This early concentration of energy is the condition that pulse-train waveform be minimum-delay. Each deep reflection horizon contributes a minimum-delay reverberation pulse-train waveform to a seismic trace. If we let a spike series represent the deep horizons in the sense that the timing of a spike represents the direct arrival time of a reflection and the amplitude of the spike represents the strength of the reflection, the seismic trace may be considered as the convolution of the spike series with the reverberation pulse-train waveform. Because the reverberation pulse-train waveform is minimum-delay, and because at least approximately the deep horizon spike series represents a statistically uncorrelated series, the two conditions required for the application of the method of predictive deconvolution are met, and hence this method can be used as a practical digital data processing method to eliminate water reverberations on field seismic traces. The concept of minimum delay is therefore an important link in chaining together the deterministic approach and the statistical approach to seismic record analysis in the single-channel case.

Introduction

Many successful digital seismic data processing methods can be developed from time-series theory. In this chapter we will concentrate our attention on the fundamental concepts involved. It is possible to process seismic data successfully without knowledge of many of these underlying geophysical and mathematical concepts. However, if one is interested in getting the ultimate information out of seismic data at a reasonable cost, or is interested in extreme accuracy in the time domain, or is concerned about the difference between a satisfactory deconvolution method and a better than satisfactory deconvolution method, then basic concepts become important and are worth learning about.

Inverse filtering or deconvolution has proved to be an effective way to increase the resolving power of the seismograph. Since the advent of the "digital revolution" in exploration seismology during the early and mid-1960s, virtually all seismograms recorded in the exploration for petroleum and natural gas have been processed by use of these methods. In this chapter we want both to go back and review the method of predictive deconvolution in terms of the basic concepts involved, and apply the method to the removal of ghosts and water reverberations from the seismic trace. The main concept we will make use of is that of minimum delay; the main mathematical tool we will use is the z transform. We want to show why the concept of minimum delay is important in geophysical model building and seismic data processing.

There are two basic approaches to seismology, the deterministic approach and the statistical approach. The deterministic approach is concerned with the building of mathematical and physical models of the layered earth in order to better understand seismic wave propagation. These models involve no random elements; they are completely deterministic. The statistical approach is concerned with the building of seismic models involving random components. For example, in the statistical model that we will discuss, the depths and reflectivity of the deep reflecting horizons are considered to have a random distribution. A major justification for using the statistical approach in seismology is due to the fact that large amounts of data must be processed; any data in large enough quantities take on a statistical character, even if each individual piece of data is of a deterministic nature.

The model required for the application of the predictive deconvolution is a statistical model. This model depends upon two basic hypotheses: (1) the statistical hypothesis that the strengths and arrival times of the information-bearing events on a seismic trace can be represented as a random spike series, and (2) the deterministic hypothesis that the basic waveform associated with each of these events is minimum-delay. There are various ways of checking a model to see if it conforms with the physical situation. The ultimate test of this model occurs in its day-to-day usage in processing seismic data.

From a theoretical point of view, the concept of minimum delay provides much fertile ground for mathematical research. We recall that minimum phase lag was introduced as a frequency-domain concept by Bode (1945). As we have seen in Chapter 2, a system is called "minimum phase lag" provided that it has the minimum possible phase-lag spectrum in the class of all systems with the same magnitude spectrum. This concept has proven to be vital in the design of feedback control systems. In this book we have taken the concept of minimum phase lag out of the frequency domain and transferred it into the time domain; by so doing we have arrived at the identical concept in the time domain, which we call "minimum delay." (That is, the terms "minimum phase lag" and "minimum delay" are synonymous, except that the former connotes the frequency domain and the latter the time domain.) A minimum-delay waveform is defined as the waveform that has the largest concentration of energy in the early part of the waveform in the class of all waveforms with the same magnitude spectrum. In this chapter we will give two fundamental examples of minimum-delay pulse-train waveforms, those occurring in the standard models of seismic ghost generation and water reverberation generation. The reason we can often expect minimum-delay pulse-train waveforms in the layered earth situation is that reflection coefficients are less than unity in magnitude. Hence, the more times a seismic pulse is reflected and transmitted, the more it is delayed and attenuated. Thus, the concentration of energy in a seismic pulse-train waveform appears at its beginning rather than at its end; this is the condition that the pulse-train waveform be minimum-delay.

The standard ghost and reverberation models that we have mentioned above are special cases of much more general deterministic layered-earth models. These models are treated in Chapter 13, where we will see that transmitted seismic pulse-train waveforms generated by such models are minimum-delay. Work on layered-earth models has done much from a theoretical point of view to justify the deterministic hypothesis required in our statistical model, namely that the basic waveform is minimum-delay (i.e., minimum-phase-lag).

With the advent of routine digital field recording and extensive digital computer processing of seismic data, it has been possible to test statistical models on a large-scale basis with high-quality data. Because a seismic trace is made up of many overlapping wave forms, it is usually not possible to obtain a direct measurement of the individual waveform shape. Hence, the minimum-delay nature of a pulse-train waveform must be verified indirectly. This indirect verification can be carried out by applying the method of predictive deconvolution to seismic data; because the method works satisfactorily as a routine seismic data processing method, we can conclude that the minimum-delay hypothesis is generally upheld.

The method of predictive deconvolution is used successfully in deconvolving field records, for both marine and land records with reverberations. The general success of the method shows that the minimum-delay hypothesis is valid under a wide range of field situations. The next step is therefore to categorize these various field situations so as to be able to predict those cases where the minimum-delay hypothesis does not apply. For such non-minimum-delay cases, adaptive procedures must be devised whereby some measure of the deviations from minimum-delay conditions can be determined. With such information, it is then possible to design the necessary deconvolution operators.

Minimum Delay

The concept of minimum delay is identical to the concept of minimum phase lag, except that term "minimum delay" connotes the time domain, whereas the term "minimum phase lag" connotes the frequency domain. An extensive discussion of these matters has already been given in Chapter 3. Here we review some of the salient features that are required by the discussion that follows.

Each of the three-point wavelets (a_0, a_1, a_2) in the set

$$\text{wavelet } a: (4, 0, -1)$$
$$\text{wavelet } b: (2, 3, -2)$$
$$\text{wavelet } c: (-2, 3, 2)$$
$$\text{wavelet } d: (-1, 0, 4)$$

has the same autocorrelation:

$$r_0 = a_0^2 + a_1^2 + a_2^3 = 17$$
$$r_1 = r_{-1} = a_0 a_1 + a_1 a_2 = 0$$
$$r_2 = r_{-2} = a_0 a_2 = -4$$
$$r_t = r_{-t} = 0 \qquad \text{for } t > 2$$

The cumulative energy buildup, or partial energy curve $(a_0^2, a_0^2 + a_1^2, a_0^2 + a_1^2 + a_2^2)$, of each of these wavelets is:

$$\text{wavelet } a: \ (16, 16, 17)$$
$$\text{wavelet } b: \ (4, 13, 17)$$
$$\text{wavelet } c: \ (4, 13, 17)$$
$$\text{wavelet } d: \ (1, 1, 17)$$

We see that wavelet a has the quickest energy buildup; this is the minimum-delay wavelet. Why do we say minimum-delay? Because the energy delay of wavelet a, as we progress from the beginning of the wavelet to the end, is a minimum when compared to that of any other wavelet in the given set. Hence, the minimum-delay wavelet is the one with most of its energy concentrated at the beginning. More specifically, we may say "minimum energy delay" instead of just "minimum delay."

At the other extreme is the maximum-delay wavelet. This wavelet has the slowest energy buildup. Most of its energy is concentrated at the end. Wavelet d is the maximum-delay wavelet of the given set.

Between the two extremes of minimum delay and maximum delay are the mixed-delay wavelets. The energy concentration of a mixed-delay wavelet is intermediate relative to the time scale of the wavelet. Wavelets b and c are mixed-delay wavelets.

In terms of the frequency domain, the condition that each wavelet in the set has the same autocorrelation becomes the condition that each wavelet in the set has the same magnitude spectrum. Figure 11-1 shows the magnitude spectrum that wavelets a, b, c, and d have in common.

The minimum-delay condition becomes the condition that the phase-lag spectrum is a minimum. Thus, the term "minimum energy delay" is equivalent to "minimum phase lag." The maximum-delay condition becomes the condition that the phase-lag spectrum is a maximum. Figure 11-2 shows the phase-lag spectra for the given wavelets a, b, c, and d.

At this point let us recall that the convolution of two minimum-delay signals yields a minimum-delay signal and that the inverse of a mini-

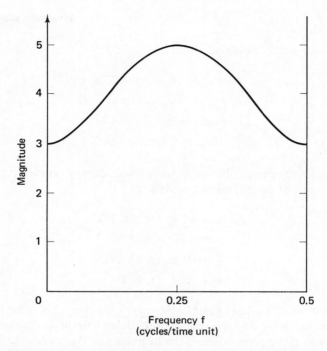

Figure 11-1. Magnitude spectrum for each of the wavelets a, b, c, and d.

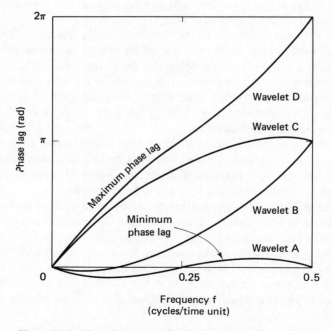

Figure 11-2. Phase-lag spectra for the wavelets a, b, c, and d.

mum-delay signal is a minimum-delay signal; that is, the property of minimum delay is preserved under convolution and inversion.

We have thus looked at the concepts of minimum delay and the associated concepts of maximum delay and mixed delay. The key word here is "delay"; why is this word important? In order to bring out the physical meaning of delay, let us look at two examples taken from exploration seismology: the elimination of seismic ghost reflections and the elimination of water reverberations. These two examples will serve to illustrate the concept of minimum delay as it applies to seismic data processing.

The Elimination of Seismic Ghost Reflections

The first example we will discuss is that of the elimination of seismic ghost reflections. In seismic exploration the explosion is set off in a shot hole drilled a hundred meters more or less below the surface of the ground. Now the existence of a large velocity discontinuity above the seismic shot can act as the source of *ghost reflections* in the following way. The *primary reflections* on the seismogram are caused by the reflection from deep strata of the energy moving directly downward from the shotpoint. Meanwhile the energy moving directly upward from the shotpoint is reflected from the overlying discontinuity, and thus there is a source of "secondary" energy moving directly downward. This secondary energy, in turn, is reflected from the deep strata, thereby producing corresponding ghost reflections on the seismogram. Thus, a given deep reflecting horizon appears on the seismogram as two reflection wavelets, displaced in time by twice the travel time from the shot to the overlying discontinuity. This relationship is illustrated in Figure 11-3.

Any differences in shape between the primary and ghost reflections can be attributed to various causes. However, in many instances the primary and ghost have approximately the same shape, so the two most important parameters become c and n, where c is a constant (of magnitude less than unity) repersenting the relative amplitude of the ghost with respect to the primary, and n is a constant (assumed to be an integer) representing the time delay (in discrete time units) of the ghost with respect to the primary. Hence, the primary and ghost reflection wavelets from a given deep reflecting strarum may be represented as the couplet

$$(1, 0, 0, \ldots, 0, c)$$
$$\text{time:} \quad 0, 1, 2 \qquad n$$

where the coefficient 1 represents the primary (at relative time 0) and c the ghost (at relative time n). This couplet is minimum-delay because its initial coefficient 1 is greater in magnitude than its final coefficient c, so more of the

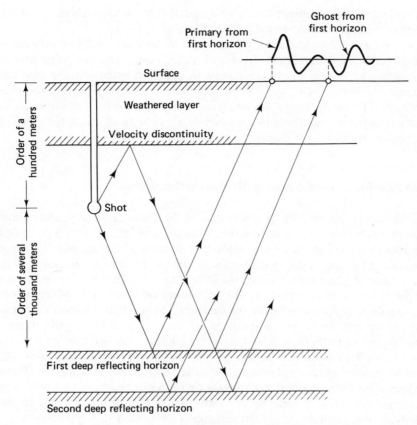

Figure 11-3. Ray paths of primary and ghost reflections from a deep reflecting horizon. Here the base of the weathering represents the ghost-producing discontinuity.

energy is concentrated at the front of the couplet than at its end. Alternatively, we may use the polynomial (or z-transform) representation given by

$$1 + cz^n$$

A filter to eliminate ghost reflections would be one that converts a primary/ghost reflection couplet at the input into a primary reflection alone at the output. If we designate the z transform of the filter by $F(z)$, the required filter may be specified mathematically as

(z transform of input)(z transform of filter) = (z transform of output)

or

$$(1 + cz^n)F(z) = 1$$

Solving for $F(z)$, we obtain the expression

$$F(z) = \frac{1}{1 + cz^n}$$

for the ghost elimination filter. Because of the linear property of the filter, a train of overlapping primary reflections plus their ghosts is transformed by the filter into a train of primary reflections only.

The action of the filter may be described by means of the block diagram shown in Figure 11-4. Upon entering the filter a primary wavelet appears at the output with no delay, and is simultaneously stored in the box cz^n, which produces a delay of n and attenuation of c. As its ghost wavelet enters the filter n time units later, the stored (delayed and attenuated) primary wavelet exactly cancels the ghost wavelet, so the ghost does not appear at the output of the filter.

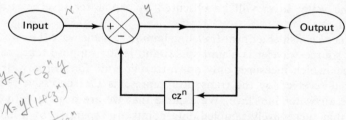

Handwritten annotations:
x
y
$y = x - cz^n y$
$x = y(1 + cz^n)$
$\frac{y}{x} = \frac{1}{1 + cz^n}$
$H = \frac{y}{x}$

Figure 11-4. Filter for removing ghost reflections.

There are various ways of estimating the required parameters from the seismic data for the design of the prototype ghost elmination filter described above. Also, more refined designs of ghost elimination filters may be used. For example, a ghost elimination filter may be designed as the least-squares shaping filter (see Chapter 6), where the input is the waveform made up of the primary with its ghost, and the desired output is the waveform of the primary only.

The Elimination of Water Reverberations

The second example is that of the elimination of water reverberations; that is, the ringing or "singing" that appears on a marine seismic trace. The water reverberation problem in marine seismic operations may be described as follows. The water/air interface is a strong reflector, with reflection coefficient nearly equal to -1. If the water/bottom interface is also a strong reflector, the water layer represents a nonattenuating medium bounded by two strong reflecting interfaces, and hence represents an energy trap. A seismic pulse generated in this energy trap will be successively reflected

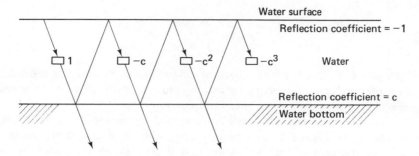

Figure 11-5. Water-confined reverberation. For clarity the ray paths have been drawn as slanting lines, although in the model they are perpendicular (or normal) to the two interfaces.

between the two interfaces. Consequently, reflections from deep horizons below the water layer will be obscured by the water reverberations (see Figure 11-5).

Let us consider a down-traveling impulsive plane wave (i.e., we assume that the source wavelet is a unit spike) and let us suppose that we have a transducer which measures only down-traveling motion in the water layer. The signal received by the transducer represents the successive reflections from the air/water interface. We assume that we are dealing with discrete time so that successively sampled data points are separated by a fixed sampling-time increment; for example, the sampling-time unit might be 4 ms. We let the integer n represent the two-way travel time in the water measured in terms of the sampling-time unit; for example, if the water depth is 100 ft and the water velocity is 5000 ft/s, the two-way travel time would be 40 ms, so for a 4-ms sampling increment the value of n would be 10.

Referring to Figure 11-5, we see that the coefficient 1 occurs at time 0 and represents the initial downgoing spike. The coefficient $-c$ occurs at discrete time n and represents the second downgoing spike, which has suffered a reflection at the bottom (reflection coefficient c) and a reflection at the surface (reflection coefficient -1). The coefficient c^2 occurs at discrete time $2n$ and represents the third downgoing spike, which has suffered two reflections at the bottom (c^2) and two reflections at the surface ($-1)^2$. The coefficient $-c^3$ occurs at discrete time $3n$ and represents the fourth downgoing spike, which has suffered three reflections at the bottom (c^3) and three reflections at the surface ($-1)^3$, and so on. Thus, the *water-confined reverberation spike train* is of the form

$$(1, 0, 0, \ldots, 0, -c, 0, 0, \ldots, 0, c^2, 0, 0, \ldots, 0, -c^3, 0, 0, \ldots)$$

$$\text{time:} \quad 0, 1, 2, \ldots \qquad n \qquad\qquad 2n \qquad\qquad 3n \qquad \text{etc.} \tag{11-1}$$

The reflection cofficient c cannot exceed unity in magnitude; generally, c will be less than unity in magnitude, for otherwise the water layer would represent a perfect energy trap and no energy would penetrate to the deeper layers. Thus, in the case where $|c| < 1$, the water-confined reverberation spike train (11-1) represents a convergent sequence, and is a wavelet (according to the mathematical definition of a discrete wavelet as being a one-sided stable time sequence; see Chapter 3). The z transform of the water-confined reverberation spike train is

$$1 - cz^n + c^2z^{2n} - c^3z^{3n} + \cdots$$

which is in the form of a geometric series, which can be summed to give

$$\frac{1}{1 + cz^n}$$

[*Note*: The sum of the geometric series $1 - x + x^2 - x^3 + x^4 - \cdots$, where $|x| < 1$ is $1/(1 + x)$.]

We thus see that the present water reverberation case is exactly reciprocal to the seismic ghost reflection case treated in the preceding section, for we recall that the z transform of the primary/ghost complex was of the form $1 + cz^n$ (where, of course, the parameters c and n represent different physical constants in the case of ghosts than in the case of water reverberations). The z transform of the water-confined reverberation elimination filter would therefore have the form

$$1 + cz^n$$

so that the impulse response function would be the couplet

$$
\begin{array}{cccc}
(1, & 0, & 0, \ldots, & c) \\
\uparrow & \uparrow & \uparrow & \uparrow \\
\text{time:} \quad 0, & 1, & 2 & n
\end{array}
\tag{11-2}
$$

Because $|c| < 1$, this couplet is minimum-delay, and it also follows that the water-confined reverberation spike train (11-1) is minimum-delay because the inverse of a minimum-delay wavelet is also minimum-delay. To verify that the water-confined reverberation elimination filter (11-2) does indeed convert the water-confined reverberation spike train (11-1) into a spike (and thereby eliminates the reverberation), we may convolve (11-1) with (11-2) and thereby obtain the spike $(1, 0, 0, 0, \ldots)$, which represents the impulsive plane wave without the water reverberation.

In our analysis so far we have only considered the water reverberation effect itself, but we have not yet included the effect of the water layer on deep subsurface reflections. To do so, we must take into account that the water layer acts as a filter, and that the seismic energy passes through this filter once as it goes down to the deep reflecting horizon, and again as it returns to the surface. That is, the water layer acts on the deep reflection data twice, once going down to the deep strata and once again coming back to the surface. Hence, we have, in effect, a situation where the deep reflection data pass through two cascaded sections of the water layer, so that the overall z transform of the filtering introduced by the water layer is the square of the z transform due to a single pass, as given in the foregoing paragraph (see Figure 11-6). That is, the z transform of the reverberation spike train resulting from a deep reflecting horizon is

$$(1 - cz^n + c^2z^{2n} - c^3z^{3n} + \cdots)^2$$

or

$$\frac{1}{(1 + cz^n)^2}$$

In the time domain, we see that the *deep-reflection reverberation spike train* is given by

$$(1, 0, 0, \ldots, 0, -2c, 0, 0, \ldots, 0, 3c^2, 0, 0, \ldots, 0, -4c^3, 0, 0, \ldots)$$

$$\text{time:} \quad 0, 1, 2, \ldots \qquad n \qquad \qquad 2n \qquad \qquad 3n$$

$$(11\text{-}3)$$

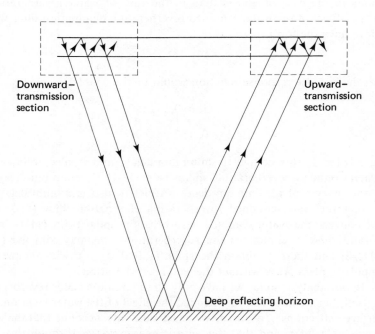

Downward–
transmission
section

Upward–
transmission
section

Deep reflecting horizon

Figure 11-6. Deep reflection reverberation-producing filter as two cascaded sections.

This expression may be obtained by convolving the water-confined reverberation spike train (11-1) with itself. Since spike train (11-1) is minimum-delay, it follows that spike train (11-3) is also minimum-delay. The reverberation spike train (11-3) may be interpreted as consisting of the deep reflection spike at time 0, and the attached reverberation spikes at times n, $2n$, $3n$, $4n$, ... To eliminate the reverberation part of spike train (11-3), we must pass spike train (11-3) through the inverse filter whose z transform is given by

$$(1 + cz^n)^2$$

or

$$1 + 2cz^n + c^2 z^{2n}$$

The impulse response function of this inverse filter is

$$(1, 0, 0, \ldots, 0, 2c, 0, 0, \ldots, 0, c^2)$$
$$\begin{array}{ccc} \uparrow \uparrow \uparrow & \uparrow & \uparrow \\ \text{time:} \quad 0, 1, 2 & n & 2n \end{array} \qquad (11\text{-}4)$$

Convolving spike train (11-3) with spike train (11-4), we see that the output is indeed a spike $(1, 0, 0, 0, \ldots)$ which represents the direct deep reflection spike (at time 0) with the attached reverberation spikes eliminated.

The Method of Predictive Deconvolution

In practice, the reverberations are often generated by a more complicated physical situation than the one we have described, so the reverberation elimination operator (11-4) is not adequate in many field applications. However, there is an important principle which we can extract from the foregoing deterministic reverberation model—the fact that the deep-reflection reverberation spike train (11-3) is minimum-delay. This minimum-delay property of reverberation spike trains is also valid in other more complicated reverberation models (e.g., see Chapter 13), and moreover it is upheld on recorded data from various field situations. As a result, we would want to use a deconvolution method that exploits this minimum-delay property.

Up to now, for simplicity, we have assumed that the seismic source pulse is a spike. In practice, it will not be a spike, but will have some more complicated shape. Let us designate the pulse-train waveform due to a deep reflection by the symbol b_t. The pulse-train waveform b_t is given by the convolution of the deep-reflection reverberation spike train (whatever it is) with the seismic source wavelet (whatever it is). For example, in the reverberation model of the preceding section, the deep-reflection reverberation spike train was given by (11-3). As we have just noted, the deep-reflection reverberation spike train in many instances is minimum-delay, or at least approximately so. Hence, if the seismic source is sufficiently sharp so that the source wavelet is also approximately minimum-delay, it follows that the waveform b_t is also approximately minimum-delay (this follows from the fact that the signal resulting from the convolution of two minimum-delay signals is minimum-delay).

Let us designate the received seismic trace by x_t (where it is assumed that the filtering effect of the recording instruments has already been removed from x_t by some restoring filtering operation). The received seismic trace x_t is the resultant of many deep reflections, each of which contributes a pulse-train waveform of shape b_t. However, because of the fact that all the pulse-train waveforms are overlapping with each other to various degrees, it is not possible to obtain a direct measurement of the individual waveform shape b_t. In such a physical situation, the method of predictive deconvolution of seismic traces can be used to eliminate the reverberations, and thus better bring out the true reflection pulses.

The predictive deconvolution method is based on the following statistical model. The received seismic trace x_t (included within an appropriately chosen time gate) is considered to be the result of convolving the waveform b_t with a random spike series ϵ_t, that is, $x_t = b_t * \epsilon_t$. The spike series ϵ_t represents the reflections from the deep reflecting horizons in the sense that the timing of a spike represents the direct arrival time of a reflection and the amplitude of the spike represents the strength of the reflection. For data processing purposes, this spike series ϵ_t is considered as a random uncorrelated (i.e., white noise) series. Hence, the autocorrelation of the received seismic trace x_t is the same as the autocorrelation of the individual waveform b_t, except for a constant scale factor. This scale factor will not affect the final results, so it may be neglected. We can therefore compute the autocorrelation of the waveform b_t from the received seismic trace x_t. From the autocorrelation we can compute the prediction operator for the waveform b_t. Because the waveform b_t is minimum-delay, a prediction operator for prediction distance n ($=$ two-way travel time in the water) will predict the reverberation component of the waveform b_t. (This would not be true if the waveform were not minimum-delay. For an illustration of this principle, see Chapter 8.) A delay of n time units will line up this predicted reverberation with the reverberation portion of the waveform b_t; by subtracting this delayed predicted reverberation from the waveform b_t we obtain the prediction error. Because the delayed predicted reverberation cancels the reverberation part of the waveform b_t, it follows that the prediction error represents the nonreverberation part of the waveform b_t. Hence, the prediction error operator eliminates the reverberation part of the waveform b_t. Since the prediction error operator is linear, we can apply it to the received seismic trace $x_t = b_t * \epsilon_t$ (which represents many overlapping waveforms b_t with arrival times and strengths given according to the spike series ϵ_t). In so doing, we eliminate the reverberations from each of the wave-forms b_t but leave intact the initial nonreverberation portions, thereby increasing seismic resolution. If more resolution is desired, the prediction distance can be lessened, which will have the effect of further compressing the energy in the waveform. Furthermore, some type of bandpass or shaping filter can be applied as a postfiltering operation to yield a smoother decon-volved seismic trace. As an extra result, we can obtain the shape of the wave-form b_t, as described in Chapter 10.

Computation Steps for Predictive Deconvolution

Let us now summarize the computational steps required for predictive deconvolution. The sampled values of the received seismic trace are denoted by x_i. The subscript i denotes the discrete-time index; for example, if the sampling were at 4-ms intervals, the trace reading x_2 would follow the read-ing x_1 by 4 ms in seismic time, and so on.

The first step is to compute the autocorrelation function of that por-tion of the sampled trace within a specified time gate. The autocorrelation coefficients ϕ_j are computed from the sampled trace by means of the formula

$$\phi_j = \sum x_{i+j} x_i$$

where the summation runs over all the time indices i within the time gate. Because $\phi_{-j} = \phi_j$ the autocorrelation coefficients need only be computed for nonnegative values of the time-shift index j.

Often it is useful to weight the autocorrelation by some set of tapered weighting factors w_j in order to obtain the weighted autocorrelation

$$r_j = w_j \phi_j$$

A typical set of weighting factors would be the triangular weghting factors given by

$$w_j = \begin{cases} 1 - \dfrac{|j|}{N} & \text{for } |j| = 0, 1, 2, 3, \ldots, N \\ 0 & \text{for } |j| = N+1, N-2, \ldots \end{cases}$$

Here N represents the time index at which the autocorrelation is truncated; the value of the parameter N must be specified.

The second step is to compute the coefficients of the prediction op-erator, which we designated by

$$(k_0, k_1, k_2, \ldots, k_m)$$
$$\uparrow \ \ \uparrow \ \ \uparrow \qquad\qquad \uparrow$$
$$\text{time:} \ \ 0, \ \ 1, \ \ 2, \ \ldots, \ m$$

There are various methods of computing this operator; for numerical work the Gauss method of least squares has certain advantages. According to the method of least squares the predicton operator is determined by minimizing the mean square prediction error. The minimization leads to the set of simultaneous linear equations called the *normal equations*, which involve the auto-correlation coefficients (that we have just computed) as the knowns, and the prediction operator coefficients as the unknowns. The normal equations are (see also Chapters 6 and 8)

$$k_0 r_0 + k_1 r_1 + k_2 r_2 + \cdots + k_m r_m = r_\alpha$$

$$k_0 r_1 + k_1 r_0 + k_2 r_1 + \cdots + k_m r_{m-1} = r_{\alpha+1}$$

$$\vdots$$

$$k_0 r_m + k_1 r_{m-1} + k_2 r_{m-2} + \cdots + k_m r_0 = r_{\alpha+m}$$

[handwritten annotations: "Prediction error operator" and "Prediction operator"]

In this set of equations, the positive integer α denotes the prediction distance or prediction span; we must specify some value for α. Clearly, we may set α equal to the two-way travel time n, or equal to unity, or to some other value, depending on the purpose at hand. More on the choice of α will be said in Chapter 12.

These equations may be solved by the Toeplitz recursive procedure (described in Appendix 6-2). Using this procedure, the machine time required to solve the normal equations for a digital filter with m coefficients is proportional to m^2, as compared to m^3 for the conventional methods of solving simultaneous equations. Another advantage of using this recursive method is that it requires computer storage space proportional to m, rather than m^2 as in the case of the conventional methods.

Once we have computed the prediction operator coefficients $(k_0, k_1, k_2, \ldots, k_m)$, we then know, in effect, the coefficients of the prediction error operator, for the prediction error operator coefficients are given by (for prediction span α)

$$(1, 0, 0, \ldots, 0, \qquad -k_0, -k_1, -k_2, \ldots, -k_m)$$

$$\uparrow \uparrow \uparrow \qquad \uparrow \qquad \uparrow \quad \uparrow \quad \uparrow \qquad \uparrow$$

$$\text{time:} \quad 0, 1, 2, \ldots, \alpha - 1, \quad \alpha, \quad \alpha + 1, \alpha + 2, \ldots, \alpha + m$$

For a further discussion of this relationship between the prediction operator and the prediction error operator, see Chapter 12. It is the prediction error operator that is the required inverse operator for deconvolving the seismic trace x_i. Accordingly, the third step is to convolve the prediction error operator with the seismic trace x_i; this computation is carried out according to the discrete convolution formula:

$$y_i = x_i - k_0 x_{i-\alpha} - k_1 x_{i-\alpha-1} - k_2 x_{i-\alpha-2} - \cdots - k_m x_{i-\alpha-m}$$

The result y_t is the prediction error series (for prediction span α), which represents the deconvolved (but unsmoothed) seismic trace. Note that the "deconvolution" of the trace is accomplished by "convolving" the trace with the "inverse" operator (i.e., with the prediction error operation).

In the special case when the prediction distance α is chosen to be unity, the prediction error operator is the least-squares inverse of the (minimum-delay) waveform b_t, and the prediction error series represents the random spike series ϵ_t (i.e., the series designating the strengths and arrival times of the deep reflections), as described in Chapter 10.

In the special case when the prediction distance α is chosen equal to the two-way water travel time n, the prediction error series represents the seismic trace made up of only the reflection pulses; that is, the initial portion of each minimum-delay waveform b_t from its onset to a time duration of n units, for the reverberation pulse train has been removed (i.e., the tail of each waveform b_t from time n on has been eliminated).

All of the above, of course, holds within the limitations of statistical errors imposed by noise, computational approximation, and the finiteness of the data, and within the limitations of specification errors imposed by the model. The success of the method of predictive deconvolution, as we have discussed, depends largely upon the validity of the basic hypotheses regarding the minimum-delay nature of the waveform b_t and the random nature of the spikes series ϵ_t. The beauty of the predictive method is that the only data required to perform the deconvolution is the received seismic trace (and the water depth in the case when we choose $\alpha = n$).

In order to smooth the prediction error series, we can apply a postfiltering operation to it. This postfilter can be some type of digital bandpass filter or a digital shaping filter, as described in Chapter 6. Instead of applying the postfilter to the prediction error series, we can instead first cascade the prediction error filter with the postfilter, and apply the cascaded filter to the received seismic trace. The final output will be the same in either case, namely the smoothed deconvolved seismic trace without reverberations.

Concluding Remarks

One of the most gratifying successes of the method of predictive deconvolution in seismic exploration has been its ability to attenuate ghost reflections and reverberations, both on land and at sea. The model for its implementation is very simple. It depends on the statistical hypothesis that the strengths and arrival times of the information bearing events are representable as a random spike series. Further, it depends on the deterministic hypothesis that the basic seismic waveform associated with each of these events is minimum delay.

deconvolution for the elimination of short-period and long-period multiple reflections

Summary

Least-squares inverse filters have found widespread use in the deconvolution of seismograms. The least-squares prediction error filter with unit prediction distance is equivalent within a scale factor to the least-squares, zero-lag inverse filter. The use of least-squares prediction filters with prediction distances greater than unity leads to the method of predictive deconvolution, which represents a more generalized approach to this subject.

The predictive technique allows one to control the length of the desired output wavelet, and hence to specify the desired degree of resolution. Multiple reflections are events that are periodic within given repetition ranges; they can be attenuated selectively by deconvolution. Deconvolution is effective in the suppression of rather complex multiple-reflection patterns.

Introduction

Deconvolution is one of the most effective tools for the digital reduction of seismic traces. It constitutes the keystone of many current seismic processing methods. One can deconvolve a reverberating pulse train into an approximation of a zero-delay unit impulse. More generally, it is possible to arrive at deconvolution filters which remove repetitive events having specified periodicities. Thus, the deconvolution filter is better viewed as an operator that removes coherent energy by its predictability than merely as a spiker of "leggy" wave trains. In this chapter we continue the treatment of deconvolution begun in the last chapter, but with particular emphasis on the design of deconvolution filters for prediction distances greater than unity.

Basic Concepts

The digital filtering process is described by the discrete convolution formula

$$y_t = \sum_\tau x_\tau a_{t-\tau}$$

where x_t is the input, a_t the filter, and y_t the output. In the sequel, t and τ are discrete-time variables and the sampling increment is $\Delta t = 1$ unless otherwise specified.

If a_t is a prediction operator with prediction distance α, the output y_t will be an estimate of the input x_t at some future time $t + \alpha$. We thus write

$$y_t = \sum_\tau x_\tau a_{t-\tau} = \hat{x}_{t+\alpha} \tag{12-1}$$

where $\hat{x}_{t+\alpha}$ is an estimate of $x_{t+\alpha}$.

An error series may be defined as the difference between the true values $x_{t+\alpha}$ and the estimated or predicted value $\hat{x}_{t+\alpha}$,

$$\epsilon_{t+\alpha} = x_{t+\alpha} - \hat{x}_{t+\alpha} \tag{12-2}$$

Thus, ϵ_t is an output series that represents the nonpredictable part of x_t.

Replacement of the term $\hat{x}_{t+\alpha}$ in equation (12-2) with its equivalent as defined by equation (12-1) results in

$$\epsilon_{t+\alpha} = x_{t+\alpha} - \sum_\tau x_\tau a_{t-\tau} \tag{12-3}$$

The z transform of equation (12-3) is

$$z^{-\alpha}E(z) = z^{-\alpha}X(z) - X(z)A(z) \tag{12-4}$$

Multiplication of both sides of equation (12-4) by z^α yields

$$E(z) = X(z) - z^\alpha X(z)A(z)$$
$$= X(z)[1 - z^\alpha A(z)] \tag{12-5}$$

The quantity $[1 - z^\alpha A(z)]$ is the z transform of the prediction error operator. It is seen to be the difference between the zero-delay unit spike and the prediction operator $A(z)$ delayed by the prediction distance α. Thus, one may calculate the error series ϵ_t by computing $\hat{x}_{t+\alpha}$ from equation (12-1), and follow with a subtraction as defined by equation (12-2). Alternatively, one may compute the error series in a single step by use of the prediction error operator. Let us assume that a seismic trace is represented by the convolution of an uncorrelated reflection coefficient series with a reverberating pulse train, which by its nature is rich in repetitive energy (see Chapter 11). The prediction error filter will then remove the predictable portion of such a trace, which to a good approximation will be given by the repetitive energy in the reverberations. The output of this filtering operation is the error series ϵ_t, which within the framework of this model constitutes the estimate of the reflection coefficient series of the layered subsurface.

Suppose that the prediction operator a_t is given by the n-length series

$$a_0, a_1, \ldots, a_{n-1}$$

Then the corresponding prediction error operator f_t with prediction distance α is

$$\underbrace{1, 0, 0, \ldots, 0}_{\alpha - 1 \text{ zeros}}, -a_0, -a_1, \ldots, -a_{n-1}$$

We must next deal with the explicit design of the prediction operator.

The Least-Squares Predictive Filtering Model

A general least-squares filter model involves the three signals illustrated in Figure 12-1: (1) the input signal x_t, (2) the desired output signal z_t, and (3) the actual output signal y_t. Minimization of the energy existing in the difference between the desired output z_t and the actual output y_t; that is, minimization of the expression

$$I = \sum_t (z_t - y_t)^2$$

results in the least-squares filter described in Chapter 6.

The n-length least-squares filter results from the solution of the normal equations with matrix representation

$$
\begin{bmatrix}
r_0 & r_1 & \cdots & r_{n-1} \\
r_1 & r_0 & \cdots & r_{n-2} \\
 & & \vdots & \\
 & & \vdots & \\
r_{n-1} & r_{n-2} & \cdots & r_0
\end{bmatrix}
\begin{bmatrix}
f_0 \\
f_1 \\
\cdot \\
\cdot \\
f_{n-1}
\end{bmatrix}
=
\begin{bmatrix}
g_0 \\
g_1 \\
\cdot \\
\cdot \\
g_{n-1}
\end{bmatrix}
\qquad (12\text{-}6)
$$

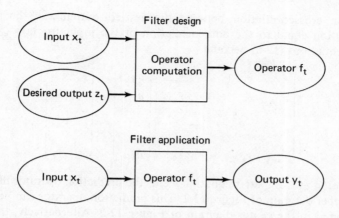

Figure 12-1. General model illustrating the design and application of the least-squares filter.

where r_t is the autocorrelation of the input, g_t the crosscorrelation between the desired output and the input, and f_t the filter coefficients.

We have seen that the prediction operator is that filter which acts on an input trace up to time t and estimates the trace amplitude at some future time $t + \alpha$. Thus, it is reasonable to define the desired output for the predictive filter as a time-advanced version of the input x_t. We can now express the prediction filter in terms of a particular least-squares filter, the one for which the desired output trace is simply a time-advanced version of the input trace.

In order to solve equations (12-6), we must know the autocorrelation of the input and the positive lag coefficients of the crosscorrelation between the desired output and the input. The autocorrelation of the input trace is

$$r_\tau = \sum_t x_t x_{t-\tau}$$

while the crosscorrelation between the desired output and input traces is

$$g_\tau = \sum_t z_t x_{t-\tau} \tag{12-7}$$

Since the desired output for the prediction operator is a time-advanced version of the input; that is, since

$$z_t = x_{t+\alpha}$$

equation (12-7) becomes

$$g_\tau = \sum_t x_{t+\alpha} x_{t-\tau} = \sum_t x_t x_{t-(\tau+\alpha)} = r_{\tau+\alpha}$$

Thus, the crosscorrelation between the desired output and the input is by definition equal to the autocorrelation of the input for lags $\geq \alpha$. The normal equations (12-6) become

$$
\begin{bmatrix}
r_0 & r_1 & \cdots & r_{n-1} \\
r_1 & r_0 & \cdots & r_{n-2} \\
& & \cdot & \\
& & \cdot & \\
& & \cdot & \\
r_{n-1} & r_{n-2} & \cdots & r_0
\end{bmatrix}
\begin{bmatrix}
a_0 \\
a_1 \\
\cdot \\
\cdot \\
\cdot \\
a_{n-1}
\end{bmatrix}
=
\begin{bmatrix}
r_\alpha \\
r_{\alpha+1} \\
\cdot \\
\cdot \\
\cdot \\
r_{\alpha+n-1}
\end{bmatrix}
\qquad (12\text{-}8)
$$

The solution to this matrix equation yields the prediction operator illustrated in the upper diagram of Figure 12-2. This prediction operator can be utilized as indicated in the center diagram of Figure 12-2. Alternatively, the corresponding prediction error operator can be formed from the prediction operator and utilized as indicated in the lower diagram of Figure 12-2.

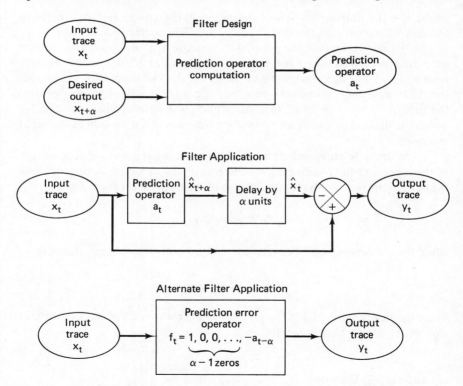

Figure 12-2. Predictive filter model which illustrates prediction operator design (upper diagram) and application (center diagram). Alternately, the prediction error operator may be formed and utilized as illustrated in the lower diagram.

We have thus shown how one can use the least-squares error criterion to generate the least-squares prediction operator, and how the prediction error operator is derived from its corresponding prediction operator. Our aim in the next section is to indicate how the prediction error operator can also be expressed in the form of a particular least-squares filter.

Predictive Filtering and Deconvolution

We shall demonstrate that the least-squares deconvolution filter which ideally transforms an unknown signal to an impulse at zero delay is equivalent to the prediction error filter for which the prediction distance α is unity. The matrix relation for the prediction operator a_t with prediction distance unity ($\alpha = 1$) and length n is obtained by setting $\alpha = 1$ in equation (12-8),

$$
\begin{bmatrix}
r_0 & r_1 & \cdots & r_{n-1} \\
r_1 & r_0 & \cdots & r_{n-2} \\
& & \cdot & \\
& & \cdot & \\
& & \cdot & \\
r_{n-1} & r_{n-2} & \cdots & r_0
\end{bmatrix}
\begin{bmatrix}
a_0 \\
a_1 \\
\cdot \\
\cdot \\
\cdot \\
a_{n-1}
\end{bmatrix}
=
\begin{bmatrix}
r_1 \\
r_2 \\
\cdot \\
\cdot \\
\cdot \\
r_n
\end{bmatrix}
\qquad (12\text{-}9)
$$

The system above may be written in the form of the n simultaneous linear equations,

$$r_0 a_0 + r_1 a_1 + \cdots + r_{n-1} a_{n-1} = r_1$$
$$r_1 a_0 + r_0 a_1 + \cdots + r_{n-2} a_{n-1} = r_2$$
$$\vdots$$
$$r_{n-1} a_0 + r_{n-2} a_1 + \cdots + r_0 a_{n-1} = r_n$$

Let us modify the system above by first subtracting the coefficient r_i from both sides of the ith row of each equation such that the right-hand side vanishes (the original system is shown within the box):

$$
\begin{aligned}
-r_1 + & \boxed{r_0 a_0 + r_1 a_1 + \cdots + r_{n-1} a_{n-1} = r_1} & -r_1 \\
-r_2 + & \boxed{r_1 a_0 + r_0 a_1 + \cdots + r_{n-2} a_{n-1} = r_2} & -r_2 \\
& \qquad\qquad\qquad \vdots & \\
-r_n + & \boxed{r_{n-1} a_0 + r_{n-2} a_1 + \cdots + r_0 a_{n-1} = r_n} & -r_n
\end{aligned}
$$

Let us next augment this system in the form

$$-r_0 + \boxed{\begin{matrix} r_1 a_0 & + r_2 a_1 & + \cdots + r_n a_{n-1} \end{matrix}} = -\beta$$

$$\begin{array}{l} -r_1 + \\ -r_2 + \\ \\ \\ \\ -r_n + \end{array} \boxed{\begin{matrix} r_0 a_0 & + r_1 a_1 & + \cdots + r_{n-1} a_{n-1} = r_1 \\ r_1 a_0 & + r_0 a_1 & + \cdots + r_{n-2} a_{n-1} = r_2 \\ & & \cdot \\ & & \cdot \\ & & \cdot \\ r_{n-1} a_0 + r_{n-2} a_1 + \cdots + r_0 a_{n-1} & = r_n \end{matrix}} \begin{array}{l} -r_1 \\ -r_2 \\ \\ \\ \\ -r_n \end{array}$$

This system may be written

$$
\begin{aligned}
r_0 - r_1 a_0 &\; - r_2 a_1 \; - \cdots - r_n a_{n-1} &= \beta \\
r_1 - r_0 a_0 &\; - r_1 a_1 \; - \cdots - r_{n-1} a_{n-1} &= 0 \\
&\qquad\qquad\cdot \\
&\qquad\qquad\cdot \\
r_n - r_{n-1} a_0 &- r_{n-2} a_1 - \cdots - r_0 a_{n-1} &= 0
\end{aligned}
$$

for which the associated matrix equation is

$$
\begin{bmatrix}
r_0 & r_1 & \cdots & r_n \\
r_1 & r_0 & \cdots & r_{n-1} \\
& & \cdot & \\
& & \cdot & \\
& & \cdot & \\
r_n & r_{n-1} & \cdots & r_0
\end{bmatrix}
\begin{bmatrix}
1 \\
-a_0 \\
\cdot \\
\cdot \\
\cdot \\
-a_{n-1}
\end{bmatrix}
=
\begin{bmatrix}
\beta \\
0 \\
\cdot \\
\cdot \\
\cdot \\
0
\end{bmatrix}
\qquad (12\text{-}10)
$$

We now see that the least-squares filter of equation (12-10) can be identified as the unit prediction error operator associated with the prediction operator of equation (12-9). Let us rewrite equation (12-10) in the form

$$
\begin{bmatrix}
r_0 & r_1 & \cdots & r_n \\
r_1 & r_0 & \cdots & r_{n-1} \\
& & \cdot & \\
& & \cdot & \\
& & \cdot & \\
r_n & r_{n-1} & \cdots & r_0
\end{bmatrix}
\begin{bmatrix}
b_0 \\
b_1 \\
\cdot \\
\cdot \\
\cdot \\
b_n
\end{bmatrix}
=
\begin{bmatrix}
\beta \\
0 \\
\cdot \\
\cdot \\
\cdot \\
0
\end{bmatrix}
\qquad (12\text{-}11)
$$

where

$$
\begin{aligned}
b_0 &= 1 \\
b_i &= -a_{i-1} \qquad (i = 1, \ldots, n) \\
\beta &= \sum_{i=0}^{n} b_i r_i
\end{aligned}
$$

In Appendix 12-1 we describe the standard deconvolution method, which is based on the use of the least-squares, zero-delay inverse filter. We note that the system of normal equations for the inverse filter given by equation (12-14) is identical to the system of equation (12-11), except for a scale factor β. Thus, the $(n + 1)$-length prediction error operator with prediction distance unity is identical to the zero-delay inverse filter of length $(n + 1)$, except for a scale factor.

We shall now show that the prediction error filter with prediction distance greater than unity can also serve as a deconvolution operator; it turns out that such a prediction-error filtering technique constitutes a more generalized approach to deconvolution. We remark that under certain assumptions described in Appendix 12-3, the autocorrelation of an input seismic signal can be identified with the autocorrelation of the source wavelet. The source wavelet is here meant to be the shot pulse modified by near-surface reverberations.

The inverse filter described in Appendix 12-1 shapes the unknown source wavelet to an impulse at zero lag time. We will show here that the prediction error filter shapes the unknown source wavelet of length $\alpha + n$ to another unknown wavelet of length α. Thus, by having control of the desired output wavelet length, one may specify the desired degree of resolution.

The prediction filter matrix equation for filter length n and prediction distance α is given by equation (12-8),

$$
\begin{bmatrix}
r_0 & r_1 & \cdots & r_{n-1} \\
r_1 & r_0 & \cdots & r_{n-2} \\
& & \cdot & \\
& & \cdot & \\
& & \cdot & \\
r_{n-1} & r_{n-2} & \cdots & r_0
\end{bmatrix}
\begin{bmatrix}
a_0 \\
a_1 \\
\cdot \\
\cdot \\
\cdot \\
a_{n-1}
\end{bmatrix}
=
\begin{bmatrix}
r_\alpha \\
r_{\alpha+1} \\
\cdot \\
\cdot \\
\cdot \\
r_{\alpha+n-1}
\end{bmatrix}
$$

or

$$
\begin{aligned}
r_0 a_0 + r_1 a_1 &+ \cdots + r_{n-1} a_{n-1} = r_\alpha \\
r_1 a_0 + r_0 a_1 &+ \cdots + r_{n-2} a_{n-1} = r_{\alpha+1} \\
&\cdot \\
&\cdot \\
&\cdot \\
r_{n-1} a_0 + r_{n-2} a_1 + \cdots + \quad r_0 a_{n-1} &= r_{\alpha+n-1}
\end{aligned}
\qquad (12\text{-}12)
$$

The system above can be augmented in such a way that the prediction operator is converted into the corresponding prediction error operator. This is accomplished by the addition of suitable terms to both sides of the equations (12-12). Proceeding as in the case of the unit prediction error filter [equation (12-9) et seq.], one obtains

$$-r_0 1 - \quad r_1 0 - \cdots - r_{\alpha-1} 0 + \quad r_\alpha a_0 + r_{\alpha+1} a_1 + \cdots + r_{\alpha+n-1} a_{n-1} = -p_0$$
$$-r_1 1 - \quad r_0 0 - \cdots - r_{\alpha-2} 0 + r_{\alpha-1} a_0 + \quad r_\alpha a_1 + \cdots + r_{\alpha+n-2} a_{n-1} = -p_1$$
$$\vdots$$
$$-r_{\alpha-1} 1 - \quad r_{\alpha-2} 0 - \cdots - \quad r_0 0 + \quad r_1 a_0 + \quad r_2 a_1 + \cdots + \quad r_n a_{n-1} = -p_{\alpha-1}$$
$$\vdots$$
$$-r_\alpha 1 - \quad r_{\alpha-1} 0 - \cdots - \quad r_1 0 + \boxed{\; r_0 a_0 + \quad r_1 a_1 + \cdots + \quad r_{n-1} a_{n-1} = r_\alpha \;} - r_\alpha$$
$$-r_{\alpha+1} 1 - \quad r_\alpha 0 - \cdots - \quad r_2 0 + \boxed{\; r_1 a_0 + \quad r_0 a_1 + \cdots + \quad r_{n-2} a_{n-1} = r_{\alpha+1} \;} - r_{\alpha+1}$$
$$\vdots$$
$$-r_{\alpha+n-1} 1 - r_{\alpha+n-2} 0 - \cdots - \quad r_n 0 + \boxed{\; r_{n-1} a_0 + r_{n-2} a_1 + \cdots + \quad r_0 a_{n-1} = r_{\alpha+n-1} \;} - r_{\alpha+n-1}$$

where the original set (12-12) is enclosed in the box. The associated matrix equation is

$$
\begin{bmatrix}
r_0 & r_1 & \cdots & r_{\alpha+n-1} \\
r_1 & r_0 & \cdots & r_{\alpha+n-2} \\
 & & \vdots & \\
r_{\alpha-1} & r_{\alpha-2} & \cdots & r_n \\
r_\alpha & r_{\alpha-1} & \cdots & r_{n-1} \\
 & & \vdots & \\
r_{\alpha+n-1} & r_{\alpha+n-2} & \cdots & r_0
\end{bmatrix}
\begin{bmatrix}
1 \\
0 \\
\vdots \\
0 \\
-a_0 \\
\vdots \\
-a_{n-1}
\end{bmatrix}
=
\begin{bmatrix}
p_0 \\
p_1 \\
\vdots \\
p_{\alpha-1} \\
0 \\
\vdots \\
0
\end{bmatrix}
\qquad (12\text{-}13)
$$

where

$$p_0 = r_0 - (r_\alpha a_0 + r_{\alpha+1} a_1 + \cdots + r_{\alpha+n-1} a_{n-1})$$
$$p_1 = r_1 - (r_{\alpha-1} a_0 + r_\alpha a_1 + \cdots + r_{\alpha+n-2} a_{n-1})$$
$$\vdots$$
$$p_{\alpha-1} = r_{\alpha-1} - (r_1 a_0 + r_2 a_1 + \cdots + r_n a_{n-1})$$

The solution of the foregoing matrix equation yields the prediction error operator with prediction distance α. Let us interpret this equation in terms of the least-squares filter model, where the left-hand matrix is the input auto-correlation matrix, and where the elements of the right-hand column vector constitute the positive lag values of the crosscorrelation between the desired output and the input. Subject to the assumptions given in Appendix 12-3, the autocorrelation function $r_0, r_1, \ldots, r_{\alpha+n-1}$ can be identified with the auto-corrleation of a source, or characteristic, wavelet, of length $\alpha + n$. However, we still require an interpretation of the crosscorrelation g_τ for $\tau = 0, 1, \ldots,$ $\alpha + n - 1$ given by

$$\underbrace{\rho_0, \rho_1, \ldots, \rho_{\alpha-1}}_{\alpha \text{ terms}}, \underbrace{0, \ldots, 0}_{n \text{ zeros}}$$

Although the crosscorrelation function is complicated, we can make one important observation. Since the crosscorrelation vanishes for lags greater than $\alpha - 1$, the length of the implied desired output wavelet cannot be greater than α. In other words, the input wavelet is of length $\alpha + n$, while the implied desired output wavelet is of length α, and hence the prediction error operator shortens an input wavelet of length $\alpha + n$ to output wavelet of length α. Since α is an independent variable, we are free to select whatever length we choose for the desired output wavelet. We conclude that the prediction error filter with prediction distance α leads to a more generalized approach to deconvolution, in which one may control the desired degree of resolution or wavelet contraction.

We have shown earlier that zero-delay least-squares inverse filter is equal within a scale factor to the prediction error filter with prediction distance unity ($\alpha = 1$). Experience has taught us that the output from these filters cannot, in general, be interpreted with ease. This is due to the presence of high-frequency components in the deconvolved trace, which result from the fact that this kind of deconvolution makes use of inverse, or "spiking" filters. One improves this condition by passing the raw deconvolved trace through suitable low-pass filters, by smoothing the autocorrelation function, or by other related means. In summary, the use of prediction error filters with an arbitrary prediction distance α leads to a deconvolution method in which one has more effective control on the desired degree of resolution.

It is of interest to establish how the prediction error filters perform on an idealized, noise-free reverberation model. These matters are discussed in Appendix 12-2, where we also shown that under appropriate simplifying conditions the prediction error filter becomes identical to the exact three-point filter for a single-layer system (see Chapter 11).

Applications of Predictive Deconvolution

The concept that permits resolution control by means of the prediction distance parameter has been introduced previously. We have seen that the crosscorrelation between the desired output and the input is zero between lag positions α and $\alpha + n - 1$, and we were thus able to deduce that the implied desired output pulse cannot be of length greater than α. We note from the autocorrelation matrix of equation (12-13) that the prediction error filter does not utilize any autocorrelation coefficient beyond lag position $\alpha + n - 1$. Our model thus implies that the source wavelet is of length $\alpha + n$.

Since this filter attempts to shape the input into some desired output, we can argue that the autocorrelation of the actual output data will tend to vanish between lag positions α and $\alpha + n - 1$. This is because the autocorrelation of the implied desired output wavelet vanishes for lags greater than $\alpha - 1$. The prediction error filter will thus modify the input in such a way that the autocorrelation of the actual output will tend to vanish between α and $\alpha + n - 1$. We cannot expect the autocorrelation to be 0 everywhere beyond lag $= \alpha + n - 1$, since the filter computation makes use of no autocorrelation coefficients for lags greater than $\alpha + n - 1$.

Experience has shown that the autocorrelation can be an interpretative aid in the analysis of reverberatory problems. When we consider that the prediction error filter is designed only from knowledge of the input autocorrelation and that the magnitude of this input autocorrelation at a particular lag is an indication of the degree of predictability at that lag, we see that the autocorrelation function is a very important entity to gauge the effectiveness of dereverberation by means of predictive deconvolution.

Thus, we set our parameters α (the prediction distance) and n (the prediction operator length) such that predictable (i.e., repetitive) energy having periods between α and $\alpha + n - 1$ time units will tend to be removed, and hence the autocorrelation of the output will tend to vanish between lags α and $\alpha + n - 1$.

Another means to measure the effectiveness of the predictive deconvolution process has already been described in Chapter 10, where it was pointed out that the reduction in energy content of the output trace relative to the input trace gives a measure of the predictable energy removed by the filtering operation.

A given reverberation may be characterized as either short-period or long-period. *Long-period reverberations* appear on the autocorrelation function as distinct waveforms which are separated by quiet zones. *Short-period reverberations* appear on the autocorrelation function in the form of decaying waveforms which are not separated by any noticeable quiet intervals.

The upper diagram of Figure 12-3 illustrates the autocorrelation of a typical trace which exhibits a moderate degree of short-period reverberation. The prediction distance α is chosen to specify the degree of wavelet contraction desired. As α approaches unity, more contraction and consequently more high-frequency noise is introduced. Thus, we choose α so that we may obtain a compromise between wavelet contraction and signal-to-noise

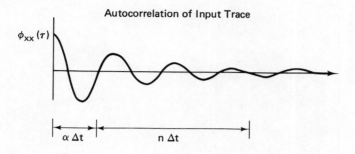

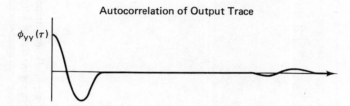

Figure 12-3. Typical autocorrelation of an input trace with a moderate amount of short-period reverberation is illustrated by the upper diagram. The lower diagram illustrates the appearance of the output autocorrelation after application of the prediction error operator with prediction distance α and length n, as established in the upper diagram. If the reverberating pattern on the autocorrelation is highly regular, one need set n such that only the first full cycle is spanned. If the pattern is irregular, the significant portion of the reverberation should be spanned.

ratio in the output trace. Empirical studies indicate that α should be set roughly equal to the lag that corresponds to the second zero crossing of the autocorrelation function. The lower diagram of Figure 12-3 illustrates the fact that the autocorrelation of the output signal trace tends to zero between lags α and $\alpha + n - 1$.

Figure 12-4 shows three different predictive deconvolution runs on offshore traces having reverberations with characteristics somewhat in between our definitions of the short-period and long-period types. The product $\alpha \, \Delta t$ has been given the values 32, 16, and 4 ms. Since these data have a sampling interval of 4 ms, the third run actually corresponds to deconvolution by the zero-delay least-squares inverse filter. We see that for $\alpha \, \Delta t = 32$ ms ($\alpha = 8$), we obtain a good dereverberation which does not exhibit the noise buildup associated with the smaller prediction distances.

The upper diagram of Figure 12-5 illustrates the autocorrelation of a typical trace with long-period reverberations. We define the appropriate prediction distance α such that the window to be deleted on the autocorrelogram begins just before the onset of the first multiple indication. Depending

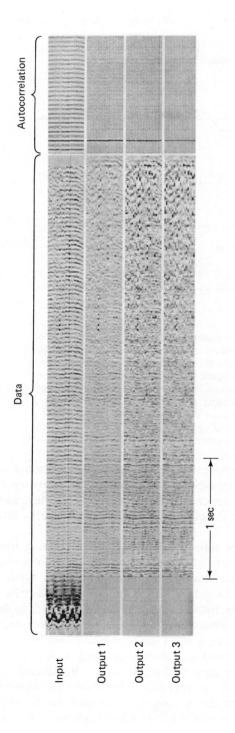

Figure 12-4. Effect that a variation in the α, the prediction distance, has on the output data. There appears to be some value of α that gives the best compromise between deringing and signal-to-noise ratio. Seismograms—top to bottom: input, outputs 1, 2, 3.

Run	$\alpha \, \Delta t$, (ms)	$n \, \Delta t$, (ms)
1	32	200
2	16	216
3	4	228

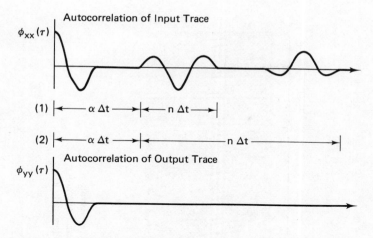

Figure 12-5. Typical autocorrelation of an input trace with a moderate amount of long-period reverberation is illustrated by the upper diagram. The lower diagram illustrates the appearance of the output autocorrelation after application of the prediction error operator with prediction distance α and length n as established in the upper diagram. The parameters should be defined as indicated in (1) if the ringing is of a first-order nature, or as indicated in (2) if the ringing is of a second-order nature.

upon the nature and period of the reverberation, we define the filter length n such that the window to be deleted spans one, two, or more orders of the multiple pattern. The autocorrelation of the resulting output trace will show very little energy between lags α and $\alpha + n - 1$. In addition, further repetitions of the waveform centered at multiples of $\alpha + n/2$ will be attenuated.

We note that the prediction error filter enables us to suppress selected waveform portions of the autocorrelation function. This is highly advantageous, since some waveforms on the autocorrelation might be due to accidentally strong correlations between certain primary reflections, and in this case we would choose not to suppress them. We may avoid their suppression by the proper selection of the prediction distance α, and we then concentrate on those waveforms associated with reverberations. We remark that we can often successfully remove the long-period multiple indication from the autocorrelation by means of the prediction error filter. Even so, the net change on the record section itself is not always significant, and the long-period dereverberation problem is by no means a solved one.

Figure 12-6 illustrates a predictive deconvolution run on a record with short-period reverberations. The data and associated autocorrelation functions show the pulse compression that has been obtained.

Figure 12-7 pictures a predictive deconvolution run on marine data

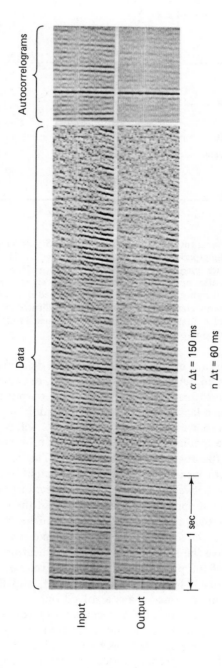

Figure 12-6. Input record that has a short-period reverberation, and output record after processing with prediction error operator. $\alpha \Delta t = 50$ ms, $n \Delta t = 150$ ms; input above, output below; data left, autocorrelation right.

Autocorrelograms

Data

Input

Output

$\alpha \Delta t = 150$ ms

$n \Delta t = 60$ ms

1 sec

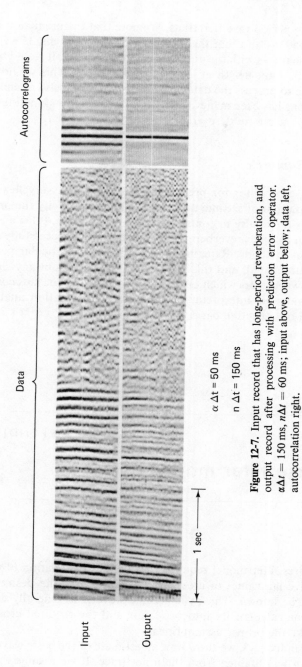

Figure 12-7. Input record that has long-period reverberation, and output record after processing with prediction error operator. $\alpha \Delta t = 150$ ms, $n\Delta t = 60$ ms; input above, output below; data left, autocorrelation right.

$\alpha \Delta t = 50$ ms

$n \Delta t = 150$ ms

1 sec

Input

Output

Autocorrelograms

Data

that exhibit long-period reverberations. We note that the prediction distance in this case is 150 ms and that the filter length is only 60 ms. If we were to deconvolve these traces with the unit prediction error filter, it would be necessary to make the filter length at least equal to 210 ms. This would require much more time to process the data. We note that a successful attenuation of the reverberations has been achieved on the autocorrelation functions as well as a moderately successful dereverberation of the data itself.

Concluding Remarks

The prediction error filter for prediction distance α is a very flexible tool for the deconvolution of seismic traces. The ability to specify the prediction distance implies the ability to control output resolution, and this means that a broad range of complex reverberatory problems can be successfully attacked with the present methods. Repetitive waveforms of a particular period can be selectively attenuated, and this is accomplishable without any significant disturbance of waveforms which one may wish to retain. The autocorrelation function is a valuable interpretative device for reverberation analysis and should be used on a routine basis.

APPENDIX 12-1

the inverse filter model

The least-squares filter model requires that the autocorrelation of the input and the positive lag values of crosscorrelation between the desired output and the input be known. The basic seismic wavelet is generally unknown; however, we can calculate its autocorrelation and the required crosscorrelation if we make the proper assumptions.

In Appendix 12-3, we show how an estimate of the basic wavelet autocorrelation can be obtained from the input trace. If we assume the desired output to be an impulse at zero lag time, the crosscorrelation between desired output and input also becomes an impulse at zero lag time. In other words, since the crosscorrelation is given by

$$\phi_{db}(\tau) = \sum_t d_t b_{t-\tau} \qquad \text{for } \tau = 0, 1, \ldots, n - 1$$

where $1, 0, 0, \ldots$ is the desired output signal d_t and $b_0, b_1, b_2, \ldots$ is the basic wavelet or input signal, b_t, we see that

$$\phi_{db}(\tau) = \begin{cases} b_0 & \text{for } \tau = 0 \\ 0 & \text{for } \tau = 1, 2, \ldots, n - 1 \end{cases}$$

which can be scaled in the form

$$\phi_{db}(\tau) = \begin{cases} 1 & \text{for } \tau = 0 \\ 0 & \text{for } \tau = 1, 2, \ldots, n - 1 \end{cases}$$

The matrix equation for the least-squares filter (see Chapter 6) then becomes

$$\begin{bmatrix} r_0 & r_1 & \cdots & r_{n-1} \\ r_1 & r_0 & \cdots & r_{n-2} \\ & & \cdot & \\ & & \cdot & \\ & & \cdot & \\ r_{n-1} & r_{n-2} & \cdots & r_0 \end{bmatrix} \begin{bmatrix} f_0 \\ f_1 \\ \cdot \\ \cdot \\ \cdot \\ f_{n-1} \end{bmatrix} = \begin{bmatrix} 1 \\ 0 \\ \cdot \\ \cdot \\ \cdot \\ 0 \end{bmatrix} \qquad (12\text{-}14)$$

where the f_t are the n coefficients which shape the basic, or characteristic, wavelet b_t to an approximation of the impulse at zero lag time.

We have thus assumed a model of the form

$$x_t = b_t * \epsilon_t$$

where x_t is the signal trace and ϵ_t is an uncorrelated series which represents the reflection coefficients of the layered subsurface. Since the desired output is an impulse at zero lag time, we see that the model requires a filter f_t such that

$$f_t * b_t \doteq 1$$

or

$$f_t \doteq b_t^{-1}$$

where the symbol $\doteq$ means "approximately equal to." The filter f_t then deconvolves the input trace as follows:

$$y_t = x_t * f_t \doteq b_t * b_t^{-1} * \epsilon_t$$
$$y_t \doteq \delta_t * \epsilon_t = \epsilon_t$$

where δ_t is the unit impulse function (or Kronecker delta),

$$\delta_t = \begin{cases} 1 & t = 0 \\ 0 & t \neq 0 \end{cases}$$

and in this sense the output trace tends to approximate the subsurface reflection coefficient series.

a study of a two-layer marine reverberation model

Impulse Response of First-Order Component

Figure 12-8 represents an idealized noise-free model of an offshore seismic situation. Reflector 1 is the water surface, reflector 2 is the water bottom, reflector 3 is some strong interface beneath the water bottom, and S is the source location just beneath the water surface. The associated normal incidence reflection coefficients are 1, c_1, and c_2, respectively, while the transmission coefficient across reflector 2 is $t_1 = 1 + c_1$. If c is the downward reflection coefficient, the corresponding upward reflection coefficient is $-c$. From physical considerations, we know that the magnitudes of all reflection coefficients are less than unity. A detailed discussion of reflection and transmission for normal incidence will be found in Chapter 13. The two-way travel time through the water layer is τ_1.

Let us compute the first-order reverberation portion of the two-layer impulse response as indicated in Figure 12-8. This response can be expressed in terms of the following z transform:

$$R_1(z) = 1 - c_1 z^{\tau_1} + c_1^2 z^{2\tau_1} + \cdots \qquad (12\text{-}15)$$

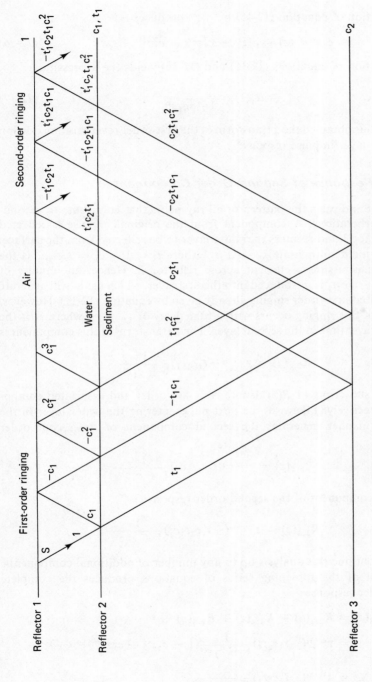

Figure 12-8. First- and second-order ringing in a two-layer marine model.

Multiplication of equation (12-15) by $c_1 z^{\tau_1}$ produces

$$c_1 z^{\tau_1} R_1(z) = c_1 z^{\tau_1} - c_1^2 z^{2\tau_1} + c_1^3 z^{3\tau_1} + \cdots \qquad (12\text{-}16)$$

while addition of equations (12-15) and (12-16) yields the expression

$$R_1(z) = \frac{1}{1 + c_1 z^{\tau_1}} \qquad (12\text{-}17)$$

We have thus obtained the z transform of the first-order reverberation portion of the two-layer impulse response.

Impulse Response of Second-Order Component

Figure 12-8 indicates the pattern of all ray paths that contribute to second-order reverberations. A component from the original impulse is reflected from reflector 3 and reenters the water layer to be reflected from the surface. At this point its amplitude is $-t_1' c_2 t_1$, where $t_1' = 1 + c_1' = 1 - c_1$ is the upward transmission coefficient across reflector 2. Hence, an impulse of amplitude $-t_1' c_2 t_1$ is introduced into the first layer, which again will generate the associated first-order ringing already given by equation (12-17). However, the onset of this ringing occurs with a time delay of $\tau_1 + \tau_2$, where τ_2 is the two-way travel time in the second layer. The z transform of this component is

$$R_{2,1}(z) = z^{\tau_1 + \tau_2}(-t_1' c_2 t_1)\frac{1}{1 + c_1 z^{\tau_1}}$$

where the subscripts of $R(z)$ denote response order and associated components, respectively. Likewise, the next pulse entering the water layer in the foregoing manner generates the second component of the second-order response

$$R_{2,2}(z) = z^{2\tau_1 + \tau_2}(t_1' c_2 t_1 c_1)\frac{1}{1 + c_1 z^{\tau_1}}$$

The third component of the second-order response is

$$R_{2,3}(z) = z^{3\tau_1 + \tau_2}(-t_1' c_2 t_1 c_1^2)\frac{1}{1 + c_1 z^{\tau_1}}$$

One can continue this analysis up to any number of additional components. Summation of the foregoing series of equations produces the complete second-order response

$$R_2(z) = R_{2,1}(z) + R_{2,2}(z) + R_{2,3}(z) + \cdots$$

$$= z^{\tau_1 + \tau_2}(-t_1' c_2 t_1)\frac{1}{1 + c_1 z^{\tau_1}}(1 - c_1 z^{\tau_1} + c_1^2 z^{2\tau_1} + \cdots)$$

$$= z^{\tau_1 + \tau_2}(-t_1' c_2 t_1)\frac{1}{(1 + c_1 z^{\tau_1})^2}$$

where we recall that $|c_1| < 1$. Since $z^{\tau_1+\tau_2}$ is simply a delay factor and $-t_1' c_2 t_1$ is a constant, we may shift the time origin and normalize the second-order response. This yields

$$R_2(z) = \frac{1}{(1 + c_1 z^{\tau_1})^2} \tag{12-18}$$

an expression that we see to be the square of the first-order response given by equation (12-17).

Removal of First-Order Ringing

Let us incorporate the first-order impulse response into the predictive deconvolution model. We will assume that the two-way travel time through the water layer is τ_1 sample units. Thus, our impulse response $x_1(t)$ for $t = 0, 1, 2, 3, \ldots$ becomes

$$1, \underbrace{0, 0, \ldots, 0}_{\tau_1 - 1 \text{ zeros}}, -c_1, \underbrace{0, 0, \ldots, 0}_{\tau_1 - 1 \text{ zeros}}, c_1^2, \ldots$$

In order to compute the prediction error filter, we require the autocorrelation of $x_1(t)$, which is

$$
\begin{aligned}
r_\tau &= 1 + c_1^2 + c_1^4 + \cdots & \text{for } \tau = 0 \\
r_\tau &= 0 & \text{for } 0 < \tau < \tau_1 \\
r_\tau &= -c_1(1 + c_1^2 + c_1^4 + \cdots) = -c_1 r_0 & \text{for } \tau = \tau_1
\end{aligned}
$$

and so on. Thus, the autocorrelation of $x_1(t)$ can be written

$$r_\tau = E_x, \underbrace{0, 0, \ldots, 0}_{\tau_1 - 1 \text{ zeros}}, -c E_x, \ldots$$

where E_x is the energy in $x_1(t)$. Let the filter length n be less than τ_1, and let the prediction distance be $\alpha = \tau_1$. Then the normal equations become

$$
\begin{bmatrix}
r_0 & 0 & \cdots & 0 \\
0 & r_0 & \cdots & 0 \\
 & & \cdot & \\
 & & \cdot & \\
 & & \cdot & \\
0 & 0 & \cdots & r_0
\end{bmatrix}
\begin{bmatrix}
a_0 \\
a_1 \\
\cdot \\
\cdot \\
\cdot \\
a_{n-1}
\end{bmatrix}
=
\begin{bmatrix}
r_{\tau_1} \\
0 \\
\cdot \\
\cdot \\
\cdot \\
0
\end{bmatrix}
$$

The only member of this system whose right side does not vanish is

$$r_0 a_0 = r_{\tau_1}$$

and thus

$$a_0 = \frac{r_{\tau_1}}{r_0} = -\frac{c_1 r_0}{r_0} = -c_1$$

The associated prediction error operator $f_2(t)$ for $t = 0, 1, 2, \ldots, \tau_1$ is

$$1, \underbrace{0, 0, \ldots, 0}_{\tau_1 - 1 \text{ zeros}}, c_1 \qquad (12\text{-}19)$$

In practice, it is not necessary to set the prediction distance α exactly to τ_1. The present model permits α to take on any value as long as it is less than or equal to τ_1. Furthermore, the filter length must be such that the inequality $\alpha + n > \tau_1$ holds true.

We note that the z tranform of the prediction error operator of equation (12-19) is $1 + c_1 z^{\tau_1}$, which is the inverse of the first-order impulse response given by equation (12-17). If this first-order impulse response is convolved with the foregoing prediction error filter, the output will be 1; in other words, we will have deconvolved the ringing signal. Figure 12-9 illustrates the input signal $x_1(t)$, the prediction error operator $f_1(t)$, and the output signal $y_1(t)$ for this situation.

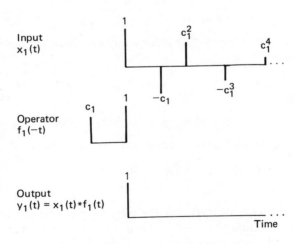

Figure 12-9. Deconvolution of a first-order ringing system. The operator is shown in time-reversed form.

Removal of Second-Order Ringing

Let us use the predictive deconvolution method on the second-order portion of the two-layer impulse response given by equation (12-18). This response $x_2(t)$ for $t = 0, 1, 2, \ldots$ can be written

$$1, \underbrace{0, 0, \ldots, 0}_{\tau_1 - 1 \text{ zeros}}, -2c_1, \underbrace{0, 0, \ldots, 0}_{\tau_1 - 1 \text{ zeros}}, 3c_1^2, \underbrace{0, 0, \ldots, 0}_{\tau_1 - 1 \text{ zeros}}, -4c_1^3, \ldots$$

The autocorrelation of $x_2(t)$ is

$$r_\tau = 1 + 4c_1^2 + 9c_1^4 + 16c_1^6 + \cdots$$

$$= \frac{1 + c_1^2}{(1 - c_1^2)^3} \qquad \text{for } \tau = 0$$

$$r_\tau = 0 \qquad \text{for } 0 < \tau < \tau_1$$

$$r_\tau = -2c_1 - 6c_1^3 - 12c_1^5 - 20c_1^7 + \cdots$$

$$= \frac{-2c_1}{(1 - c_1^2)^3} \qquad \text{for } \tau = \tau_1$$

$$r_\tau = 0 \qquad \text{for } \tau_1 < \tau < 2\tau_1$$

$$r_\tau = 3c_1^2 + 3c_1^4 + 15c_1^6 + 24c_1^8 + \cdots$$

$$= \frac{-c_1^4 + 3c_1^2}{(1 - c_1^2)^3} \qquad \text{for } \tau = 2\tau_1$$

and so on. Thus, the normalized autocorrelation of x_t, namely r_τ for $\tau = 0, 1, 2, \ldots$, becomes

$$1 + c_1^2, \underbrace{0, 0, \ldots, 0}_{\tau_1 - 1 \text{ zeros}}, -2c_1, \underbrace{0, 0, \ldots, 0}_{\tau_1 - 1 \text{ zeros}}, 3c_1^2 - c_1^4, \ldots$$

If the filter length is $n = \tau_1 + 1$ and the prediction distance is $\alpha = \tau_1$, the normal equations become

$$\tau_1 - 1 \text{ rows} \left\{ \overset{\displaystyle \tau_1 - 1 \text{ columns}}{\begin{bmatrix} 1 + c_1^2 & 0 & 0 & \cdots & 0 & -2c_1 \\ 0 & 0 & 0 & \cdots & 0 & 0 \\ & & & \vdots & & \\ & & & \vdots & & \\ 0 & 0 & 0 & \cdots & 0 & 0 \\ -2c_1 & 0 & 0 & \cdots & 0 & 1 + c_1^2 \end{bmatrix}} \begin{bmatrix} a_0 \\ a_1 \\ \vdots \\ \vdots \\ a_{\tau_1 - 1} \\ a_{\tau_1} \end{bmatrix} = \begin{bmatrix} -2c_1 \\ 0 \\ \vdots \\ \vdots \\ 0 \\ 3c_1^2 - c_1^4 \end{bmatrix} \right.$$

The two nonvanishing equations of this system yield the solution

$$a_0 = -2c_1$$

and

$$a_{\tau_1} = -c_1^2$$

Hence, the associated prediction error operator $f_2(t)$ for $t = 0, 1, 2, \ldots, 2\tau_1$ is

$$1, \underbrace{0, 0, \ldots, 0}_{\tau_1 - 1 \text{ zeros}}, 2c_1, \underbrace{0, 0, \ldots, 0}_{\tau_1 - 1 \text{ zeros}}, c_1^2 \qquad (12\text{-}20)$$

This particular prediction error operator is the exact three-point filter, so we see that in the noise-free case the present predictive deconvolution model yields the classical results obtained on the basis of strictly deterministic considerations. The predictive deconvolution scheme allows a more general attack on the dereverberation problem, as the present treatment has sought to demonstrate.

It is not necessary to set the prediction distance exactly equal to τ_1, nor is it necessary to set the filter length exactly equal to $\tau_1 + 1$. However, these parameters must be set such that $\alpha \leq \tau_1, n \geq \tau_1$, and $\alpha + n \geq 2\tau_1$. The z transform of equation (12-20) is

$$F_2(z) = 1 + 2c_1 z^{\tau_1} + c_1^2 z^{2\tau_1} = (1 + c_1 z^{\tau_1})^2$$

which is the inverse of the z transform of the second-order impulse response given by equation (12-18). Thus, convolution of the second-order impulse response $x_2(t)$ with the prediction error operator $f_2(t)$ produces a zero delay spike (Figure 12-10). In other words, the second-order ringing system has been deconvolved by means of the prediction error operator.

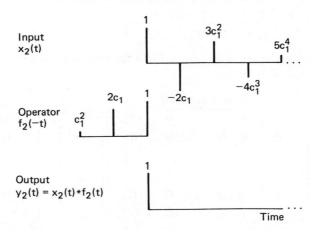

Figure 12-10. Deconvolution of a second-order ringing system. The operator is shown in time-reversed form.

the autocorrelation of a seismic trace

Under the proper assumptions the autocorrelation of a seismic trace is an estimate of the autocorrelation of the "basic" seismic wavelet. The basic seismic wavelet is assumed to be either the initial shot pulse, or the initial shot pulse modified by near-surface reverberations. The derivation presented here is similar to one given in Appendix 6-1.

Suppose we have a signal x_t, which results from the convolution of a basic wavelet b_t with an uncorrelated series ϵ_t, where we assume that ϵ_t can be identified with the reflection coefficient series of a layered medium (Chapter 10), that is,

$$x_t = b_t * \epsilon_t.$$

The z-transform of the autocorrelation of x_t is given by

$$\Phi_{xx}(z) = [B(z)E(z)][B(1/z)E(1/z)]$$

where $B(z)$ and $E(z)$ are the z-transforms of b_t and ϵ_t respectively. The above equation can be rewritten in the form

$$\Phi_{xx}(z) = [B(z)B(1/z)][E(z)E(1/z)],$$

which is the z-transform of

$$\phi_{xx}(\tau) = \phi_{bb}(\tau) * \phi_{\epsilon\epsilon}(\tau). \qquad (12\text{-}21)$$

Therefore the autocorrelation of x_t is equal to the convolution of the auto-correlation of b_t with the autocorrelation of ϵ_t. Since ϵ_t is an uncorrelated series, we obtain

$$\phi_{\epsilon\epsilon}(\tau) = E_\epsilon \qquad \text{for } \tau = 0$$

and

$$\phi_{\epsilon\epsilon}(\tau) = 0 \qquad \text{for } \tau \neq 0,$$

where E_ϵ is the energy in ϵ_t. Thus equation (12.21) reduces to

$$\phi_{xx}(\tau) = \sum_t \phi_{\epsilon\epsilon}(t)\phi_{bb}(\tau - t) = E_\epsilon\phi_{bb}(\tau),$$

and we see that the autocorrelation of x_t is simply a scaled version of the autocorrelation of b_t. This means that subject to the above assumptions, we can obtain an estimate of ϕ_{bb} even though we do not know b_t itself. The consistency of the autocorrelation estimates can be improved through use of suitable weighting functions.

seismic wave
propagation in a layered system

Summary

The problem of a normally incident plane P wave propagating in a system of horizontally stratified homogeneous perfectly elastic layers is reformulated in terms of concepts drawn from communication theory. We show how both the reflected and transmitted responses of such a system can be expressed as a z transform which is the ratio of two polynomials in z. Since this response must be stable, the denominators of the z transforms describing the reflected and transmitted motion are minimum delay (i.e., minimum phase lag). If the layered medium is bounded at depth by a perfect reflector, the reflected impulse response recorded at the surface is in the form of a dispersive all-pass z transform. A dispersive all-pass system is one whose z transform is the ratio of the z transform of a maximum-delay wavelet to that of its corresponding minimum-delay wavelet; hence, the magnitude spectrum of a dispersive all-pass system is unity for all frequencies. This means that the magnitude spectrum of the reflected response is identical to the magnitude spectrum of the input wavelet used to excite the system. More specifically, all the energy put in is returned with the same frequency content, but is differentially delayed. The phase-lag spectrum of the reflected response lies everywhere above the phase-lag spectrum of the input wavelet. Thus, the all-pass situation implies that the layered-earth model with a perfect reflector at depth, while not able to alter the magnitude of the frequency components of the input wavelet, will introduce differential time delays with certain properties into each such component. Finally, since the reflected impulse response of such a perfectly reflecting model is an all-pass wavelet, its auto-correlation is a spike of unit magnitude at $\tau = 0$, and zero for all other lags.

Introduction

Much of the existing work on the reflection and transmission of seismic waves through stratified media can be recast in the framework of communication theory. Such an approach not only has appeal because of the simplification and unification it makes possible, but also because it opens up new directions of research into this important area. In this chapter we are concerned with the normal-incidence-plane compressional-wave synthetic seismogram produced by a stratified system consisting of perfectly elastic horizontal layers. Our treatment relies heavily on the work of Wuenschel (1960), Goupillaud (1961), Kunetz and D'Erceville (1962), and Sherwood and Trorey (1965).

Wave Propagation across a Boundary

Let us review the mechanics of seismic wave propagation across a boundary between two media. For simplicity, we confine our discussion to a horizontal plane interface and to compressional (P-wave) propagation along the vertical (y) axis, as shown in Figure 13-1. Medium 1 lies everywhere above the origin

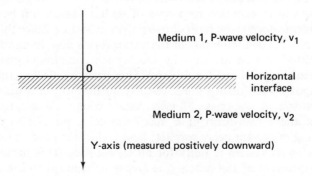

Figure 13-1. Normal-incidence P-wave propagation across an interface.

($y = 0$), and medium 2 lies everywhere below the origin. In media 1 and 2 the P-wave velocities are v_1 and v_2, respectively.

Let us imagine that a plane harmonic wave of frequency f and amplitude A_i travels downward in medium 1. The incident disturbance, assumed to be the elastic particle displacement of the medium from equilibrium, is then the real part of

$$g_i = A_i e^{2\pi i f(\tau - y/v_1)}$$

where τ is the time variable and the symbol i in the exponent is $\sqrt{-1}$. The elastic displacement of the wave reflected back into medium 1 can be represented by

$$g_c = A_c e^{2\pi i f(\tau + y/v_1)}$$

while the elastic displacement of the wave transmitted into medium 2 is represented by

$$g_t = A_t e^{2\pi i f(\tau - y/v_2)}$$

We note that the incident, reflected, and transmitted waves all have the same frequency, f. A_i, A_c, and A_t are, respectively, the real-valued amplitude coefficients of the incident, reflected, and transmitted plane harmonic waves.

Let us now look at the nature of the conditions that hold at the boundary; these conditions give the relationships connecting the elastic displacements g_i, g_c, and g_t. In order to obtain A_c and A_t in terms of A_i, we need two, and only two, conditions. The first condition requires that the elastic displacement g be continuous at the boundary for all time τ. The second condition requires that the normal elastic stress be continuous at the boundary for all time τ. This stress is given by

$$E \frac{\partial g}{\partial y}$$

where E is an elastic modulus that is generally different for the two media in question.

Let us now apply these two boundary conditions. The continuity of displacement g leads to the equation

$$A_i + A_c = A_t$$

while the continuity of stress $E(\partial g/\partial y)$ leads to

$$A_i - A_c = \frac{E_2/v_2}{E_1/v_1} A_t$$

where E_1 and E_2 are the elastic moduli of media 1 and 2, respectively. For a perfectly elastic medium $E = \rho v^2$, where ρ is the mass density and v the elastic P-wave velocity. The quantity ρv is called the *acoustic impedance* of the medium, and it is denoted by Z. Thus, the equation of continuity of stress becomes

$$A_i - A_c = \frac{Z_2}{Z_1} A_t$$

Solving the equations of stress and displacement continuity for A_c and A_t in terms of A_i we have

$$A_c = \frac{1 - Z_2/Z_1}{1 + Z_2/Z_1} A_i$$

and

$$A_t = \frac{2}{1 + Z_2/Z_1} A_i$$

For example, in the limiting case when $Z_1 = Z_2$, the equations above give $A_c = 0$ and $A_t = A_i$, which means that the boundary between media 1 and 2 ceases to be a reflector. In the case when $Z_2 \gg Z_1$, they give

$$A_c \doteq -A_i \quad \text{and} \quad A_t \doteq 0$$

which means that the incident wave is almost entirely reflected, but with a change in phase of π, as indicated by the minus sign. On the other hand, in the case when $Z_1 \gg Z_2$, they give

$$A_c \doteq A_i \quad \text{and} \quad A_t \doteq 2A_i$$

which means that the reflected amplitude is approximately equal to the incident amplitude, and the transmitted amplitude is approximately twice the incident amplitude. This paradox can be resolved by introducing energy considerations.

The power per unit area, P, conveyed by the radiation is proportional to the product of the partial time derivative of g and the partial space derivative of g; that is,

$$P = E \frac{\partial g}{\partial \tau} \frac{\partial g}{\partial y}$$

Letting ρ_1 and ρ_2 be the densities of media 1 and 2, respectively, the incident average power per unit area, $\langle P_i \rangle$, is (Lindsay, 1960, p. 77),

$$\langle P_i \rangle = \tfrac{1}{2} \rho_1 v_1 \omega^2 A_i^2 \qquad (\omega = 2\pi f)$$

while the transmitted average power per unit area, $\langle P_t \rangle$, is

$$\langle P_t \rangle = \tfrac{1}{2} \rho_2 v_2 \omega^2 A_t^2$$

Therefore,

$$\langle P_t \rangle = \frac{4(Z_2/Z_1)}{(1 + Z_2/Z_1)^2} \langle P_i \rangle$$

Returning to the case $Z_1 \gg Z_2$, we see that $\langle P_t \rangle \doteq 0$, so practically no energy is transmitted across the boundary, despite the fact that $A_t \doteq 2A_i$.

In a similar way we can find the expression for $\langle P_c \rangle$, which is

$$\langle P_c \rangle = \left[\frac{1 - Z_2/Z_1}{1 + Z_2/Z_1} \right]^2 \langle P_i \rangle$$

We see that in either the case when $Z_1 \gg Z_2$ or the case when $Z_2 \gg Z_1$, we have $\langle P_c \rangle \doteq \langle P_i \rangle$, or nearly all the energy is reflected. Thus, if the Z values are nearly equal, the media are well matched and considerable energy transmission takes place; whereas, if the Z values differ greatly, the media are badly matched and little energy transmission takes place.

Reflection and Transmission Coefficients

As we have seen, the amplitudes of the reflected and transmitted waves resulting from an incident wave traveling downward in medium 1 to the interface are

$$A_c = \frac{Z_1 - Z_2}{Z_1 + Z_2} A_i$$

and

$$A_t = \frac{2Z_1}{Z_1 + Z_2} A_i$$

It is convenient to denote the coefficients of A_i in these equations by c and t, respectively; that is,

$$A_c = cA_i \qquad \text{where } c = \frac{Z_1 - Z_2}{Z_1 + Z_2}$$

and

$$A_t = tA_i \qquad \text{where } t = \frac{2Z_1}{Z_1 + Z_2}$$

The coefficient c is called the *particle displacement reflection coefficient* and

t the *particle displacement transmission coefficient*. These coefficients satisfy the relation

$$t = 1 + c$$

Similarly, the amplitudes A'_c of the reflected wave and A'_t of the transmitted wave resulting from an incident wave A'_i traveling upward in medium 2 to the interface are

$$A'_c = \frac{Z_2 - Z_1}{Z_2 + Z_1} A'_i$$

and

$$A'_t = \frac{2Z_2}{Z_2 + Z_1} A'_i$$

Let us define the coefficients of A'_i in these equations by c' and t', respectively; that is

$$A'_c = c' A'_i \qquad \text{where } c' = \frac{Z_2 - Z_1}{Z_2 + Z_1}$$

and

$$A'_t = t' A'_i \qquad \text{where } t' = \frac{2Z_2}{Z_2 + Z_1}$$

We see that these coefficients satisfy the relation

$$t' = 1 + c'$$

The four coefficients c, t, c', t' are depicted in Figure 13-2 (where, for clarity, we have displaced the rays slightly from normal incidence). These coefficients satisfy the additional relations

$$c = -c'$$

and

$$tt' - cc' = 1$$

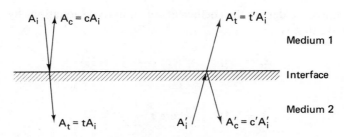

Figure 13-2. Reflection and transmission of a normally incident *P*-wave across a boundary. (*Left*) Reflection coefficient c and transmission coefficient t. (*Right*) Transmission coefficient t' and reflection coefficient c'.

The derivations above have been carried out for the particle displacement reflection and transmission coefficients. In Appendix 13-2 we derive the appropriate coefficients for stress, particle velocity, and particle acceleration.

Introduction of z-Transform Notation

Let us use the conventional model of stratified media in which the vertical travel time of an elastic wave through each medium is the same, which we designate to be one-half time unit. Although we restrict ourselves to normal incidence, it is convenient to draw the diagrams with time displacement along the horizontal axis, so that the rays appear to be at nonnormal incidence. Hence, the possible wave paths lie on a diamond-shaped grid, as depicted in Figure 13-3. Lines sloping downward to the right correspond to vertically downgoing waves, and lines sloping upward to the right correspond to vertically upgoing waves. We assume that our layered medium is excited by an impulsive downgoing source pulse at time $\tau = 0$ located just above the first layer (see Figure 13-3).

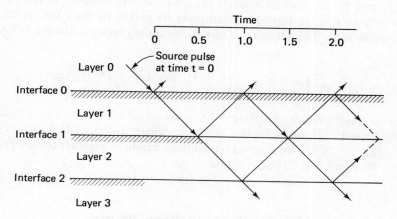

Figure 13-3. Ray paths in the layered medium.

Let us concentrate our attention on the downgoing waves. The diagram shown in Figure 13-4 exhibits all the downward waves positioned at the tops of the layers. We use the coefficient

$$d_k\left(n + \frac{k-1}{2}\right)$$

to denote the displacement amplitude of the downward wave at the *top* of layer k occurring at time instant $n + (k-1)/2$, where $k = 1, 2, 3, 4, \ldots$ and $n = 0, 1, 2, 3, \ldots$. Here k is the layer index, while n is a suitable time index. For example, the coefficient $d_1(1)$ is the displacement amplitude of the

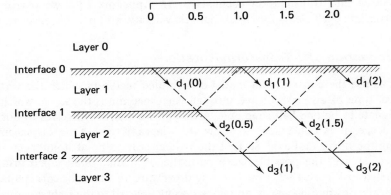

Figure 13-4. Downgoing waves in the layered medium, with the waves positioned at the layer *tops*.

downgoing wave at the top of layer $k = 1$ at time instant $n + (k - 1)/2 = 1$ (i.e., $n = 1$). We may summarize the information contained in the entire suite of waves at the top of a given layer by means of the z transform of their coefficients. The z transform of the downgoing waves at the top of layer k is

$$D_k(z) = d_k\left(\frac{k - 1}{2}\right)z^{(k-1)/2} + d_k\left(\frac{k + 1}{2}\right)z^{(k+1)/2} + d_k\left(\frac{k + 3}{2}\right)z^{(k+3)/2} + \cdots$$

Here we have taken the coefficient $d_k(j)$ associated with a given time instant j and have therefore multiplied it by z^j, where $j = (k - 1)/2, (k + 1)/2, (k + 3)/2, \ldots$. Then we have added all these products together for the given layer. We recall (see Chapter 2) that

$$z = e^{-i\omega}$$

represents a pure delay of one time unit. Hence, the term

$$d_k\left(n + \frac{k - 1}{2}\right)z^{n+[(k-1)/2]}$$

indicates that the coefficient $d_k[n + (k - 1)/2]$ is associated with a time delay of $n + (k - 1)/2$ from the instant that the impulsive source has excited our system.

Let us next consider the situation in which all the downward waves are positioned at the bottoms of the layers, as depicted in Figure 13-5. Here we have used the coefficient

$$d'_k\left(n + \frac{k}{2}\right)$$

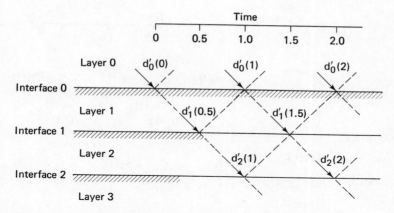

Figure 13-5. Downgoing waves in the layered medium. The waves have traveled to the layer *bottoms*.

to denote the downward wave at the *bottom* of layer k (i.e., the prime signifies the bottom of the layer) occurring at time $n + k/2$, where $k = 0, 1, 2, 3, \ldots$ and $n = 0, 1, 2, 3, \ldots$. For example, $d'_1(1.5)$ is the coefficient of the downgoing wave at the bottom of layer $k = 1$ at time $n + k/2 = 1.5$ (i.e., $n = 1$). Again we may summarize this information by use of the z transform. The z transform of the downgoing waves at the bottom of layer k is

$$D'_k(z) = d'_k\left(\frac{k}{2}\right)z^{k/2} + d'_k\left(\frac{k+2}{2}\right)z^{(k+2)/2} + d'_k\left(\frac{k+4}{2}\right)z^{(k+4)/2} + \cdots$$

What is the relationship between $D_k(z)$, the z transform of the downgoing waves at the top of layer k, and $D'_k(z)$, the z transform of the downgoing waves at the bottom of layer k? They are related by a z transform $A_k(z)$ characteristic of layer k; that is,

$$D'_k(z) = A_k(z)D_k(z)$$

In the absorption-free case, the effect of layer k is only to introduce a time delay of one-half time unit *from the top to the bottom of the layer*, so $A_k(z)$ reduces to

$$A_k(z) = z^{1/2}$$

and hence

$$D'_k(z) = z^{1/2}D_k(z)$$

In a model with absorption, the function $A_k(z)$ includes both the effects of the time delay and the absorption.

Next let us consider the upgoing waves. As in the case of downgoing waves, we must consider both the tops and bottoms of the layers. Figure

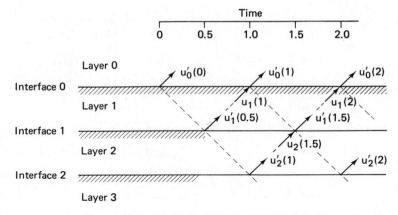

Figure 13-6. Upgoing waves in the layered medium, showing waves at the layer tops and at the layer bottoms.

13-6 shows for upgoing waves what is shown for downgoing waves in both Figures 13-4 and 13-5; that is, Figure 13-6 exhibits the upgoing waves both at the tops and bottoms of the layers.

The symbol u represents an upgoing wave. If there is no prime on the u, the wave is at the top of the layer; if there is a prime on the u (i.e., u'), the wave is at the bottom of the layer. Hence, the coefficient

$$u_k\left(n + \frac{k-1}{2}\right)$$

denotes the upward wave at the *top* of layer k occurring at time $\tau = n + (k - 1)/2$, where $k = 1, 2, 3, 4, \ldots$ and $n = 1, 2, 3, 4, \ldots$. [We note that $n = 0$ is missing here, which means that the coefficients $u_k[(k - 1)/2]$ are missing, or, equivalently, the condition $u_k[(k - 1)/2] = 0$ is required.] For example, $u_1(2)$ is the coefficient of the upgoing wave at the top of layer $k = 1$ at time $n + (k - 1)/2 = 2$ (i.e., $n = 2$). The z transform of the upgoing waves at the top of layer k is

$$U_k(z) = u_k\left(\frac{k+1}{2}\right)z^{(k+1)/2} + u_k\left(\frac{k+3}{2}\right)z^{(k+3)/2} + u_k\left(\frac{k+5}{2}\right)^{(k+5)/2} + \cdots$$

Likewise, the coefficient

$$u'_k\left(n + \frac{k}{2}\right)$$

signifies the upward wave at the *bottom* of layer k occurring at time $n + k/2$, where $k = 0, 1, 2, 3, \ldots$ and $n = 0, 1, 2, 3, \ldots$. For example, $u'_1(1.5)$ is the coefficient of the upward wave at the bottom of layer $k = 1$ at time $n + (k/2)$

$= 1.5$ (i.e., $n = 1$). The z transform of the upgoing waves at the bottom of layer k is

$$U'_k(z) = u'_k\left(\frac{k}{2}\right) z^{k/2} + u'_k\left(\frac{k+2}{2}\right) z^{(k+2)/2} + u'_k\left(\frac{k+4}{2}\right) z^{(k+4)/2} + \cdots$$

In the absorption-free model, the effect of layer k is to introduce a delay of one-half time unit from the bottom to the top of the layer, so

$$U_k(z) = z^{1/2} U'_k(z)$$

That is, for upgoing waves the delay is measured at the top of a layer with respect to the bottom of this layer; whereas, for downgoing waves, the delay is measured at the bottom of a layer with respect to the top of this layer.

Relationship between the Wave Coefficients

Let us now derive the relationship between the wave coefficients by making use of the reflection and transmission coefficients: c, c' and t, t'. Figure 13-7 depicts the situation at the interface between layers k and $k + 1$ at time instant i.

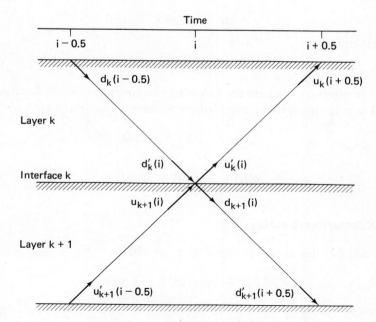

Figure 13-7. Closeup look at the interface between layers k and $k + 1$ at time i.

The coefficient $u'_k(i)$ is made up of two parts: the part due to the reflection of $d'_k(i)$ and the part due to the transmission of $u_{k+1}(i)$. Hence, we have the equation

$$u'_k(i) = cd'_k(i) + t'u_{k+1}(i)$$

The coefficient $d_{k+1}(i)$ is made up of a part due to the transmission of $d'_k(i)$ and a part due to the reflection of $u_{k+1}(i)$. Hence, we have the equation

$$d_{k+1}(i) = td'_k(i) + c'u_{k+1}(i)$$

We may rearrange this equation as

$$td'_k(i) = d_{k+1}(i) - c'u_{k+1}(i)$$

Substituting this expression for $td'_k(i)$ into the foregoing expression for $u'_k(i)$, we find that

$$u'_k(i) = \frac{c}{t}[d_{k+1}(i) - c'u_{k+1}(i)] + t'u_{k+1}(i)$$

which gives

$$tu'_k(i) = cd_{k+1}(i) + (tt' - cc')u_{k+1}(i)$$

Since

$$tt' - cc' = 1$$

this last equation is

$$tu'_k(i) = cd_{k+1}(i) + u_{k+1}(i)$$

In summary, we have the following two equations, which describe the situation at the interface at time i (where we have used $c = -c'$):

$$td'_k(i) = d_{k+1}(i) + cu_{k+1}(i)$$

and

$$tu'_k(i) = cd_{k+1}(i) + u_{k+1}(i)$$

The Scattering Matrix

If we multiply the equations above by z^i, we obtain

$$td'_k(i)z^i = d_{k+1}(i)z^i + cu_{k+1}(i)z^i$$

and

$$tu'_k(i)z^i = cd_{k+1}(i)z^i + u_{k+1}(i)z^i$$

Next, we sum these equations over $i = k/2, (k + 2)/2, (k + 4)/2, \ldots$, and thereby obtain

$$t \sum_i d'_k(i)z^i = \sum_i d_{k+1}(i)z^i + c \sum_i u_{k+1}(i)z^i$$

and

$$t \sum_i u'_k(i)z^i = c \sum_i d_{k+1}(i)z^i + \sum_i u_{k+1}(i)z^i$$

where we observe the condition that $u_{k+1}(k/2) = 0$. We recognize the summations in these equations to be z transforms; that is, they may be written as

$$tD'_k(z) = D_{k+1}(z) + cU_{k+1}(z)$$

and

$$tU'_k(z) = cD_{k+1}(z) + U_{k+1}(z)$$

In the absorption-free model, recalling that

$$D'_k(z) = z^{1/2}D_k(z)$$

and

$$U_k(z) = z^{1/2}U'_k(z)$$

we have the equations

$$tD_k(z) = z^{-1/2}D_{k+1}(z) + cz^{-1/2}U_{k+1}(z)$$

and

$$tU_k(z) = cz^{1/2}D_{k+1}(z) + z^{1/2}U_{k+1}(z)$$

These equations are called the *scattering equations*. They may be written in matrix notation as

$$\begin{bmatrix} D_k(z) \\ U_k(z) \end{bmatrix} = \mathbf{M}_k \begin{bmatrix} D_{k+1}(z) \\ U_{k+1}(z) \end{bmatrix}$$

where the matrix $\mathbf{M}_k$ is defined as

$$\mathbf{M}_k = \begin{bmatrix} \dfrac{z^{-1/2}}{t_k} & \dfrac{c_k z^{-1/2}}{t_k} \\ \dfrac{c_k z^{1/2}}{t_k} & \dfrac{z^{1/2}}{t_k} \end{bmatrix}$$

We note that we have now attached the subscript k to the reflection coefficient $c (= c_k)$ and to the transmission coefficient $t (= t_k)$ in order to explicitly indicate that they are the coefficients for the interface between layers k and $k + 1$. The matrix $\mathbf{M}_k$ is called the *scattering matrix* for the interface between these two layers.

The scattering matrix $\mathbf{M}_k$ is defined for $k = 1, 2, 3, \ldots$. For the interface between layer 0 and layer 1, we use the relations

$$t_0 D_0'(z) = D_1(z) + c_0 U_1(z)$$

and

$$t_0 U_0'(z) = c_0 D_1(z) + U_1(z)$$

which may be written as

$$\begin{bmatrix} D_0'(z) \\ U_0'(z) \end{bmatrix} = \mathbf{M}_0 \begin{bmatrix} D_1(z) \\ U_1(z) \end{bmatrix}$$

where the scattering matrix $\mathbf{M}_0$ is defined as

$$\mathbf{M}_0 = \begin{bmatrix} \dfrac{1}{t_0} & \dfrac{c_0}{t_0} \\ \dfrac{c_0}{t_0} & \dfrac{1}{t_0} \end{bmatrix}$$

An attractive feature of the scattering matrix approach is that a series of chain relationships can be written down among the z transforms of the downgoing and upgoing waves in the various layers. Thus, we have

$$\begin{aligned} \begin{bmatrix} D_0'(z) \\ U_0'(z) \end{bmatrix} &= \mathbf{M}_0 \begin{bmatrix} D_1(z) \\ U_1(z) \end{bmatrix} \\ &= \mathbf{M}_0 \mathbf{M}_1 \begin{bmatrix} D_2(z) \\ U_2(z) \end{bmatrix} \\ &= \mathbf{M}_0 \mathbf{M}_1 \mathbf{M}_2 \begin{bmatrix} D_3(z) \\ U_3(z) \end{bmatrix} \\ &= \mathbf{M}_0 \mathbf{M}_1 \mathbf{M}_2 \mathbf{M}_3 \begin{bmatrix} D_4(z) \\ U_4(z) \end{bmatrix} \end{aligned}$$

and so on. The description of the wave propagation process in terms of such matrices has been known for many years, and here we have reformulated this method with the aid of z-transform theory. Let us now proceed to give some examples.

A One-Finite-Thickness-Layer Model with a Discussion of the All-Pass Theorem

In this example we assume that there is one finite-thickness layer (= layer 1). A unit spike is set off as a downgoing wave at the bottom of layer 0; thus, this boundary condition is

$$d'_0(0) = 1, \quad d'_0(1) = 0, \quad d'_0(2) = 0, \quad \ldots$$

The z transform of the input is therefore unity; that is,

$$D'_0(z) = d'_0(0) + d'_0(1)z + d'_0(2)z^2 + \cdots = 1$$

Because layer 1 is the only finite-thickness layer, layer 2 becomes a half-space, and no energy is reflected up again after the wave passes into layer 2. Thus, we have the boundary condition

$$u_2(1.5) = 0, \quad u_2(2.5) = 0, \quad u_2(3.5) = 0, \quad \ldots$$

so that the z transform $U_2(z)$ is equal to zero. Thus, for this example we have the equation

$$\begin{bmatrix} 1 \\ U'_0(z) \end{bmatrix} = \mathbf{M}_0 \mathbf{M}_1 \begin{bmatrix} D_2(z) \\ 0 \end{bmatrix}$$

which we can solve for $U'_0(z)$ and $D_2(z)$. First, we evaluate the matrix product

$$\mathbf{M}_0 \mathbf{M}_1 = \frac{1}{t_0 t_1} \begin{bmatrix} 1 & c_0 \\ c_0 & 1 \end{bmatrix} \begin{bmatrix} z^{-1/2} & c_1 z^{-1/2} \\ c_1 z^{1/2} & z^{1/2} \end{bmatrix}$$

$$= \frac{1}{t_0 t_1} \begin{bmatrix} z^{-1/2} + c_0 c_1 z^{1/2} & c_1 z^{-1/2} + c_0 z^{1/2} \\ c_0 z^{-1/2} + c_1 z^{1/2} & c_0 c_1 z^{-1/2} + z^{1/2} \end{bmatrix}$$

Then we have

$$1 = \frac{1}{t_0 t_1} (z^{-1/2} + c_0 c_1 z^{1/2}) D_2(z)$$

and

$$U'_0(z) = \frac{1}{t_0 t_1} (c_0 z^{-1/2} + c_1 z^{1/2}) D_2(z)$$

Hence,

$$D_2(z) = \frac{t_0 t_1}{z^{-1/2} + c_0 c_1 z^{1/2}} = \frac{t_0 t_1 z^{1/2}}{1 + c_0 c_1 z} \tag{13-1}$$

and

$$U'_0(z) = \frac{c_0 z^{-1/2} + c_1 z^{1/2}}{z^{-1/2} + c_0 c_1 z^{1/2}} = \frac{c_0 + c_1 z}{1 + c_0 c_1 z} \tag{13-2}$$

Unless both c_0 and c_1 are of unit magnitude (clearly an extreme case), we have, in general, that $|c_0 c_1| < 1$. Therefore, the denominator of the rightmost members of equations (13-1) and (13-2), namely $1 + c_0 c_1 z$, is a (strictly) minimum-delay z transform, and hence $D_2(z)$ and $U_0'(z)$ are stable. The inverse of a minimum-delay function is also minimum-delay. It follows from the minimum-delay property of the denominator of $D_2(z)$ that $D_2(z)$ itself is minimum-delay, except for the pure delay of one-half time unit [indicated by the fact that we can factor $z^{1/2}$ from the expression for $D_2(z)$]. More explicitly, $D_2(z)$, given by equation (13-1), may then be expanded in the convergent series

$$D_2(z) = t_0 t_1 z^{1/2}(1 - c_0 c_1 z + c_0^2 c_1^2 z^2 - \cdots)$$
$$= t_0 t_1 z^{1/2} - t_0 t_1 c_0 c_1 z^{3/2} + t_0 t_1 c_0^2 c_1^2 z^{5/2} - \cdots$$

Thus, the downgoing wave at the top of medium 2 has coefficients

$$d_2(0.5) = t_0 t_1, \quad d_2(1.5) = -t_0 t_1 c_0 c_1, \quad d_2(2.5) = t_0 t_1 c_0^2 c_1^2, \quad \ldots$$

Furthermore, $U_0'(z)$, given by equation (13-2), may be expanded in the convergent series

$$U_0'(z) = (c_0 + c_1 z)\cdot(1 - c_0 c_1 z + c_0^2 c_1^2 z^2 - \cdots)$$
$$= c_0 + (c_1 - c_0^2 c_1)z + (c_0^3 c_1^2 - c_0 c_1^2)z^2 + \cdots$$

so the upgoing wave at the bottom of layer 0 has coefficients

$$u_0'(0) = c_0, \quad u_0'(1) = c_1 - c_0^2 c_1, \quad u_0'(2) = c_0^3 c_1^2 - c_0 c_1^2, \quad \ldots$$

These represent the waveform that would be detected by a geophone on the ground surface if layer 0 represents the air. Let us again consider the right member of equation (13-2). Since $|c_0 c_1| < 1$, the denominator is minimum-delay. On the other hand, the numerator is minimum-delay only if $|c_0| > |c_1|$. Hence, the upgoing response $U_0'(z)$ is, in general, not minimum-delay. However, let us consider the function

$$1 + U_0'(z)$$

which for the one-layer case becomes, with the aid of equation (13-2),

$$1 + U_0'(z) = \frac{(c_0 + 1) + c_1(c_0 + 1)z}{1 + c_0 c_1 z}$$

Since $|c_1| < 1$, it follows that $|c_0 + 1| > |c_1(c_0 + 1)|$, so that the numerator of the above expression is minimum-delay, while the denominator, being identical to that of $U_0'(z)$ alone, is also minimum-delay. Therefore, the function $1 + U_0'(z)$ is minimum-delay whenever both $|c_0|$ and $|c_1| < 1$.

The unit spike added to the upgoing response can be given a physical interpretation, since we may consider it to represent that part of the source excitation which travels directly upward from the interface between layers 0 and 1 at time $\tau = 0$, without suffering any prior reflection. This upgoing pulse exists in addition to the signal $u_0'(0)$ (see Figure 13-6), which is reflected from this interface at $\tau = 0$. More generally, the function $1 + U_0'(z)$ is minimum-delay for an arbitrary number of layers.

In the case when the lower interface is a perfect reflector (i.e., when $c_1 = -1$), we see that $t_1 = 1 + c_1 = 0$, so

$$D_2(z) = 0$$

and

$$U_0'(z) = \frac{c_0 - z}{1 - c_0 z} = -1 \cdot \frac{z - c_0}{1 - c_0 z}$$

Thus, $U_0'(z)$ is equal to a constant factor, -1, times the quotient of $(z - c_0)$ divided by $(1 - c_0 z)$. Since the reflection coefficient c_0 is less than one in magnitude, the z transform $1 - c_0 z$ is minimum delay, and the z transform $z - c_0$ is maximum delay. Moreover, since $z - c_0$ is the time reverse of $1 - c_0 z$, it follows that $U_0'(z)$ is a *dispersive all-pass z-transform* (see Chapter 5). Specifically, a dispersive all-pass z transform is one given by the ratio of the z transform of a maximum-delay wavelet to the z transform of the corresponding minimum-delay wavelet. Because a maximum-delay wavelet and its corresponding minimum-delay wavelet have the same magnitude spectrum, we recall from the results of Chapter 5 that a dispersive all-pass system has a magnitude spectrum equal to unity for all frequencies.

Because the condition $c_1 = -1$ means that the interface beween layer 1 and layer 2 is a perfect reflector, no energy is transmitted out of layer 1 into layer 2. Hence, it follows that all the energy is ultimately fed back into layer 0. The elastic displacement at the bottom of layer 0 is described by the function $U_0'(z)$. The total energy that enters the system is due to the unit spike source function and so is unity; that is, the input energy is

$$[d_0'(0)]^2 + [d_0'(1)]^2 + [d_0'(2)]^2 + \cdots = 1 + 0 + 0 + \cdots = 1$$

The energy that exits from the system is all contained in the upgoing waves at the bottom of medium 0; this energy is

$$[(u_0'(0)]^2 + [u_0'(1)]^2 + [u_0'(2)]^2 + \cdots$$

We could evaluate this infinite sum directly, but it is easier to evaluate it via the Fourier transformation. The z transform $U'_0(z)$ becomes a Fourier transform by the substitution

$$z = e^{-i\omega} = e^{-2\pi i f}$$

where ω is angular frequency and f is cyclical frequency. Hence, the Fourier transform is

$$U'_0(e^{-2\pi i f})$$

and since this represents an all-pass system it follows that the magnitude of $U'_0(e^{-2\pi i f})$ is unity for all values of frequency f; that is,

$$|U'_0(e^{-2\pi i f})| = 1$$

Hence, the required energy is

$$[u'_0(0)]^2 + [u'_0(1)]^2 + [u'_0(2)]^2 + \cdots = \int_{-0.5}^{0.5} |U'_0(e^{-2\pi i f})|^2 \, df$$

where the limits -0.5 and 0.5 on the integration sign are the Nyquist frequencies $\pm 1/(2\Delta\tau) = \pm\frac{1}{2}$, since the time increment $\Delta\tau$ was chosen to be one unit (here we have made use of Parseval's theorem, e.g., Robinson, 1967a, p. 58). In the case of a discrete function of time a_t [time increment $\Delta\tau$, spectrum $A(f)$ periodic with period $(1/\Delta\tau)$], Parseval's theorem takes the form

$$\sum_i a_i^2 = \frac{1}{1/\Delta\tau} \int_{-1/2\Delta\tau}^{1/2\Delta\tau} |A(f)|^2 \, df$$

which is perfectly analogous to the form for a periodic function of time, $a(t)$ (period T; spectrum A_t digitized at the interval $1/T$),

$$\sum_i |A_i|^2 = \frac{1}{T} \int_{-T/2}^{T/2} a^2(t) \, dt$$

We may thus write

$$[u'_0(0)]^2 + [u'_0(1)]^2 + \cdots$$
$$= \int_{-0.5}^{+0.5} |U'_0(e^{-2\pi i f})|^2 \, df$$
$$= \int_{-0.5}^{+0.5} 1 \, df = 1$$

That is, the output energy of the system is also unity. This is the result that we expected, in agreement with energy conservation.

Let us now explain why the case $c_1 = -1$ (i.e., when the lower interface is a perfect reflector) leads to a response $U'_0(z)$, which is all-pass. To do so, we return to the general case of a one-finite-thickness layer for which c_1 is not necessarily -1. Before we had assumed that the source was a unit spike in the downgoing direction. Now we want to allow the source to produce an arbitrarily shaped wavelet in the downgoing direction; that is, the source wavelet is

$$\hat{d}_0'(0), \quad \hat{d}_0'(1), \quad \hat{d}_0'(2), \quad \ldots$$

where the numerical values of these coefficients are arbitrary. The caret symbol, $\wedge$, is used to distinguish the case of an arbitrary source excitation from that of a unit spike, or impulsive source. The z transform of the source wavelet is

$$\hat{D}_0'(z) = \hat{d}_0'(0) + \hat{d}_0'(1)z + \hat{d}_0'(2)z^2 + \cdots$$

$\hat{D}_0'(z)$ then appears in the equation

$$\begin{bmatrix} \hat{D}_0'(z) \\ \hat{U}_0'(z) \end{bmatrix} = \mathbf{M}_0 \mathbf{M}_1 \begin{bmatrix} \hat{D}_2(z) \\ 0 \end{bmatrix}$$

As before, we solve this equation for $\hat{U}_0'(z)$ and $\hat{D}_2(z)$. We obtain

$$\hat{D}_2(z) = \frac{t_0 t_1 z^{1/2}}{1 + c_0 c_1 z} \hat{D}_0'(z)$$

$$\hat{U}_0'(z) = \frac{c_0 + c_1 z}{1 + c_0 c_1 z} \hat{D}_0'(z)$$

The function $\hat{D}_0'(z)$ is the z transform of the input; the function $\hat{D}_2(z)$ is the z transform of one output that we are interested in, and the function $\hat{U}_0'(z)$ is the z transform of the other output. Hence, each of these equations represents the input/output relations of a time-invariant linear digital filter, as depicted in Figure 13-8. The coefficients associated with the z transform are the weighting coefficients of the digital filter. But we have already seen that the z transforms of these two filters are, respectively, $D_2(z)$ and $U_0'(z)$, which represent the responses of the system to a unit spike input. This is in perfect

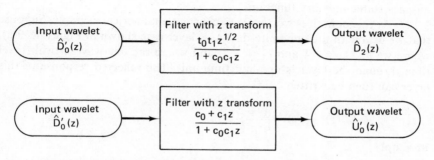

Figure 13-8. Input/output relations for the one-finite-thickness-layer model.

agreement with the interpretation of the weighting coefficients of a digital filter as the time response of the filter to a unit spike input. Thus, the output downgoing wavelet $[\hat{d}_2(0.5), \hat{d}_2(1.5), \hat{d}_2(2.5), \ldots]$ is the convolution of the input wavelet $[\hat{d}_0'(0), \hat{d}_0'(1), \hat{d}_0'(2), \ldots]$ with the filter weighting function

$$d_2(0.5) = t_0 t_1, \quad d_2(1.5) = -t_0 t_1 c_0 c_1, \quad d_2(2.5) = t_0 t_1 c_0^2 c_1^2, \quad \ldots$$

(which we obtained previously as the downgoing wave due to a spike input). Likewise, the output upgoing wavelet $[\hat{u}_0'(0), \hat{u}_0'(1), \hat{u}_0'(2), \ldots]$ is the convolution of the input wavelet $[\hat{d}_0'(0), \hat{d}_0'(1), \hat{d}_0'(2), \ldots]$ with the filter weighting function

$$u_0'(0) = c_0, \quad u_0'(1) = c_1 - c_0^2 c_1, \quad u_0'(2) = c_0^3 c_1^2 - c_0 c_1^2, \quad \ldots$$

(which we obtained previously as the upgoing wave due to a spike input).

Let us now return to the case where the bottom interface is a perfect reflector (i.e., $c_1 = -1$). In this case the first filter discussed above is an *all-stop filter*, so the output wavelet $[\hat{d}_2(0.5), \hat{d}_2(1.5), \hat{d}_2(2.5), \ldots]$ is identically zero; the second filter is an *all-pass filter*, so the output wavelet $[\hat{u}_0'(0), \hat{u}_0'(1), \hat{u}_0'(2), \ldots]$ has exactly the same magnitude spectrum as the input wavelet $[\hat{d}_0'(0), \hat{d}_0'(1), \hat{d}_0'(2), \ldots]$ but has a phase-lag spectrum that lies above the phase-lag spectrum of the input wavelet. In other words, the all-pass situation means that the layered-earth model is not able to alter the frequency makeup of the reflected wavelet, but is able to introduce differential delays into the frequency components. This delaying effect is best illustrated by the energy-buildup curves: although both input and output have the same total energy, the partial energy of the input exceeds at every time instant the partial energy of the output. The concept of partial energy was discussed in Chapter 5. The all-pass property of the filter relating the output $\hat{U}_0'(z)$ to the input $\hat{D}_0'(z)$ in the case of a perfect reflector at the bottom follows also from the general consideration that the model used is linear and that all energy enters the system as $\hat{D}_0'(z)$ and leaves it as $\hat{U}_0'(z)$. As regards phase, it was shown in Chapter 5 that the positive phase lag is a general property of an all-pass stable memory function.

At this point it is worthwhile to consider a numerical example. Suppose that we deal with the one-finite-thickness-layer case shown in Figure 13-9(a), where we let $c_0 = -\frac{1}{2}$ and $c_1 = -1$, and where the one-way vertical travel time through the layer is one-half time unit. The reflected response of this layer can then be written

$$U_0'(z) = -1 \cdot \frac{\frac{1}{2} + z}{1 + \frac{1}{2}z}$$

or simply

$$-U_0'(z) = \frac{\frac{1}{2} + z}{1 + \frac{1}{2}z} = \frac{1}{2} + \frac{3}{4}z - \frac{3}{8}z^2 + \frac{3}{16}z^3 - \cdots$$

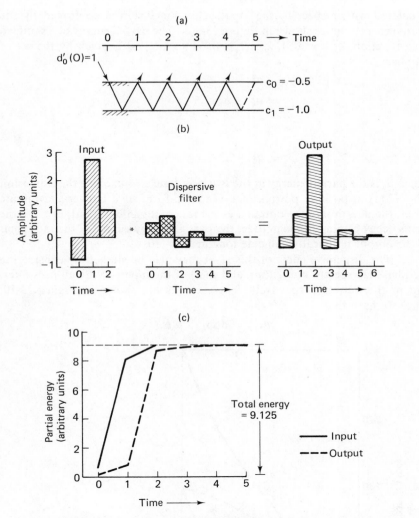

Figure 13-9. (a) Example of a one-finite-thickness-layer model. (b) Convolution of the all-pass response of part (a) with the pulse $(-0.75, +2.75, +1)$. (c) Partial energy curves of the input and output of the operation (b).

where we have switched the minus sign to the left-hand side for convenience. This response is shown in Figure 13-9(b) and is, as we have seen, the memory function of a dispersive all-pass filter. If we wish to find the response of this layer to an arbitrary excitation, such as that given by the pulse $(-0.75, 2.75, 1)$ shown on the left side of Figure 13-9(b), we convolve the dispersive memory function with this input. We then obtain the *dispersed* output wavelet as shown in the same figure. It is quite apparent that the energy of the output is *delayed* with respect to the input. This conclusion is confirmed by the partial energy curves shown in Figure 13-9(c). The partial energy of a sampled time function, a_τ, $\tau = 0, 1, 2, \ldots, n$ (n = an integer) is given by the set of numbers

$$p_0 = a_0^2$$
$$p_1 = a_0^2 + a_1^2$$
$$\cdot$$
$$\cdot$$
$$\cdot$$
$$p_n = a_0^2 + a_1^2 + a_2^2 + \cdots a_n^2$$

where p_n is the partial energy at the two-way time $\tau = n$. Since the expansion of $-U_0'(z)$ in positive powers of z is in the form of a converging infinite series, the duration of the output wavelet is actually infinite; but, because the terms decay rapidly in magnitude, the partial energy curves of the input and of the output practically coincide for time $\tau > 10$.

The dispersive effect of this layer may also be studied by considering the phase-lag spectra $\delta(\omega)$, $\alpha(\omega)$, and $\mu(\omega)$ of the input, of the all-pass filter, and of the output, respectively. These curves are shown in Figure 13-10. Since we have that

$$\mu(\omega) = \delta(\omega) + \alpha(\omega)$$

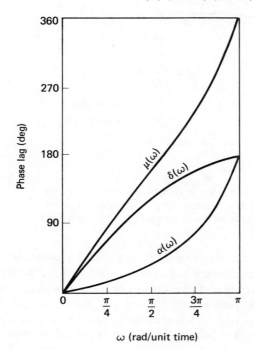

Figure 13-10. Phase-lag spectra $\delta(\omega)$, $\alpha(\omega)$, and $\mu(\omega)$ of the input, all-pass filter, and output of the one-finite-thickness-layer model, of Figure 13-9.

the phase-lag spectrum $\mu(\omega)$ of the reflected, upgoing response lies above the phase-lag spectrum $\delta(\omega)$ of the input for $0 < \omega \leq \pi$. (The Nyquist frequency ω_N is here π, since our sampling increment $\Delta\tau = 1$ time unit [i.e., $\omega_N = 2\pi f_N = 2\pi(1/2\Delta\tau) = \pi/\Delta\tau = \pi$]. On the other hand, since the layer is an all-pass filter, the magnitude spectra of both input and output are identical.

A Two-Finite-Thickness-Layer Model

Figure 13-11 is a schematic illustration of the two-finite-thickness-layer model. By solving the equation

$$\begin{bmatrix} D'_0(z) \\ U'_0(z) \end{bmatrix} = \mathbf{M}_0 \mathbf{M}_1 \mathbf{M}_2 \begin{bmatrix} D_3(z) \\ 0 \end{bmatrix}$$

we obtain the downgoing response

$$D_3(z) = \frac{t_0 t_1 t_2 z}{1 + (c_0 c_1 + c_1 c_2)z + c_0 c_2 z^2} D'_0(z) \qquad (13\text{-}3)$$

and the upgoing response

$$U'_0(z) = \frac{c_0 + (c_1 + c_0 c_1 c_2)z + c_2 z^2}{1 + (c_0 c_1 + c_1 c_2)z + c_0 c_2 z^2} D'_0(z) \qquad (13\text{-}4)$$

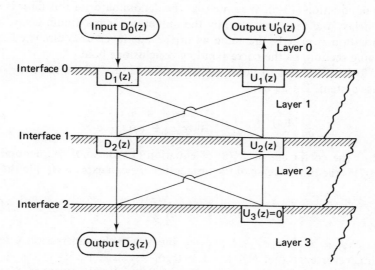

Figure 13-11. Lattice representation of the two-finite-thickness-layer model.

The coefficients of $D'_0(z)$ represent the z transforms of the two filters. Since the response of a system of layered elastic plates to a vertically incident plane compressional wave must be stable, the denominators of the filters of equations (13-3) and (13-4) must be minimum delay, for otherwise their expansion in positive powers of z would not lead to a convergent series. In other words, the zeros of these denominators must all be outside the unit circle $|z| = 1$. Hence, the coefficient of $D'_0(z)$ in equation (13-3) (i.e., the transmitted impulse response) is, except for a pure delay of one time unit, of the minimum-delay type. Just as is true for the one-finite-thickness-layer case, the weighting coefficients of these filters may be directly found from their z transforms by expansion in positive powers of z, and these weighting coefficients may then be convolved with the input to yield the respective outputs.

In the case when the lower interface is a perfect reflector (i.e., when $c_2 = -1$ and $t_2 = 0$), we find from equations (13-3) and (13-4) that

$$D_3(z) = 0 \cdot D'_0(z) \tag{13-5}$$

$$U'_0(z) = -1 \frac{z^2 + (c_0 c_1 - c_1)z - c_0}{1 + (c_0 c_1 - c_1)z - c_0 z^2} D'_0(z) \tag{13-6}$$

We observe that the downgoing filter of equation (13-3) reduces to an all-stop filter. The numerator of the coefficient of $D'_0(z)$ in equation (13-6) is now the time reverse of its denominator. Hence, the upgoing filter of equation (13-4) now becomes an all-pass filter, which is given by the coefficient of $D'_0(z)$ of equation (13-6). We note that the denominator of this filter is minimum delay, and that consequently the numerator is maximum delay. Thus, the situation is exactly the same as in the one-layer case discussed in the foregoing section, so the corresponding conclusions hold.

It is of interest to point out a further feature of the response of the all-pass system. Let us write

$$A(z) = -1 \cdot \frac{z^2 + (c_0 c_1 - c_1)z - c_0}{1 + (c_0 c_1 - c_1)z - c_0 z^2} \tag{13-7}$$

which is the coefficient of $D'_0(z)$ of equation (13-6). Since the denominator of (13-7) is the time reverse of the numerator, we can express $A(z)$ in the form

$$A(z) = -\frac{N(z)}{N^R(z)} \tag{13-8}$$

where $N(z) = z^2 + (c_0 c_1 - c_1)z - c_0$ and where the superscript R denotes the time reverse, so that $N^R(z) = 1 + (c_0 c_1 - c_1)z - c_0 z^2$.

What, now, is the autocorrelation of the all-pass response $A(z)$? We know that the convolution of two time functions is obtainable by multiplication of their respective z transforms. Since correlation involves the reversal of one time function in the convolution process, we conclude that the z transform of the autocorrelation of a time function can be found by multiplying the z transform of the original time function by the z transform of its time reverse. Thus, the z transform $\Phi_{AA}(z)$ of the autocorrelation function of the all-pass response given by equation (13-8) is

$$\Phi_{AA}(z) = \left[-\frac{N(z)}{N^R(z)} \right] \left[-\frac{N^R(z)}{N(z)} \right] = 1z^0 = 1$$

We thus see that the autocorrelation of the all-pass response is a unit spike for lag $\tau = 0$, and zero for all other lags. This interesting result is quite general and, in particular, holds for any number of layers of the model we are considering, provided that the system of layers is bounded by a perfect reflector at depth. We also observe that this result is in agreement with the energy considerations outlined previously in this chapter, and is actually an alternative proof for energy conservation in such a system. This conclusion results from the fact that the zeroth lag of an autocorrelation function is equal to the energy (or power, as the case may be) of the original function.

Our all-pass result holds also when the deepest reflection coefficient is $+1$ rather than -1. In the former case the acoustic impedance of the half-space below the system of finite-thickness layers is much smaller than that of the deepest layer of this system, while in the latter case the opposite is true; that is, the acoustic impedance of the half-space is much greater than that of the deepest finite-thickness layer immediately above it. Our discussion has been based on the choice of a deepest reflection coefficient of -1, because, in general, the acoustic impedance becomes larger as the depth increases.

The General Model and the All-Pass Theorem

Once a model has more than two or three layers, it is easier to handle it by means of a digital computer. In the examples discussed in the foregoing two sections we have been primarily interested in the outputs at the upper and lower interfaces of the layered system. In these computations, however, it is easy to take out as outputs the upgoing and downgoing wave motion at any interface or within a given layer. The delay properties of these functions provide valuable potential diagnostic information in seismic interpretation.

In our treatment, each layer is assumed to have the same one-way travel time; in order to obtain the effect of a layer with an arbitrary travel time, one can lump several successive layers together by setting the reflection coefficient to zero and the transmission coefficient to unity at the required number of interfaces existing between the constituent iso-travel-time layers.

The fact that the transmitted impulsive response of a system of layered strata is minimum delay has an interesting consequence in field seismic data processing, as we have already seen in Chapters 10 and 11. If this system represents the reverberating surface layers, the reverberations due to a deep reflection are caused by the transmission of seismic energy through the system twice: once on the way down to the deep reflecting horizon, and once again on the way up to the detector. By the reciprocity principle, the transmission impulsive response is the same for both upward and downward transmission, except for a constant factor. It follows, therefore, that the reverberation spike train due to a deep reflection is the minimum-delay response obtained by convolving the transmission impulse response of the surface layers with itself. Because the reverberation trains originating from deep reflections are minimum delay, the method of predictive deconvolution can be used to eliminate them from the received seismic trace. These questions have already been treated in ample detail in earlier chapters of this book. A detailed discussion of the minimum-delay property of the reverberation spike train can be found in Appendix 13-1, which follows directly.

As in the special cases that we have treated, whenever an interface is a perfect reflector all the input is ultimately returned to the surface, and the corresponding upgoing filter is an all-pass filter. This result may be called the "all-pass theorem." One of the most convincing arguments in support of this theorem comes from energy-delay relationships, as we have indicated in the one-finite-thickness-layer case.

Concluding Remarks

The response of a layered medium of homogeneous, perfectly elastic layers to a normally incident plane P wave may be conveniently expressed in terms of z-transform theory. This formulation leads to a degree of insight that would otherwise be difficult to achieve. Such a model exhibits a number of interesting properties, chief among which is the fact that a layered medium bounded by a perfect reflector at depth is representable as a dispersive, all-pass system with positive phase lag.

the minimum-delay property
of the reverberation spike train

In the chapter we stated that the reverberation spike train due to a deep reflection is minimum delay. In this appendix we want to give the derivation in the case of two reverberating layers, say a water layer over a mud layer; we will show that the reverberation spike train resulting from a deep reflection is minimum delay. As we will see, this derivation is general in the sense that it does not depend upon the fact that we are considering only two reverberating layers; hence it follows that the minimum-delay property of reverberations holds in the case of an arbitrary number of reverberating surface layers.

We use exactly the same definitions, notations, and assumptions as given in the chapter, except that we will now allow each layer to have an arbitrary one-way travel time. (In the chapter each layer was assumed to have the same one-way travel time; in order to obtain the effect of a layer with an arbitrary travel time, one had to lump several successive layers together with zero reflection coefficients and unit transmission coefficients at their interfaces.) We now assume that the air half-space is layer 0; the water layer is layer 1; the underlying mud layer is layer 2; and the layer under the mud is layer 3. The one-way travel times in layers 1 and 2 are designated by τ_1 and τ_2, respectively. We assume that the shot and the transducer are in the air (i.e., layer 0) just above the air/water interface; by dividing the results by t_0 and t_0', respectively, we in effect move the shot and the transducer into the water layer (i.e., layer 1) just below the air/water interface.

Our derivation is based on the fact that the water and mud layers (i.e., layers 1 and 2) may be regarded as a reverberation-producing filter. The energy passes through the water–mud filter as it goes down to the deep reflecting horizon and passes through the mud–water filter as it returns to the surface. Hence, we have in effect a situation where the deep reflection energy passes through two cascaded filters, the water–mud filter and the mud–water filter. We must therefore compute the z transform of each of

321

these two filters; the z transform of the overall reverberation-producing filter is the product of these two z transforms.

We note that there is a pure time delay of $\tau_1 + \tau_2$ (i.e., the travel time of the direct energy) in the water–mud filter; this delay will appear in its z transform as the factor $z^{\tau_1+\tau_2}$. Likewise, there is the pure time delay of $\tau_2 + \tau_1$ in the mud–water filter which appears as the factor $z^{\tau_2+\tau_1}$ in its z transform. We will want to divide out these factors, as these time delays are accounted for as part of the travel time of the direct reflected energy on the seismogram. This matter has already been brought out in Appendix 12-2 in our discussion of the properties of the ideal predictive deconvolution filter.

The basic equation that we will use to find each of these two z transforms is

$$\begin{bmatrix} D_0'(z) \\ U_0'(z) \end{bmatrix} = \mathbf{M_0 M_1 M_2} \begin{bmatrix} D_3(z) \\ U_3(z) \end{bmatrix} \qquad (13\text{-}9)$$

which is derived in the present chapter, except that now the scattering matrices $\mathbf{M_1}$ and $\mathbf{M_2}$ are defined as

$$\mathbf{M_1} = \frac{1}{t_1} \begin{bmatrix} z^{-\tau_1} & c_1 z^{-\tau_1} \\ c_1 z^{\tau_1} & z^{\tau_1} \end{bmatrix}$$

$$\mathbf{M_2} = \frac{1}{t_2} \begin{bmatrix} z^{-\tau_2} & c_2 z^{-\tau_2} \\ c_2 z^{\tau_2} & z^{\tau_2} \end{bmatrix}$$

in order to take into account the different travel times τ_1 and τ_2. It is convenient to designate the matrix product $\mathbf{M_0 M_1 M_2}$ in the form of a single matrix; that is, we define the elements $A(z)$, $B(z)$, $C(z)$, and $D(z)$ by the equation

$$\mathbf{M_0 M_1 M_2} = \frac{1}{t_0 t_1 t_2} \begin{bmatrix} A(z) & B(z) \\ C(z) & D(z) \end{bmatrix}$$

For example, by direct matrix multiplication of $\mathbf{M_0 M_1 M_2}$ we see that the element $A(z)$ is given by

$$A(z) = z^{-(\tau_1+\tau_2)}[1 + c_0 c_1 z^{2\tau_1} + c_1 c_2 z^{2\tau_2} + c_0 c_2 z^{2(\tau_1+\tau_2)}]$$

Let us define the polynomial $P(z)$ by the equation

$$A(z) = z^{-(\tau_1+\tau_2)} P(z)$$

Hence we see that $P(z)$ is explicitly given by the expression

$$P(z) = 1 + c_0 c_1 z^{2\tau_1} + c_1 c_2 z^{2\tau_2} + c_0 c_2 z^{2(\tau_1+\tau_2)}$$

[*Note:* In the case of N reverberating surface layers each with one-way travel times $\tau_1, \tau_2, \tau_3, \ldots, \tau_N$, respectively, the element $A(z)$ would be defined as the upper left-hand element in the matrix $\mathbf{M}_0\mathbf{M}_1\mathbf{M}_2\mathbf{M}_3 \ldots \mathbf{M}_N$, and $P(z)$ would be defined as the polynomial that satisfies $A(z) = z^{-(\tau_1+\tau_2+\tau_3+\cdots\tau_N)}P(z)$. All of the following arguments can be so modified for this more general case.]

Let us designate the z transform of the water–mud filter by $F(z)$. We may obtain $F(z)$ from the basic equation (13-9) by specifying the following conditions:

$D_0'(z) = 1 = z$ transform of downgoing unit spike (at bottom of layer 0)

$U_0'(z)$ (arbitrary for this derivation)

$D_3(z) = F(z) = z$ transform of downgoing spike train (at the top of layer 3) produced in response to unit spike = z transform of water–mud filter

$U_3(z) = 0$ (i.e., no upgoing energy in layer 3)

Equation (13-9) therefore becomes

$$\begin{bmatrix} 1 \\ U_0'(z) \end{bmatrix} = \frac{1}{t_0 t_1 t_2}\begin{bmatrix} A(z) & B(z) \\ C(z) & D(z) \end{bmatrix}\begin{bmatrix} F(z) \\ 0 \end{bmatrix}$$

which yields

$$1 = \frac{A(z)F(z)}{t_0 t_1 t_2}$$

Hence, the z transform of the water–mud filter is

$$F(z) = \frac{t_0 t_1 t_2}{A(z)} = z^{\tau_1+\tau_2}\frac{t_0 t_1 t_2}{P(z)}$$

Given $F(z)$, we want to divide out the factor t_0 (which corresponds to moving the source from just above to just below the air/water interface) and to divide out the factor $z^{\tau_1+\tau_2}$ (which corresponds to the pure time delay due to the direct travel time through the water and mud); we therefore obtain the modified water–mud filter

$$F_M(z) = z^{-(\tau_1+\tau_2)}t_0^{-1}F(z) = \frac{t_1 t_2}{P(z)}$$

Now as a physical fact we know that the (modified) water–mud filter is stable (i.e., the filter cannot produce an unbounded response to a bounded input) and is causal (i.e., the filter cannot respond to an input before the input occurs); hence it follows that the polynomial $P(z)$ which occurs in the denominator of $F_M(z)$ must be minimum delay. Because the inverse of a minimum-delay filter is minimum delay, it follows that the reciprocal of this polynomial is minimum delay. That is, the modified water–mud filter $F_M(z)$ is minimum delay; it has a causal and stable impulse response whose coefficients can be found by polynomial division in which we expand in terms of a series of nonnegative powers of z.

Let us designate the z transform of the mud–water filter by $G(z)$. We may obtain $G(z)$ from the basic equation (13-9) by specifying the following conditions:

$D_0'(z) = 0$ (i.e., no downgoing energy in layer 0)

$U_0'(z) = G(z) = z$ transform of upgoing spike train (at the bottom of layer 0) produced in response to unit spike $= z$ transform of mud–water filter

$D_3(z)$ (of no interest for this derivation)

$U_3(z) = 1 = z$ transform of upgoing unit spike (at top of layer 3)

Equation (13-9) therefore becomes

$$\begin{bmatrix} 0 \\ G(z) \end{bmatrix} = \frac{1}{t_0 t_1 t_2} \begin{bmatrix} A(z) & B(z) \\ C(z) & D(z) \end{bmatrix} \begin{bmatrix} D_3(z) \\ 1 \end{bmatrix}$$

To solve for $G(z)$, let us invert the equation. We obtain

$$\begin{bmatrix} D_3(z) \\ 1 \end{bmatrix} = \frac{1}{t_0' t_1' t_2'} \begin{bmatrix} D(z) & -B(z) \\ -C(z) & A(z) \end{bmatrix} \begin{bmatrix} 0 \\ G(z) \end{bmatrix}$$

where we have used the fact that

$$\det(\mathbf{M_0 M_1 M_2}) = (\det \mathbf{M_0})(\det \mathbf{M_1})(\det \mathbf{M_2})$$
$$= (1 - c_0^2)(1 - c_1^2)(1 - c_2^2)$$
$$= (t_0 t_0')(t_1 t_1')(t_2 t_2')$$

where $\det(\cdot)$ is the determinant of the argument.

Solving for $G(z)$ we find the following expression for the z transform of the mud–water filter:

$$G(z) = \frac{t_0' t_1' t_2'}{A(z)} = z^{\tau_1 + \tau_2} \frac{t_0' t_1' t_2'}{P(z)}$$

Thus, the modified mud–water filter is

$$G_M(z) = z^{-(\tau_1 + \tau_2)}(t_0')^{-1} G(z) = \frac{t_1' t_2'}{P(z)}$$

Comparing $F_M(z)$ with $G_M(z)$ we see that the modified water–mud filter is the same as the modified mud–water filter, except that the former has the (constant) factor $t_1 t_2$, indicating downward transmission, and the latter has the (constant) factor $t_1' t_2'$, indicating upward transmission. The equivalence of these two filters could have been established by the principle of reciprocity.

Cascading these two filters we have the overall (modified) reverberation-producing filter; its z transform is

$$F_M(z) G_M(z) = \frac{t_1 t_1' t_2 t_2'}{P(z) P(z)}$$

Because the cascading of two minimum-delay filters results in a minimum-delay filter, it follows that the overall (modified) reverberation-producing filter is minimum delay; that is, the impulse response of $F_M(z) G_M(z)$ is minimum delay [the coefficients of this impulse response can be found either by dividing $P^2(z)$ into the constant $t_1 t_1' t_2 t_2'$ in terms of a series of nonnegative powers of z or by convolving the coefficients of $F_M(z)$ with those of $G_M(z)$]. Since the impulse response of the filter $F_M(z) G_M(z)$ represents the reverberation spike train produced by a deep reflection, it follows that the reverberation spike train is minimum delay, as we wished to show.

the reflection coefficient in terms of functions continuous at an interface

The results of this chapter have yielded the particle displacement reflection coefficient c_k,

$$c_k = \frac{Z_k - Z_{k+1}}{Z_k + Z_{k+1}} \qquad (13\text{-}10)$$

where Z_k is the acoustic impedance of layer k. We shall establish how this coefficient depends on certain other functions continuous at an interface, namely particle velocity, particle acceleration, stress, and pressure. Let the origin $y = 0$ of a downward pointing y axis be at the top of layer k. Let

$$d_k\left(t - \frac{y}{v_k}\right) \quad \text{and} \quad u_k\left(t + \frac{y}{v_k}\right)$$

be the downgoing and upgoing waves in layer k, where $v_k = P$-wave velocity in layer k. The total displacement is

$$d_k\left(t - \frac{y}{v_k}\right) + u_k\left(t + \frac{y}{v_k}\right)$$

Let a dot indicate differentiation with respect to t, and let a prime indicate differentiation of a function with respect to its argument. Thus, the particle velocity $\dot{d}_k$ is

$$\dot{d}_k\left(t - \frac{y}{v_k}\right) = d'_k\left(t - \frac{y}{v_k}\right)\frac{\partial(t - y/v_k)}{\partial t} = d'_k\left(t - \frac{y}{v_k}\right)$$

The total stress in layer k is

$$E_k \frac{\partial}{\partial y}\left[d_k\left(t - \frac{y}{v_k}\right) + u_k\left(t + \frac{y}{v_k}\right)\right]$$

or

$$E_k\left(-\frac{1}{v_k}\right)d'_k + E_k\left(\frac{1}{v_k}\right)u'_k$$

where E_k = elastic modulus in layer k. This follows because

$$\frac{\partial d_k}{\partial y} = d'_k\left(t - \frac{y}{v_k}\right)\frac{\partial(t - y/v_k)}{\partial y} = -\frac{1}{v_k}d'_k\left(t - \frac{y}{v_k}\right)$$

and

$$\frac{\partial u_k}{\partial y} = u'_k\left(t + \frac{y}{v_k}\right)\frac{\partial(t + y/v_k)}{\partial y} = +\frac{1}{v_k}u'_k\left(t + \frac{y}{v_k}\right)$$

At the interface ($y = 0$) we have an initial downgoing wave d_k, with a reflected wave u_k and a transmitted wave d_{k+1}. Here the two conditions

1. Continuity of displacement.
2. Continuity of stress.

must be satisfied. These give

$$(1) \quad d_k + u_k = d_{k+1}$$

$$(2) \quad \frac{E_k}{v_k}(-d'_k + u'_k) = \frac{E_{k+1}}{v_{k+1}}(-d'_{k+1})$$

Since

$$E_k = \rho_k v_k^2 = (\rho_k v_k)v_k = Z_k v_k$$

where $\rho_k = $ density, we have

$$(1) \quad d_k + u_k = d_{k+1}$$

$$(2) \quad (d'_k - u'_k)Z_k = d'_{k+1}Z_{k+1} \qquad\qquad (13\text{-}11)$$

Taking the derivative of the first of the equations (13-11), we obtain

$$(1) \quad d'_k + u'_k = d'_{k+1}$$

$$(2) \quad (d'_k - u'_k)Z_k = d'_{k+1}Z_{k+1}$$

which, since $d'_k = \dot{d}_k$ and $u'_k = \dot{u}_k$, are in terms of the velocities,

$$(1) \quad \dot{d}_k + \dot{u}_k = \dot{d}_{k+1}$$

$$(2) \quad (\dot{d}_k - \dot{u}_k)Z_k = \dot{d}_{k+1}Z_{k+1} \qquad\qquad (13\text{-}12)$$

Simultaneous solution of these two equations gives

$$\frac{\dot{d}_{k+1}}{\dot{d}_k} = \frac{2Z_k}{Z_k + Z_{k+1}} = t$$

and

$$\frac{\dot{u}_k}{\dot{d}_k} = \frac{Z_k - Z_{k+1}}{Z_k + Z_{k+1}} = c$$

where t and c are, respectively, the transmission and reflection coefficients for particle velocity. Differentiating equations (13-12) once more with respect to time and solving yields

$$\frac{\ddot{d}_{k+1}}{\ddot{d}_k} = \frac{2Z_k}{Z_k + Z_{k+1}} = t$$

and

$$\frac{\ddot{u}_k}{\ddot{d}_k} = \frac{Z_k - Z_{k+1}}{Z_k + Z_{k+1}} = c$$

which are seen to be the transmission and reflection coefficients for particle acceleration. Comparison of these results with equation (13-10) shows that the reflection and transmission coefficients for particle displacement, particle velocity, and particle acceleration are identical.

Now let $\hat{c}$ and $\hat{t}$ be the corresponding reflection coefficients for the stress,

$$\hat{c} = \frac{\text{reflected stress}}{\text{initial stress}}$$

$$= \frac{E_k \dfrac{\partial u_k(t + y/v_k)}{\partial y}}{E_k \dfrac{\partial d_k(t - y/v_k)}{\partial y}}$$

$$= \frac{Z_k u_k'}{-Z_k d_k'} = -\frac{u_k'}{d_k'} = -\frac{\dot{u}_k}{\dot{d}_k}$$

$$= -c = \frac{Z_{k+1} - Z_k}{Z_{k+1} + Z_k}$$

and

$$\hat{t} = \frac{\text{transmitted stress}}{\text{initial stress}}$$

$$= \frac{E_{k+1} \dfrac{\partial d_{k+1}(t - y/v_{k+1})}{\partial y}}{E_k \dfrac{\partial d_k(t - y/v_k)}{\partial y}}$$

$$= \frac{-Z_{k+1} d_{k+1}'}{-Z_k d_k'} = \frac{Z_{k+1} \dot{d}_{k+1}}{Z_k \dot{d}_k}$$

$$= \frac{Z_{k+1}}{Z_k} t = \frac{2 Z_{k+1}}{Z_k + Z_{k+1}}$$

In other words, the reflection coefficient for stress is

$$\hat{c} = -c$$

while the transmission coefficient for stress is

$$\hat{t} = \frac{Z_{k+1}}{Z_k} t$$

Furthermore, since pressure is "negative stress" in the context of our model, the reflection and transmission coefficients for stress and for pressure are seen to be identical.

In particular, at the free surface $k = 0$ we have approximately that $Z_0 = 0$, so the value of the reflection coefficient for particle displacement, particle velocity, or particle acceleration is

$$c_0 = \frac{Z_0 - Z_1}{Z_0 + Z_1} \approx -1$$

In terms of pressure (or stress), we have

$$\hat{c}_0 = \frac{Z_1 - Z_0}{Z_1 + Z_0}$$
$$\approx +1$$

signal-to-noise-ratio
enhancement filters

Summary

Up to this point, we have emphasized the usefulness of the least squares (or Wiener) filter for the solution of a broad class of geophysical signal analysis problems. We shall now devote our attention to two additional digital filters, whose design criteria are based on the explicit enhancement of the signal-to-noise ratio of the given data. First, we describe the matched, or cross-correlation filter, which has seen extensive use in the processing of seismograms recorded with vibratory sources. Second, we shall deal with the output energy filter, whose design principle rests on the maximization of the signal energy within a specified interval.

The matched filter is designed such that ideally the presence of a given signal is indicated by a single large deflection in the output. The output energy filter ideally reveals the presence of such a signal by producing a longer burst of energy in the time interval where the signal occurs. The input is assumed to be an additive mixture of signal and noise. The shape of the signal must be known in order to design the matched filter, but only its autocorrelation function need be known to obtain the output energy filter. The derivation of these filters differs according to whether the noise is white or colored. In the former case, the noise autocorrelation function consists of only a single spike at lag zero, while in the latter the shape of this noise autocorrelation function is arbitrary.

A Simplified Derivation of the Matched and Output Energy Filters

Let us assume that a 2-length signal (b_0, b_1) is fed into a filter with the 2-length response function (a_0, a_1). The output is then given by the 3-length series

$$(c_0, c_1, c_2) = (b_0, b_1) * (a_0, a_1)$$
$$= (a_0 b_0, a_0 b_1 + a_1 b_0, a_1 b_1) \qquad (14\text{-}1)$$

where the symbol $*$ denotes convolution.

Suppose that we wish to maximize the square of the central output value, c_1^2, subject to the unit energy constraint on the filter coefficients, namely $a_0^2 + a_1^2 = 1$. The operator (a_0, a_1) so chosen is called the *matched filter*.

Now suppose that we wish to maximize the sum of the squares of all output values, $c_0^2 + c_1^2 + c_2^2$, subject to the same unit energy constraint on the filter coefficients. The operator chosen in this manner is called the *output energy filter*, since $c_0^2 + c_1^2 + c_2^2$ is the energy in the output.

The solution of the matched filter problem can be obtained with the aid of Cauchy's inequality (see, e.g., Hardy et al., 1952). If a_0, a_1, b_0, b_1 are real numbers, then the inequality

$$(a_0 b_1 + a_1 b_0)^2 \leq (a_0^2 + a_1^2)(b_0^2 + b_1^2)$$

is true, and the equality holds only when $a_0 = k b_1$ and $a_1 = k b_0$, where k is any constant. But we recognize the quantity on the left side of the expression above to be c_1^2, while the two factors on the right side are the respective energies of the filter and of the input. In other words,

(Central output value)$^2 \leq$ (Energy of filter)(Energy of input)

Thus, we may maximize the square of the central output value by choosing the filter coefficients (a_0, a_1) in such a way that the expression above becomes an equality. We already know that this can be done by setting

$$(a_0, a_1) = (k b_1, k b_0)$$

We have yet to impose the unit energy constraint on the filter; if we do so, then $k = 1/\sqrt{b_0^2 + b_1^2}$ and hence the sought matched filter is

$$(a_0, a_1) = \left(\frac{b_1}{\sqrt{b_0^2 + b_1^2}}, \frac{b_0}{\sqrt{b_0^2 + b_1^2}} \right)$$

These filter coefficients are optimum in the sense that they maximize the square of the middle output value subject to the unit energy constraint. We note that this filter is obtainable by reversing the input signal (b_0, b_1) in time, and then dividing each filter coefficient by the normalization factor $\sqrt{b_0^2 + b_1^2}$. It can therefore be said that these optimum coefficients are "matched" to the input signal, and this is the reason for the term "matched filter."

The solution of the output energy filter problem can be obtained by means of the calculus of variations (see, e.g., Hildebrand, 1952). We desire to maximize the output energy $c_0^2 + c_1^2 + c_2^2$ subject to the constraint $a_0^2 + a_1^2 = 1$. Using the method of Lagrange multipliers, we must maximize the expression

$$J = c_0^2 + c_1^2 + c_2^2 - \lambda(a_0^2 + a_1^2 - 1) \qquad (14\text{-}2)$$

where λ is the Lagrange multiplier. Substituting equation (14-1) into this relation, we have

$$J = (a_0 b_0)^2 + (a_0 b_1 + a_1 b_0)^2 + (a_1 b_1)^2 - \lambda(a_0^2 + a_1^2 - 1)$$

By setting the partial derivatives of J with respect to each of the filter coefficients a_0 and a_1 equal to zero, we obtain the two linear simultaneous equations in a_0 and a_1,

$$(b_0^2 + b_1^2)a_0 + (b_0 b_1)a_1 - \lambda a_0 = 0$$
$$(b_0 b_1)a_0 + (b_0^2 + b_1^2)a_1 - \lambda a_1 = 0$$

Since $b_0^2 + b_1^2 = r_0$ and $b_0 b_1 = r_1$, where r_0 and r_1 are, respectively, the zeroth and first lags of the autocorrelation of the input (b_0, b_1), we can write

$$r_0 a_0 + r_1 a_1 - \lambda a_0 = 0$$
$$r_1 a_0 + r_0 a_1 - \lambda a_1 = 0$$

or, in matrix form,

$$\left\{ \begin{bmatrix} r_0 & r_1 \\ r_1 & r_0 \end{bmatrix} - \begin{bmatrix} \lambda & 0 \\ 0 & \lambda \end{bmatrix} \right\} \begin{bmatrix} a_0 \\ a_1 \end{bmatrix} = \begin{bmatrix} 0 \\ 0 \end{bmatrix} \qquad (14\text{-}3)$$

This is the ordinary eigenvalue problem, with two eigenvalues λ_1 and λ_2 and two associated eigenvectors. Anticipating a later result in this chapter, we state that the eigenvector associated with the larger of these two eigenvalues constitutes the optimum output energy filter.

In what follows we shall rederive our rather heuristic results on a more rigorous basis. By doing this we shall bring out some of the many interesting properties that these filters possess.

The Matched Filter in White Noise

In this section we shall assume that the noise is white, which is appropriate for many radar problems, but less so for seismic work. In the next section we then assume that the noise is stationary with an arbitrary autocorrelation function.

Suppose that the signal is the *known* finite-length wavelet

$$(b_0, b_1, b_2, \ldots, b_n)$$

Let us consider two possibilities that may arise for the received data x_t. First, x_t can be an additive mixture of the signal b_t and unwanted noise u_t,

$$x_t = b_t + u_t$$

and second, the signal may be absent, so that only the white noise u_t arrives at the receiver,

$$x_t = u_t$$

The received data x_t are fed into a filter a_t to yield the output y_t. In the presence of the signal b_t, the output y_t is

$$
\begin{aligned}
y_t = x_t * a_t &= (b_t + u_t) * a_t \\
&= b_t * a_t + u_t * a_t \\
&= c_t + v_t
\end{aligned}
$$

where c_t and v_t are the respective filter responses to a pure signal input and to a pure white noise input. This model is illustrated in Figure 14-1. In the absence of a signal, we obtain only filtered noise as output,

$$
\begin{aligned}
y_t = x_t * a_t &= u_t * a_t \\
&= v_t
\end{aligned}
$$

We shall fix the length of the filter $(a_0, a_1, \ldots, a_n)$ to be equal to the length of the input signal $(b_0, b_1, \ldots, b_n)$. Then the filtered data y_t is given by the convolution

$$y_t = x_t * a_t = a_0 x_t + a_1 x_{t-1} + \cdots + a_n x_{t-n}$$

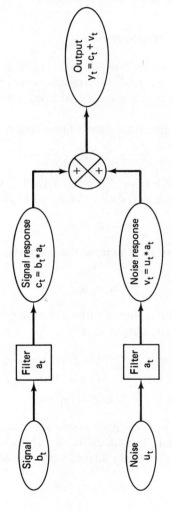

Figure 14-1. Relationships among signal, noise, and filtered output.

At this point we would like to decide whether the filtered data y_t consist of a signal plus noise, or whether they are only noise. To answer this question at a particular instant of time, say $t = t_0$, we require that the output of the filter at this time be greater when the signal is present than when it is absent. This goal can be achieved by making the instantaneous power in the filter output containing a signal at time $t = t_0$ as large as possible relative to the average power in the filtered noise at that instant. Hence, we shall define a signal-to-noise ratio at the filter output, μ, as

$$\mu = \frac{(\text{Value of filtered signal at time } t_0)^2}{\text{Average power of filtered noise at time } t_0}$$

If we convolve the $(n + 1)$-length signal $(b_0, b_1, \ldots, b_n)$ with the $(n + 1)$-length filter $(a_0, a_1, \ldots, a_n)$, we obtain a $(2n + 1)$-length output series $(c_0, c_1, \ldots, c_n, \ldots, c_{2n-1}, c_{2n})$, where c_n is the central value of this output series. It is convenient to choose the time instant t_0 to be n, as will soon become apparent. Thus, μ becomes

$$\mu = \frac{c_n^2}{E\{v_n^2\}} = \frac{(a_0 b_n + a_1 b_{n-1} + \cdots + a_n b_0)^2}{E\{v_n^2\}}$$

where $E\{v_n^2\}$ is the average value of the noise output power at the time instant $t = n$. Now we have

$$E\{v_t^2\} = E\{(v_t * a_t)^2\} = q_0(a_0^2 + a_1^2 + \cdots + a_n^2) \qquad (14\text{-}4)$$

where q_0 is the power of the white noise u_t, and where $(a_0^2 + a_1^2 + \cdots + a_n^2)$ is the energy in the filter's memory function. This expression is derived in Appendix 14-1. Because the noise u_t is stationary, its power q_0 is actually a constant independent of the time t. Thus, the output signal-to-noise ratio μ is

$$\mu = \frac{c_n^2}{E\{v_n^2\}} = \frac{(a_0 b_n + a_1 b_{n-1} + \cdots + a_n b_0)^2}{q_0(a_0^2 + a_1^2 + \cdots + a_n^2)} \qquad (14\text{-}5)$$

and it is this quantity that we wish to maximize. In other words, we want to use the instantaneous output value

$$y_n = c_n + v_n$$

to detect the presence of the signal, and we thus want the ratio of c_n^2 to $E\{v_n^2\}$ to be as large as possible. The memory function of the filter we seek can again be found by means of Cauchy's inequality, which for the present case is

$$(a_0 b_n + a_1 b_{n-1} + \cdots + a_n b_0)^2 \le (a_0^2 + a_1^2 + \cdots + a_n^2)(b_0^2 + b_1^2 + \cdots + b_n^2)$$

Dividing both sides of this inequality by $q_0(a_0^2 + a_1^2 + \cdots + a_n^2)$, and combining with equation (14-5), we obtain

$$\mu \leq \frac{1}{q_0}(b_0^2 + b_1^2 + \cdots + b_n^2)$$

The equal sign in this inequality holds when the filter is given by

$$(a_0, a_1, \ldots, a_n) = (kb_n, kb_{n-1}, \ldots, kb_0) \tag{14-6}$$

where k is an arbitrary scaling factor, which we choose to be unity for convenience. This operator is called the *matched filter* for detecting a known signal $(b_0, b_1, \ldots, b_n)$ in the presence of white noise.

Among all linear filters, the matched filter is optimum in the sense that it maximizes the signal-to-noise ratio μ. We emphasize that the signal-to-noise ratio must be defined as in equation (14-5) for this result to be true—in other words, signal-to-noise ratios defined on an alternate basis will not necessarily be at a maximum for the matched filter.

Let us now briefly turn to the frequency-domain expression for the matched filter. Since the memory function of this filter is equal to the time reverse of the signal, that is, since

$$a_t = b_{n-t}, \qquad t = 0, 1, \ldots, n \tag{14-7}$$

its Fourier transform is

$$A(f) = \sum_{t=0}^{n} a_t e^{-2\pi i f t} = \sum_{t=0}^{n} b_{n-t} e^{-2\pi i f t}$$

where f is the cyclical frequency. Letting $n - t = s$, where s is a new summation variable, we obtain

$$A(f) = \sum_{s=0}^{n} b_s e^{-2\pi i f (n-s)} = e^{-2\pi i f n} \sum_{s=0}^{n} b_s e^{2\pi i f s}$$

$$= e^{-2\pi i f n} B^*(f) \tag{14-8}$$

where $B^*(f)$ is the complex conjugate of the Fourier transform of the signal b_t. We see from the relation above, then, that the Fourier transform $A(f)$ of the matched filter is the complex conjugate of the Fourier transform of the signal multiplied by the phase factor $e^{-2\pi i f n}$.

The energy density spectrum of the matched filter is

$$P(f) = A(f)A^*(f) = B^*(f)B(f) = |B(f)|^2$$

But, since $|B(f)|^2 = R(f)$, where $R(f)$ is the energy density spectrum of the signal, we have

$$P(f) = R(f)$$

In other words, both the signal and its matched filter in white noise have identical energy density spectra.

We may summarize this treatment of the matched filter as follows. Suppose that we are given an autocorrelation function of a noise time series together with a signal wavelet. We wish to detect the presence of the signal which is immersed in the noise. The method to be used is to filter the received mixture of signal and noise and to state that the signal is present if the instantaneous output y_n is at a maximum value. How should one design such a filter? If the noise were white (as we are assuming here) and if the filter's memory function had unit energy, the power of the output with only noise as input would be identical to the power of the input, since in this case we can write (see Figure 14-1)

$$y_t = v_t = u_t * a_t$$

Therefore, we have by equation (14-4) that

$$E\{y_t^2\} = E\{v_t^2\} = q_0(a_0^2 + a_1^2 + \cdots + a_n^2) = q_0$$

where $E\{y_t^2\}$ is the output power and q_0 the white noise power. As a result, the output power is unaffected by the actual values of the filter coefficients, and hence the filter need concern itself only with the signal. This means that the simple example of the preceding section illustrates the entire design principle in the case of white noise; that is, the optimum filter is the one whose memory function is given by the reversed signal, namely, by the coefficient sequence $(b_n, b_{n-1}, \ldots, b_0)$.

The crosscorrelation of one wavelet with another can be implemented by passing the first wavelet through a filter whose memory function is the reverse of the second wavelet. Hence, the filtering operation performed by a matched filter is equivalent to the crosscorrelation of the received data with the signal, and thus the matched filter can also be called a *correlator*. If the two wavelets are identical, the filter output is then simply the auto-correlation function of the input wavelet, and the filter is said to be "matched" to this input signal. On passing the signal b_t alone through the matched filter, we obtain the output c_t,

$$c_t = a_t * b_t = \sum_{s=0}^{n} a_s b_{t-s} = \sum_{s=0}^{n} b_{n-s} b_{t-s}$$

since $a_s = b_{n-s}$ by equation (14-7) and s is a dummy summation variable. This output can be written

$$c_t = b_n b_t + b_{n-1} b_{t-1} + \cdots + b_0 b_{t-n}$$

which we recognize to be

$$c_t = r_{n-t}$$

where r_{n-t} is the autocorrelation coefficient for lag $n - t$ of the signal b_t. Therefore, the response c_t of the matched filter to the signal b_t alone is the autocorrelation function r_{n-t} of this signal. For $t = n$, we have

$$c_n = r_0$$

where $r_0 = (b_0^2 + b_1^2 + \cdots + b_n^2) = $ energy of the signal wavelet b_t. Because $r_0 > |r_s|$ for any s (since b_t is a wavelet), it follows that $c_n > |c_t|$, so that c_n is the maximum value of the output c_t. Hence, we see that this maximum value c_n is equal to the energy of the signal. The matched filter *compresses* the entire energy of the signal into the output value c_n, thereby facilitating the detection process.

The Matched Filter in Autocorrelated Noise

In the last section we derived the expression for the matched filter in the case when the noise is white. Let us now discuss the situation that arises when the unwanted noise u_t is stationary with a known autocorrelation function q_t, where the coefficients q_t are not necessarily zero for $t \neq 0$. We term this noise "autocorrelated noise," in contradistinction to pure white noise, whose only nonvanishing autocorrelation coefficient is q_0. Such noise is also sometimes known as "colored noise." Our notation will be the same as before, except that we must now bear in mind that the noise u_t is no longer white. As in the previous section, the matched filters to be discussed here are indeterminate in the sense of an arbitrary amplification factor k, which we set equal to unity for convenience.

We use the same definition for the signal-to-noise ratio μ as before,

$$\mu = \frac{c_n^2}{E\{v_n^2\}}$$

We wish to maximize μ subject to the assumption that the input noise u_t is of the autocorrelated kind. It will be convenient to introduce matrix notation at this point. Thus, we let

$$\mathbf{a} = (a_0, a_1, \ldots, a_n) = (1) \text{ by } (n + 1) \text{ row vector} = \text{filter}$$

$$\mathbf{b} = (b_n, b_{n-1}, \ldots, b_0) = (1) \text{ by } (n+1) \text{ row vector}$$
$$= \text{time } \textit{reverse} \text{ of signal}$$

$$\mathbf{q} = \begin{bmatrix} q_0 & \cdots & q_n \\ \vdots & \ddots & \vdots \\ q_n & \cdots & q_0 \end{bmatrix} = (n+1) \text{ by } (n+1) \text{ autocorrelation}$$
matrix of the noise u_t

Then it is shown in Appendix 14-1 that

$$\mu = \frac{c_n^2}{E\{v_n^2\}} = \frac{(\mathbf{ab'})^2}{\mathbf{aqa'}} \tag{14-9}$$

where the prime (') denotes the matrix transpose. Now,

$$(\mathbf{ab'})^2 = (\mathbf{ab'})(\mathbf{ab'})' = (\mathbf{ab'})(\mathbf{ba'})$$

so that

$$\mu = \frac{(\mathbf{ab'})(\mathbf{ba'})}{\mathbf{aqa'}} \tag{14-10}$$

In order to maximize μ, we differentiate this quantity with respect to the filter vector **a** and set the result equal to zero. We thus obtain

$$\frac{\partial \mu}{\partial \mathbf{a}} = \frac{\mathbf{aqa'}\dfrac{\partial}{\partial \mathbf{a}}(\mathbf{ab'ba'}) - \mathbf{ab'ba'}\dfrac{\partial}{\partial \mathbf{a}}(\mathbf{aqa'})}{(\mathbf{aqa'})^2} = 0$$

which leads to

$$[\underset{\substack{\uparrow \\ 1 \text{ by } 1}}{(\mathbf{aqa'})} \quad \underset{\substack{\uparrow \\ (n+1) \text{ by } 1}}{(2\mathbf{b'ba'})}] - [\underset{\substack{\uparrow \\ 1 \text{ by } 1}}{(\mathbf{ab'})} \quad \underset{\substack{\uparrow \\ (n+1) \text{ by } 1}}{(2\mathbf{qa'})} \quad \underset{\substack{\uparrow \\ 1 \text{ by } 1}}{(\mathbf{ba'})}] = 0 \tag{14-11}$$

This relation is satisfied identically by the matrix equation

$$\mathbf{qa'} = \mathbf{b'} \tag{14-12}$$

which can be written out in the form

$$\begin{bmatrix} q_0 & \cdots & q_n \\ \vdots & \ddots & \vdots \\ & & \vdots \\ q_n & \cdots & q_0 \end{bmatrix}\begin{bmatrix} a_0 \\ a_1 \\ \vdots \\ a_n \end{bmatrix} = \begin{bmatrix} b_n \\ b_{n-1} \\ \vdots \\ b_0 \end{bmatrix} \tag{14-13}$$

This is the matrix formulation for a set of $(n + 1)$ linear simultaneous equations in the $(n + 1)$ unknown filter coefficients $(a_0, a_1, \ldots, a_n)$. Its solution then yields the desired optimum matched filter in the presence of autocorrelated, or colored noise.

We notice that the matrix $\mathbf{q}$ of equation (14-13) is an autocorrelation matrix, which has the distinctive and useful property that all elements on any given diagonal are the same. Therefore, equation (14-13) may be solved by the Toeplitz recursion technique (see Chapter 6).

The known quantities in this calculation are the noise autocorrelation matrix $\mathbf{q}$ and the time reverse of the signal wavelet, b_{n-t}, while the unknown quantities are the filter coefficients a_t.

Let us now obtain an expression for the maximum signal-to-noise ratio, μ_{max}. We do so by substituting equation (14-12) into equation (14-10):

$$\mu_{max} = \frac{(\mathbf{ab'})(\mathbf{ba'})}{(\mathbf{ab'})} = \mathbf{ba'}$$

But since $\mathbf{a'} = \mathbf{q}^{-1}\mathbf{b'}$ by equation (14-12), where $\mathbf{q}^{-1}$ is the matrix inverse of $\mathbf{q}$, we have

$$\mu_{max} = \mathbf{bq}^{-1}\mathbf{b'} \qquad (14\text{-}14)$$

In particular, if the noise is white and of unit power, we have $q_0 = 1$, $q_t = 0$, $t \neq 0$, so that

$$\mathbf{q} = \mathbf{q}^{-1} = \mathbf{I}$$

where $\mathbf{I}$ is the identity, or unit matrix. In this case, we can then write

$$\mu_{max} = \mathbf{bIb'} = \mathbf{bb'} = (b_n, b_{n-1}, \ldots, b_0)(b_n, b_{n-1}, \ldots, b_0)'$$
$$= (b_0^2 + b_1^2 + \cdots + b_n^2)$$
$$= \text{energy of the signal } b_t$$

As we may expect, this result is identical to the one obtained in the previous section for the case of the matched filter in the presence of stationary, white noise.

We shall briefly turn to the frequency-domain expression for the matched filter in the presence of autocorrelated noise. First, we rewrite equation (14-13) in the form

$$\sum_{s=0}^{n} a_s q_{t-s} = b_{n-t}, \qquad t = 0, 1, \ldots, n$$

We wish to take the Fourier transform of both sides of the equation above, yet we must bear in mind that this relation holds only for $t = 0, 1, \ldots, n$.

But the noise autocorrelation function q_t must be defined between the limits $-\infty \leq t \leq +\infty$, so that we would actually have to find the complete convolution of a_s with q_s, and this could only be done by letting the range of t be $-\infty \leq t \leq +\infty$. As a result, the Fourier transform can only be taken of the *complete* convolution of a_s with q_s, but in real life we have at best only the preceding approximation to this convolution. Keeping this fact in mind, we can write the following *approximate* expression for the Fourier transform of equation (14-13):

$$\sum_{t=-\infty}^{+\infty} \left(\sum_{s=0}^{n} a_s q_{t-s} \right) e^{-i2\pi ft} = \sum_{t=-\infty}^{+\infty} b_{n-t} e^{-2\pi ift}$$

The left member of this equation is the Fourier transform of the convolution of the filter a_s with the noise autocorrelation function q_s, which is $A(f)Q(f)$, while the right member has already been evaluated in the preceding section. Therefore,

$$A(f)Q(f) = e^{-2\pi ifn} B^*(f)$$

where $B^*(f)$ is the Fourier transform of the time reverse of the signal. We can solve this frequency-domain equation for the matched filter by simply dividing through by $Q(f)$, where we assume, of course, that $Q(f) \neq 0$ for any f,

$$A(f) = e^{-2\pi ifn} \frac{B^*(f)}{Q(f)}$$

The physical meaning of this expression is simple. The larger the amplitude spectrum $|B(f)|$ of the signal and the smaller the power density spectrum $Q(f)$ of the noise in the interval $(f, f + df)$, the more the matched filter $A(f)$ transmits frequencies in that interval. Thus, if the power spectral density $Q(f)$ of the noise is small in some interval of the frequency band occupied with the signal, and large elsewhere, the matched filter is essentially transparent in this interval.

The matched filter in the presence of autocorrelated noise can also be called a *correlator*, just as is true for the previously discussed matched filter in the presence of white noise. The difference between the two filters is that, while the latter is simply the time reverse of the signal wavelet b_t, the former must be calculated by means of equation (14-13).

If the noise is white, the matched filter in the presence of white noise *guarantees* the optimum value of the signal-to-noise ratio μ. If the noise is not white, but has a known autocorrelation function, it is the matched filter in the presence of autocorrelated noise that *guarantees* the optimum value of μ. If we do not know the autocorrelation coefficients q_t of the noise, we may often do quite well by assuming that the noise is white, even though we then realize that the matching used is not optimum. On the other hand, if q_t is known, we can do better, which is not surprising since in this case we have more information at our disposal to design the matched filter.

The Output Energy Filter

Consider a signal that is being transmitted through a dispersive medium. This medium disperses the signal wavelet b_t without altering its magnitude spectrum appreciably.† Let us thus assume that we know roughly the shape of the received signal's magnitude spectrum, although we know nothing about its phase spectrum. We are able, however, to measure the power density spectrum of the ambient noise. We wish to design a digital filter which will increase the likelihood of detecting the signal when it arrives at the receiver. The matched filter is now no longer appropriate because we do not know the shape of the signal b_t, but only its magnitude spectrum $|B(f)|$. Knowledge of $|B(f)|$ is equivalent to knowledge of the signal's energy density spectrum $R(f)$, since

$$R(f) = B(f)B^*(f) = |B(f)|^2$$

But the energy density spectrum $R(f)$ is the Fourier cosine transform of the signal's autocorrelation function r_t, (see, e.g., Robinson, 1954, Chap. 5),

$$R(f) = \sum_{t=-n}^{+n} r_t \cos 2\pi ft$$

and therefore

$$r_t = \int_{-1/2}^{+1/2} R(f) \cos 2\pi ft \, df \qquad (14\text{-}15)$$

Here the Nyquist, or folding frequency limits, are $f = \pm 1/(2\Delta t) = \pm\frac{1}{2}$, where Δt is the time sampling increment, which we have set equal to unity for convenience. Similarly,

$$q_t = \int_{-1/2}^{+1/2} Q(f) \cos 2\pi ft \, df \qquad (14\text{-}16)$$

where $Q(f)$ is the power density spectrum and q_t the autocorrelation function of the noise u_t. Thus, knowledge of $R(f)$ and $Q(f)$ allows one to compute r_t and q_t, respectively. We shall need these two autocorrelation functions for the discussion that follows. In practice, one attempts to estimate r_t and

---

†The Fourier transform of the signal b_t is $B(f) = \sum_{t=0}^{n} b_t e^{-i2\pi ft}$. $B(f)$ is generally complex, and can be written $B(f) = |B(f)| e^{i\varphi(f)}$, where $|B(f)|$ is the magnitude spectrum and $\varphi(f)$ is the phase spectrum of b_t.

q_t directly from the received data or, if only estimates of $R(f)$ and $Q(f)$ are available, one uses equations (14-15) and (14-16) to obtain the two required autocorrelation functions.

Given r_t and q_t, let us try to design a digital filter a_t whose output energy is as large as possible with the signal alone as input, under the constraint that the output noise power be as small as possible with noise alone as input. Such a filter is called an *output energy* filter. We thus seek the filter coefficients $(a_0, a_1, \ldots, a_m)$ which maximize a signal-to-noise ratio at the filter output, λ, defined as

$$\lambda = \frac{\text{Energy of the filtered signal}}{\text{Average power of the filtered noise}} = \frac{\sum_{t=0}^{m+n} c_t^2}{E\{v_t^2\}} \qquad (14\text{-}17)$$

(see also Figure 14-1). We remark that this definition of the signal-to-noise ratio† differs from the ratio μ [equation (14-4)], which we have maximized previously in order to derive the matched filter.

At this point it will again be convenient to introduce matrix notation. We let

$$\mathbf{a} = (a_0, a_1, \ldots, a_m) = (1) \text{ by } (m + 1) \text{ row vector} = \text{filter}$$

$$\mathbf{r} = \begin{bmatrix} r_0 & \cdots & r_m \\ \vdots & \ddots & \vdots \\ r_m & \cdots & r_0 \end{bmatrix} = \begin{matrix} (m + 1) \text{ by } (m + 1) \text{ autocorrelation} \\ \text{matrix of the signal } b_t \end{matrix}$$

$$\mathbf{q} = \begin{bmatrix} q_0 & \cdots & q_m \\ \vdots & \ddots & \vdots \\ q_m & \cdots & q_0 \end{bmatrix} = \begin{matrix} (m + 1) \text{ by } (m + 1) \text{ autocorrelation} \\ \text{matrix of the noise } u_t \end{matrix}$$

It is then shown in Appendix 14-1 that the signal-to-noise ratio λ can be written in the convenient form

$$\lambda = \frac{\sum_{t=0}^{m+n} c_t^2}{E\{v_t^2\}} = \frac{\mathbf{ara}'}{\mathbf{aqa}'} \qquad (14\text{-}18)$$

where the prime (') denotes the matrix transpose. We can find the maximum,

---

†We consider the energy of a signal because our signals are transients and thus have finite energy content. Stationary noise is not a transient phenomenon and has, in general, infinite energy content, but its power (energy/unit time) remains finite.

or optimum, value of this ratio by differentiating λ with respect to the filter vector $\mathbf{a}$ and setting the result equal to zero:

$$\frac{\partial \lambda}{\partial \mathbf{a}} = \frac{(\mathbf{aqa'})\dfrac{\partial}{\partial \mathbf{a}}(\mathbf{ara'}) - (\mathbf{ara'})\dfrac{\partial}{\partial \mathbf{a}}(\mathbf{aqa'})}{(\mathbf{aqa'})^2} = 0$$

which leads to

$$\underset{\substack{\uparrow \\ \text{1 by 1}}}{[(\mathbf{aqa'})} \quad \underset{\substack{\uparrow \\ (m+1)\text{ by }1}}{(2\mathbf{ra'})]} \quad - \quad \underset{\substack{\uparrow \\ \text{1 by 1}}}{[(\mathbf{ara'})} \quad \underset{\substack{\uparrow \\ (m+1)\text{ by }1}}{(2\mathbf{qa'})]} \quad = 0 \qquad (14\text{-}19)$$

But from equation (14-18) we have

$$\mathbf{ara'} = \lambda \mathbf{aqa'}$$

Substituting this expression into equation (14-19), we obtain

$$(\mathbf{aqa'})(\mathbf{ra'}) - \lambda(\mathbf{aqa'})(\mathbf{qa'}) = 0$$

Dividing through by $\mathbf{aqa'}$ (which is a 1 by 1 matrix, or simply a scalar), we get

$$\mathbf{ra'} - \lambda \mathbf{qa'} = 0$$

or

$$(\mathbf{r} - \lambda \mathbf{q})\mathbf{a'} = 0 \qquad (14\text{-}20)$$

This matrix equation can be written out in the form

$$\left\{ \begin{bmatrix} r_0 & \cdots & r_m \\ \cdot & & \cdot \\ \cdot & \cdot & \cdot \\ \cdot & & \cdot \\ \cdot & & \cdot \\ r_m & \cdots & r_0 \end{bmatrix} - \lambda \begin{bmatrix} q_0 & \cdots & q_m \\ \cdot & & \cdot \\ \cdot & \cdot & \cdot \\ \cdot & & \cdot \\ \cdot & & \cdot \\ q_m & \cdots & q_0 \end{bmatrix} \right\} \begin{bmatrix} a_0 \\ a_1 \\ \cdot \\ \cdot \\ \cdot \\ a_m \end{bmatrix} = \begin{bmatrix} 0 \\ 0 \\ \cdot \\ \cdot \\ \cdot \\ 0 \end{bmatrix} \qquad (14\text{-}21)$$

or, using summation rather than matrix notation,

$$\sum_{s=0}^{m} (r_{t-s} - \lambda q_{t-s})a_s = 0, \qquad t = 0, 1, \ldots, m \qquad (14\text{-}22)$$

Here the signal autocorrelation function r_t and the noise autocorrelation function q_t represent known quantities, while the parameter λ and the filter coefficients a_s represent the unknowns. The set of linear simultaneous equations given by equation (14-20), (14-21), or (14-22) constitutes the "generalized eigenvalue" problem (see, e.g., Hildebrand, 1952, pp. 74 ff.), where the signal-to-noise ratio λ is the eigenvalue and the filter $\mathbf{a} = (a_0, a_1, \ldots, a_m)$ is the associated eigenvector. The matrices $\mathbf{r}$ and $\mathbf{q}$ of equation (14-20) are autocorrelation matrices and therefore positive definite (Robinson, 1954, Chap. 5). Hence, this problem has $(m + 1)$ distinct eigenvector solutions associated with $(m + 1)$ eigenvalues. These eigenvalues are real and positive by virtue of the positive definite property of the matrices $\mathbf{r}$ and $\mathbf{q}$ (Hildebrand, 1952, pp. 76, 78). This property of the eigenvalues λ follows also from the definition of the signal-to-noise ratio (14-17), since λ is the ratio of two real and positive quantities, and must therefore also be real and positive.

Now we want to obtain the maximum value of the signal-to-noise ratio, and therefore we select the eigenvector associated with the largest eigenvalue, $\lambda_{\max}$. This eigenvector will then constitute the desired memory function of the output energy filter.† Since we are only interested in $\lambda_{\max}$ and its associated eigenvector, it is in practice not necessary to calculate all the $(m + 1)$ eigenvalues. There are iterative techniques that allow one to compute $\lambda_{\max}$ and its associated eigenvector with ease and rapidity, particularly because $\mathbf{r}$ and $\mathbf{q}$, being autocorrelation matrices, are in general well conditioned (Hildebrand, 1952, pp. 68–74).

Up to this point we have not explicitly described the nature of the noise u_t, whose autocorrelation is q_t. If this noise is white and of unit power, then we have $q_0 = 1$ and $q_t = 0$ for $t \neq 0$, so that

$$\mathbf{q} = \mathbf{I}$$

where $\mathbf{I}$ is the identity, or unit matrix. In this case equation (14-20) becomes

$$(\mathbf{r} - \lambda\mathbf{I})\mathbf{a}' = 0 \qquad (14\text{-}23)$$

which now constitutes an ordinary eigenvalue problem. In particular, if $m = 1$ and $\mathbf{q} = \mathbf{I}$ in equation (14-21), we arrive at the simple introductory output energy filter example given by equation (14-3). On the other hand, if the noise is of the autocorrelated, or "colored," type, the autocorrelation function q_t must be known explicitly in order to solve equation (14-21). We have thus been able to obtain solutions for the output energy filter in white noise [equation (14-23)], and in correlated noise [equation (20)].

Let us now proceed to describe a few remarkable characteristics of

---

†The fact that an eigenvector is determined only to within a constant scale factor (Hildebrand, 1952) corresponds to the physical fact that the signal-to-noise ratio λ does not depend on the amplification of the filter.

the output energy filter. It is shown in Appendix 14-1 that the signal-to-noise ratio λ can also be written in the form

$$\lambda = \frac{\sum\limits_{t=-m}^{+m} p_t r_t}{\sum\limits_{t=-m}^{+m} p_t q_t} = \frac{p_0 r_0 + 2\sum\limits_{t=1}^{m} p_t r_t}{p_0 q_0 + 2\sum\limits_{t=1}^{m} p_t q_t} \qquad (14\text{-}24)$$

where p_t is the autocorrelation function of the filter a_t. The last step above follows from the fact that autocorrelation functions of real-valued processes are symmetric. Equation (14-24) tells us that λ depends upon the filter coefficients $(a_0, a_1, \ldots, a_m)$ *only* through this filter's autocorrelation function p_t. Hence, if we were able to find another filter having the same autocorrelation function as this optimum output energy filter, this new filter would also yield the optimum signal-to-noise ratio $\lambda_{\max}$. Now it was shown in Chapter 5 that there can exist many finite-length wavelets which have the identical autocorrelation function p_t. These different wavelets are obtainable by factoring the z transform of p_t, which we call $P(z)$,

$$P(z) = \sum_{t=-m}^{m} p_t z^t \qquad (14\text{-}25)$$

The factors of $P(z)$ occur in pairs of the form

$$(z - z_i) \quad \text{and} \quad \left(z - \frac{1}{z_i^*}\right) \qquad (i = 1, 2, \ldots, m)$$

(Robinson, 1954, Chap. 2), where z_i^* is the complex conjugate of z_i. These two factors correspond respectively to the roots $z = z_i$ and $z = 1/z_i^*$ of $P(z)$. There are m such root pairs, and by choosing one root from each pair we can generate a wavelet having the given autocorrelation function p_t. Thus, there can be as many as 2^m different† wavelets of length $(m + 1)$, including the original optimum filter memory function $(a_0, a_1, \ldots, a_m)$, all of which have the same autocorrelation function p_t, and consequently all of which yield the same optimum value of the signal-to-noise ratio $\lambda_{\max}$. For simplicity let us assume that the maximum eigenvalue $\lambda_{\max}$ has unit multiplicity (i.e., it only occurs once), which will generally be the case in applied work. It then follows that the associated eigenvector is unique within a constant scale factor. Therefore, the memory function $(a_0, a_1, \ldots, a_m)$ represents the unique eigenvector solution corresponding to the eigenvalue $\lambda_{\max}$, and as a result the 2^m possible wavelets must indeed be all identical, except again for scale

---

†By "different" we imply that these wavelets may vary by more than a constant scale factor. In general, not all of these 2^m possible wavelets will be different. This problem was treated in greater detail in Chapter 5.

factors of unit magnitude. Returning now to the factorization of (14-25), it follows that the two roots in each pair, z_i and $1/z_i^*$, must be the same, for otherwise the 2^m choices of one root from each pair would not all generate the same wavelet. Now a moment's thought tells us that a root z_i can be equal to its conjugate-complex reciprocal $1/z_i^*$ only if both of these roots have unit magnitude. Hence, we have established the important result that *all* the roots of the z transform $P(z)$ of the autocorrelation function ρ_t of the output energy filter a_t must lie on the unit circle in the complex z plane. Furthermore, we have

$$P(z) = A(z^{-1})A(z) \qquad (14\text{-}26)$$

where $A(z)$ is the z transform of the filter a_t. But since all the roots of $P(z)$ lie on the unit circle, all the roots of $A(z)$ and of $A(z^{-1})$ must lie on the unit circle also. We thus see that all the roots of the transfer function $A(z)$ of the output energy filter lie on the unit circle and are of the form

$$z_i = \frac{1}{z_i^*} = e^{-2\pi i f}$$

This result has the interesting consequence that the response function $(a_0, a_1, \ldots, a_m)$ of the output energy filter is either symmetric or antisymmetric. To see why this is true, consider again the roots of $A(z)$, all of which have unit magnitude. The root $z = -1$ would come from the factor $(z + 1)$, which is the z transform of the symmetric 2-length wavelet $(1, 1)$. The root $z = +1$ would come from the factor $(z - 1)$, which is the z transform of the antisymmetric 2-length wavelet $(1, -1)$. The complex root $z = e^{-i2\pi f}$ would have to occur together with its complex conjugate $z = e^{+i2\pi f}$, since we are here restricting ourselves to filters with real coefficients. But these two complex roots would come from the factor

$$(z - e^{-i2\pi f})(z - e^{+i2\pi f}) = 1 - 2(\cos 2\pi f) z + z^2$$

which is the z transform of the symmetric 3-length wavelet $(1, -2\cos 2\pi f, 1)$. The convolution of any number of symmetric wavelets with each other yields a symmetric wavelet. The convolution of any number of antisymmetric wavelets with each other yields either a symmetric or an antisymmetric wavelet according to whether the number of wavelets convolved is even or odd, respectively. Therefore, the memory function of an output energy filter is either symmetrical (when its z transform has an even number of $+1$ roots) or antisymmetrical (when its z transform has an odd number of $+1$ roots). The wavelets $(1, 1)$ and $(1, -1)$ are of the "equi-delay" type (Robinson, 1962); that is, they may be considered as minimum-delay or maximum-delay, as they are on the borderline between these two types. All roots of the z

transform of an equi-delay wavelet lie on the unit circle; hence, the $(m + 1)$-length response function of the output energy filter is an $(m + 1)$-length equi-delay wavelet.

A Simple Comparative Study of Matched, Output Energy, and Least-Squares Filtering

It now becomes desirable to study the comparative performance of the filters we have been discussing thus far. This can perhaps best be done by first restricting ourselves to some very simple numerical examples. More elaborate examples will then follow.

We have shown that the matched filter is optimum in the sense that it is the only linear filter which maximizes the signal-to-noise ratio μ. Similarly, the output energy filter is optimum in the sense that it is the only linear filter which maximizes the signal-to-noise ratio λ. In this connection we should also consider the least-squares filter introduced in Chapter 6. The least-squares filter is optimum in the sense that it is the only linear filter that minimizes the mean square difference between a desired output d_t and an actual output c_t. We shall thus deal here with three types of linear digital filters, each of which is optimum in a *particular sense*. It is therefore clear that the term "optimum filter" is by itself ambiguous; instead, we should always be careful to state exactly which filter performance criterion we are optimizing. In order to bring this point into proper focus, we shall compute *both* μ and λ for all the examples of this section, since we may then observe how each filter optimizes the particular performance criterion upon which its design has been based.†

Suppose that we are given the 2-length input signal wavelet

$$b_t = (b_0, b_1) = (3, 1)$$

which is immersed in white noise of unit power, so that we have

$$q_t = \begin{cases} 1 & \text{for } t = 0 \\ 0 & \text{for } t \neq 0 \end{cases}$$

We wish to compute the following filters for this model:

1. Matched filter in white noise of unit power.

†The performance criterion for the least-squares filter is the value of the mean square difference between the desired output d_t and the actual output c_t. However, since neither the matched filter nor the output energy filter allows us to control the *shape* of the actual output, there is no convenient way to define such a mean-square-difference criterion for these filters. Therefore, we consider here only the performance criteria μ and λ.

2. Output energy filter in white noise of unit power.
3. Least-squares spiking filter in white noise of unit power.

Let us fix the filter memory function length to be $(n + 1) = (m + 1) = 2$ so that $n = m = 1$. For the matched filter this is necessary from its definition, since the length of the filter memory function must equal the length of the signal wavelet; for the other two filters above we choose to deal with 2-length filter memory functions so that we may have a proper basis for comparison.

Thus, we may write for the foregoing three cases:

(a_0, a_1) = 2-length filter memory function

$(c_0, c_1, c_2) = (b_0, b_1) * (a_0, a_1)$ = 3-length output series

(d_0, d_1, d_2) = 3-length desired output series (for the least-squares filter only)

$$\mu = \frac{(a_0 b_1 + a_1 b_0)^2}{p_0 q_0} = \text{signal-to-noise ratio which is optimum for the matched filter}$$

$$\lambda = \frac{p_0 r_0 + 2 p_1 r_1}{p_0 q_0} = \text{signal-to-noise ratio which is optimum for the output energy filter}$$

The foregoing expressions for μ and λ are obtainable from equation (14-5) for $n = 1$ and from equation (14-24) for $m = 1$, respectively. We recall that

$(p_0, p_1) = (a_0^2 + a_1^2, a_0 a_1)$ = autocorrelation coefficients of filter memory function

$(r_0, r_1) = (b_0^2 + b_1^2, b_0 b_1)$ = autocorrelation coefficients of signal wavelet
= (10, 3)

In order to compare the performance of the various filters to be discussed here, it will be convenient to have all memory functions normalized so that the unit energy constraint $a_0^2 + a_1^2 = 1$ be satisfied.

We shall now go through the calculations necessary to obtain the filters for the three cases listed above.

Matched Filter in White Noise of Unit Power

From equation (14-6) we have for $k = 1$ and $n = 1$,

$$(a_0, a_1) = (b_1, b_0) = (1, 3)$$

which, when normalized to unit energy, becomes

$$(a_0, a_1) = (0.316, 0.948)$$

and, therefore,

$$(p_0, p_1) = (1.0, 0.3)$$

The output is

$$(c_0, c_1, c_2) = (0.948, 3.160, 0.948)$$

while we obtain for the signal-to-noise ratios μ and λ,

$$\mu = \mu_{max} = 10.00$$
$$\lambda = 11.80$$

Here we write $\mu = \mu_{max}$, since we know that the matched filter maximizes μ.

Output Energy Filter in White Noise of Unit Power

Equation (14-21) yields for $(r_0, r_1) = (10, 3)$ and $(q_0, q_1) = (1, 0)$,

$$\begin{bmatrix} 10 - \lambda & 3 \\ 3 & 10 - \lambda \end{bmatrix} \begin{bmatrix} a_0 \\ a_1 \end{bmatrix} = \begin{bmatrix} 0 \\ 0 \end{bmatrix} \tag{14-27}$$

This system will have a nontrivial solution if and only if the determinant of the coefficient matrix vanishes, that is, if

$$(10 - \lambda)^2 - 9 = 0$$

The roots, or eigenvalues, of this characteristic equation are

$$\lambda_1 = 7 \quad \text{and} \quad \lambda_2 = 13$$

Thus, the optimum signal-to-noise ratio λ is

$$\lambda = \lambda_{max} = 13.00$$

where we write $\lambda = \lambda_{max}$ since we know that the output energy filter maximizes λ.

The memory function of this filter is given by the eigenvector associated with λ_{mnx}, and is obtainable from the adjoint $F(\lambda_{max})$ of the coefficient matrix of equation (14-27). This gives

$$\mathbf{F}(13) = \begin{bmatrix} -3 & -3 \\ -3 & -3 \end{bmatrix} = \begin{bmatrix} -3 \\ -3 \end{bmatrix} \begin{bmatrix} 1 & 1 \end{bmatrix}$$

from whence†

$$(a_0, a_1) = (1, 1)$$

†Readers unfamiliar with the details of eigenvalue problem theory are referred to Hildebrand (1952, Chap. 1).

Normalizing to unit energy, we have

$$(a_0, a_1) = (0.707, 0.707)$$

and therefore

$$(\rho_0, \rho_1) = (1, 0.5)$$

The output is

$$(c_0, c_1, c_2) = (2.121, 2.828, 0.707)$$

Finally,

$$\mu = 8.00$$

Least-Squares Spiking Filter in White Noise of Unit Power

Let us attempt to shape the signal wavelet $(3, 1)$ into a spike at $t = 1$, that is,

$$(d_0, d_1, d_2) = (0, 1, 0)$$

The normal equations for this least-squares filter are obtained by suitably specializing equation (14-3) of Chapter 6. We have

$$\begin{bmatrix} 11 & 3 \\ 3 & 11 \end{bmatrix} \begin{bmatrix} a_0 \\ a_1 \end{bmatrix} = \begin{bmatrix} 1 \\ 3 \end{bmatrix}$$

from which we can compute the normalized memory function

$$(a_0, a_1) = (0.067, 0.998)$$

and therefore

$$(\rho_0, \rho_1) = (1.000, 0.067)$$

The output is

$$(c_0, c_1, c_2) = (0.201, 3.061, 0.998)$$

while we obtain for the signal-to-noise ratios μ and λ,

$$\mu = 9.37$$

$$\lambda = 10.40$$

The numerical results for these three cases are summarized in Table 14-1. We observe that the matched filter maximizes μ, while the output energy filter maximizes λ. The matched filter has been designed to maximize the signal-to-noise ratio at a given instant, in this case $t = 1$. Hence, c_1 is greater for the matched filter than for the remaining two filters. On the other hand, the output energy filter has been designed to maximize the energy in

TABLE 14-1. Summary of Results for the Detection of the Input Signal
Wavelet $(b_0, b_1) = (3, 1)$ Immersed in White Noise of Unit Power

Operation	Filter		Output			μ	λ
	a_0	a_1	c_0	c_1	c_2		
1. Matched	0.316	0.948	0.948	3.160	0.948	10.00	11.80
2. Output energy	0.707	0.707	2.121	2.828	0.707	8.00	13.00
3. Least-squares spiking	0.067	0.998	0.201	3.061	0.998	9.37	10.40

the entire output, namely $c_0^2 + c_1^2 + c_2^2$. Hence, this quantity is greater for
the output energy filter than for the other two filters, as we can readily
determine from Table 14-1. The least-squares spiking filter has been designed
to condense the input signal wavelet into a spike at time $t = 1$. Therefore,
c_1 is large while c_0 and c_2 are small. However, we notice that the matched
filter is attempting only to make c_1 large; on the other hand, the least-squares
spiking filter is attempting to make c_1 large while at the same time making
c_0 and c_2 small. Hence, it is reasonable to expect that c_1 will be larger for
the matched filter than for the least squares spiking filter.

It is also instructive to compare the energy density spectra of these
three filters, where the filter energy density spectrum $P(f)$ is

$$P(f) = \sum_{t=-n}^{+n} p_t \cos 2\pi ft$$

Since $n = 1$ here, we have

$$P(f) = p_0 + 2p_1 \cos 2\pi f, \qquad |f| \leq f_N$$

where f_N is the folding frequency for any particular choice of the sampling
increment. The normalized spectra are plotted in Figure 14-2. The signal
wavelet and the matched filter have identical energy density spectra. The
power density spectrum of white noise is uniform for all f, and this is the
reason why the frequency response of the output energy filter is relatively
small in the spectral band in which there is relatively little signal energy.
The least-squares spiking filter needs the high frequencies in order to shape
the input into a spike, and therefore it must respond more at these higher
frequencies than either the matched filter or the output energy filter. As the
white noise power increases, therefore, one must expect progressively poorer
performance for the least-squares spiking filter.

Thus far our numerical examples have dealt only with signals
immersed in white noise. Let us now consider a parallel development for the
autocorrelated noise situation. We again assume that we are given the 2-

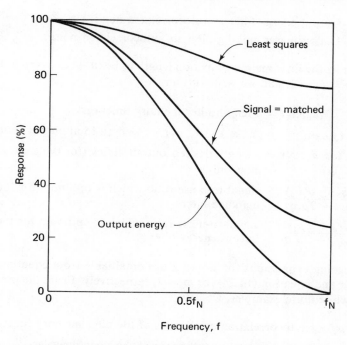

Figure 14-2. Normalized energy-density spectra for the case of white noise.

length input signal wavelet $b_t = (b_0, b_1) = (3, 1)$, but this time we assume further that we are also given a 2-length additive noise wavelet u_t,

$$u_t = (u_0, u_1) = (1, 2)$$

whose autocorrelation coefficients are

$$q_t = (q_0, q_1) = (u_0^2 + u_1^2, u_0 u_1) = (5, 2)$$

In general, we will not know u_t explicitly, but only the autocorrelation function q_t. Nevertheless, we shall assume u_t is known here in order to better illustrate the action of our filters in the presence of autocorrelated noise. Thus, the total input $x_t = b_t + u_t$ is

$$\begin{aligned} x_t = (x_0, x_1) &= (b_0, b_1) + (u_0, u_1) \\ &= (3, 1) + (1, 2) = (4, 3) \end{aligned}$$

We now wish to compute the following filters for this model:

1. Matched filter in autocorrelated noise.

2. Output energy filter in autocorrelated noise.

3. Least-squares spiking filter in autocorrelated noise.

We again fix the filter memory function length to be $(n + 1) = (m + 1) = 2$, so that $n = m = 1$, and we write further:

$$(a_0, a_1) = \text{2-length filter memory function}$$

$$(y_0, y_1, y_2) = (x_0, x_1) * (a_0, a_1) = \text{3-length total output series}$$

$$(d_0, d_1, d_2) = \text{3-length desired output series (for the least-squares filter only)}$$

$$\mu = \frac{(a_0 b_1 + a_1 b_0)^2}{p_0 q_0 + 2p_1 q_1} = \begin{array}{l}\text{signal-to-noise ratio which is optimum for the} \\ \text{matched filter}\end{array}$$

$$\lambda = \frac{p_0 r_0 + 2p_1 r_1}{p_0 q_0 + 2p_1 q_1} = \begin{array}{l}\text{signal-to-noise ratio which is optimum for the} \\ \text{output energy filter}\end{array}$$

The foregoing expressions for μ and λ are obtainable from equation (14-9) for $n = 1$ and equation (14-24) for $m = 1$, respectively. From the previously treated white noise examples, we have

$$(p_0, p_1) = \text{autocorrelation coefficients of the filter memory function}$$

$$(r_0, r_1) = \text{autocorrelation coefficients of the signal wavelet}$$

$$= (10, 3)$$

We again impose the unit energy constraint on the filter coefficients, namely that $a_0^2 + a_1^2 = 1$. Let us now go through the calculations needed in order to obtain the filters for the three cases listed above. To save space, we give only the derivation for the filter coefficients themselves, since the actual outputs y_t and the signal-to-noise ratios μ and λ are readily obtainable by application of the formulas given above. These results are then displayed in Table 14-2.

TABLE 14-2. SUMMARY OF RESULTS FOR THE DETECTION OF THE INPUT SIGNAL WAVELET $(b_0, b_1) = (3, 1)$ IMMERSED IN AUTOCORRELATED NOISE $(q_0, q_1) = (5, 2)$

	Filter		Total Output				
Operation	a_0	a_1	y_0	y_1	y_2	μ	λ
1. Matched	−0.077	0.997	−0.308	3.757	2.991	1.81	2.03
2. Output energy	0.707	−0.707	2.828	−0.707	−2.121	0.67	2.33
3. Least-squares spiking	0.000	1.000	0.000	4.000	3.000	1.80	2.00

Matched Filter in Autocorrelated Noise

From equation (14-13) we have for $n = 1$,

$$\begin{bmatrix} 5 & 2 \\ 2 & 5 \end{bmatrix}\begin{bmatrix} a_0 \\ a_1 \end{bmatrix} = \begin{bmatrix} 1 \\ 3 \end{bmatrix}$$

The normalized memory function is thus

$$(a_0, a_1) = (-0.077, 0.997)$$

Output Energy Filter in Autocorrelated Noise

Equation (14-21) yields for $(r_0, r_1) = (10, 3)$ and $(q_0, q_1) = (5, 2)$,

$$\begin{bmatrix} 10 - 5\lambda & 3 - 2\lambda \\ 3 - 2\lambda & 10 - 5\lambda \end{bmatrix}\begin{bmatrix} a_0 \\ a_1 \end{bmatrix} = \begin{bmatrix} 0 \\ 0 \end{bmatrix} \qquad (14\text{-}28)$$

This system will have a nontrivial solution if and only if the determinant of the coefficient matrix vanishes, that is, if

$$(10 - 5\lambda)^2 - (3 - 2\lambda)^2 = 0$$

The roots, or eigenvalues of this characteristic equation are

$$\lambda_1 = 1.857 \quad \text{and} \quad \lambda_2 = 2.333$$

Thus, the optimum signal-to-noise ratio λ is

$$\lambda = \lambda_{max} = 2.333$$

The memory function of the filter is given by the eigenvector associated with λ_{max}, and is obtainable from the adjoint $\mathbf{F}(\lambda_{max})$ of the coefficient matrix of equation (14-28).

$$\mathbf{F}(2.333) = \begin{bmatrix} -0.715 & +0.715 \\ +0.715 & -0.715 \end{bmatrix} = \begin{bmatrix} -0.715 \\ +0.715 \end{bmatrix}[1 \quad -1]$$

The normalized memory function is thus

$$(a_0, a_1) = (0.707, -0.707)$$

Least-Squares Spiking Filter in Autocorrelated Noise

We wish to shape the input signal $(b_0, b_1) = (3, 1)$ into a spike at $t = 1$, that is,

$$(d_0, d_1, d_2) = (0, 1, 0)$$

and this is to occur in the presence of autocorrelated noise for which $(q_0, q_1) = (5, 2)$. The normal equations for this situation are obtainable by suitably specializing equation (6-11), which here becomes

$$\begin{bmatrix} r_0 + q_0 & r_1 + q_1 \\ r_1 + q_1 & r_0 + q_0 \end{bmatrix} \begin{bmatrix} a_0 \\ a_1 \end{bmatrix} = \begin{bmatrix} b_1 \\ b_0 \end{bmatrix}$$

Since $(r_0, r_1) = (10, 3)$, we obtain

$$\begin{bmatrix} 15 & 5 \\ 5 & 15 \end{bmatrix} \begin{bmatrix} a_0 \\ a_1 \end{bmatrix} = \begin{bmatrix} 1 \\ 3 \end{bmatrix}$$

and the normalized memory function is thus $(a_0, a_1) = (0, 1)$. Hence, in this particular case the least-squares filter is merely the unit delay operator z.

The numerical results for these three filters are summarized in Table 14-2. Just as was true for the white noise model treated earlier, we again observe that the matched filter maximizes μ while the output energy filter maximizes λ. Next, we may notice the apparently disturbing fact that the energy in the total output (y_0, y_1, y_2) is actually less for the output energy filter than for the matched filter, since we have

$$y_0^2 + y_1^2 + y_2^2 = \begin{cases} 23.16 \text{ for the matched filter} \\ 13.00 \text{ for the output energy filter} \end{cases}$$

However, we recall from Figure 14-1 that

$$y_t = x_t * a_t = (b_t + u_t) * a_t = (b_t * a_t) + (u_t * a_t)$$
$$= c_t + v_t$$

where c_t is the filter response when the signal alone is the input while v_t is the filter response when the noise alone is the input. Now the design criterion for the output energy filter is *not* that $(y_0^2 + y_1^2 + y_2^2)$ be as large as possible, but that the ratio

$$\lambda = \frac{c_0^2 + c_1^2 + c_2^2}{v_0^2 + v_1^2 + v_2^2} \tag{14-29}$$

be as large as possible. This expression is a restatement of equation (14-17) for the particular example under consideration here. Thus, we have for the matched filter

$$(c_0, c_1, c_2) = (3, 1) * (-0.077, 0.997) = (-0.231, 2.914, 0.997)$$
$$(v_0, v_1, v_2) = (1, 2) * (-0.077, 0.997) = (-0.007, 0.843, 1.994)$$

while we obtain for the output energy filter

$$(c_0, c_1, c_2) = (3, 1) * (0.707, -0.707) = (2.121, -1.414, -0.707)$$
$$(v_0, v_1, v_2) = (1, 2) * (0.707, -0.707) = (0.707, 0.707, -1.414)$$

Use of equation (14-29) then yields

$$\lambda = 2.03 \text{ for the matched filter}$$

and

$$\lambda = 2.33 \text{ for the output energy filter}$$

and these values agree with the ones we obtained previously by a different route.

Finally, we show in Figure 14-3 the normalized energy density spectra of the filters tabulated in Table 14-2, together with the normalized energy density spectra of the signal $(b_0, b_1) = (3, 1)$ and of the noise wavelet $(u_0, u_1) = (1, 2)$. The noise response goes down faster than the signal response with increasing frequency, and therefore the output energy and matched

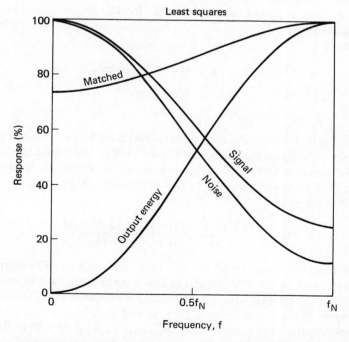

Figure 14-3. Normalized energy-density spectra for the case of autocorrelated noise.

filters become progressively more transparent at higher frequencies. This is particularly true for the output energy filter, which is quite opaque at the lower frequencies since the noise level there is too high. The least-squares filter attempts to shape the signal into a spike, which has a white light spectrum; in the present case we have seen that the memory function of this filter is given by the unit delay operator z, which also has a white light spectrum. In other words, the best least-squares approximation to the spike at $t = 1$ is here obtainable by simply delaying the total input signal by one time unit. The matched filter responds more at the lower frequencies than the output energy filter because it tries to follow or "match" the signal spectrum as much as possible, while at the same time responding as little as possible to the noise. We notice in particular that the spectrum of the matched filter in the presence of autocorrelated noise is no longer equal to the signal spectrum itself.

The Detection of a Signal Immersed in White Noise

The simple numerical examples presented thus far illustrate how one calculates matched and output energy filters in a given noise situation. Let us now consider a more involved signal detection model. We assume that a 15-length signal wavelet $b_t = (b_0, b_1, \ldots, b_{14})$ is added to a 100-length white noise sample

$$u_t = (u_0, u_1, \ldots, u_{99}) \text{ at } t = 20, \text{ so that we have}$$

$$
\begin{aligned}
x_t &= u_t + k b_{t-20} \\
&= (u_0, u_1, \ldots, u_{20} + k b_0, u_{21} + k b_1, \ldots, u_{34} + k b_{14}, u_{35}, \ldots, u_{99})
\end{aligned}
$$

Here x_t is the total input while k is a real scale factor to be described shortly. The shape of the signal b_t is shown in Figure 14-4(a). We would like to study the performance of our various filters at different amplitude levels of the signal b_t. One convenient way to do this is to *define* an *input* signal-to-noise ratio, S/N, such that

$$S/N = \frac{(kb_0)^2 + (kb_1)^2 + \cdots + (kb_{14})^2}{u_{20}^2 + u_{21}^2 + \cdots + u_{34}^2}$$

We then assume various values for the ratio S/N, and can therefore easily calculate the scale factor k by which the signal coefficients must be multiplied in order that the relation above be satisfied. We observe that this definition of the input S/N ratio is quite arbitrary, and that in particular we must know explicitly the noise sample values $(u_{20}, u_{21}, \ldots, u_{34})$, to which the signal coefficients have been added. For practical problems this definition of an input S/N ratio is unrealistic, since we will rarely know the noise sample

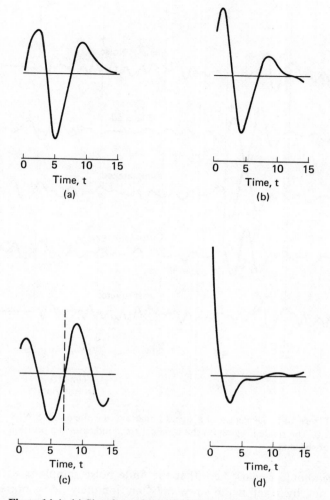

Figure 14-4. (a) Signal wavelet b_t; (b) Memory function of the mini-matched filter; (c) Memory function of the output energy filter; (d) Memory function of the least-squares filter.

values u_t explicitly, but it is quite adequate to construct our present model. In this manner five synthetic traces x_t were constructed for the S/N values,

$$2, 1, \quad 0.5, \quad 0.25, \quad \text{and} \quad 0.125.$$

These traces are shown as the topmost displays of Figures 14-5 to 14-9, and are labeled "Signal + Noise." The shaded regions on these traces indicate the position at which the signal b_t has been added to the noise u_t.

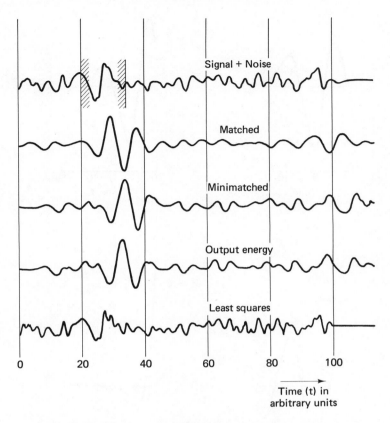

Figure 14-5. Detection of a signal immersed in white noise, $S/N =$ 2.0. The shaded region on the topmost trace indicates the position of the signal.

One should remark here that the finite noise sample u_t used in our model is not by itself white since only an infinitely long noise sample can be truly white. Nevertheless, we shall call our 100-length noise series "white" in the loose sense described above. The filters to be treated in this section are all computed under the assumption that the noise is strictly white. This is a minimum assumption for the reason that we do not need to measure the actual noise autocorrelation function q_t explicitly. If we did so, our filters would in general perform better than those to be discussed below. However, our purpose here will be to show that the white noise filters do well even if the *actual* noise on which they operate is not truly white.

Since the signal b_t is known and since the noise u_t is assumed to be white, the matched filter response to the inputs x_t is at once obtainable by

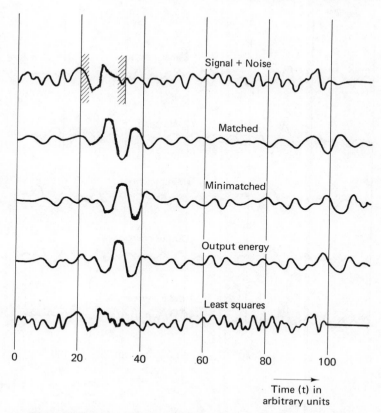

Figure 14-6. Detection of a signal immersed in white noise, $S/N = 1.0$. The shaded region on the topmost trace indicates the position of the signal.

crosscorrelating the inputs x_t with the signal wavelet b_t. The resulting outputs are shown in Figures 14-5 to 14-9 and are labeled "Matched."

The output energy filter is calculated with the aid of equation (14-21), where we assume that the noise is white and of unit power, and where we set $m = 14$. The memory function of this filter is shown in Figure 14-4(c). We shall say more about this memory function shortly. Its responses to the inputs x_t are displayed in Figures 14-5 to 14-9 and are labeled "Output Energy." For comparative purposes, we have also computed the 15-length least-squares filter under the following conditions:

$$\text{Input} = (b_0, b_1, \ldots, b_{14}) + (\text{white noise of unit power})$$
$$\text{Desired output} = (b_0, b_1, \ldots, b_{14}) \text{ at zero delay}$$

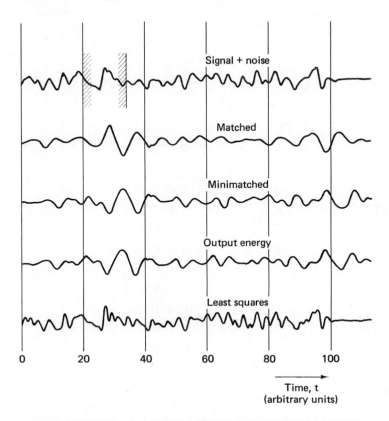

Figure 14-7. Detection of a signal immersed in white noise, $S/N = 0.5$. The shaded region on the topmost trace indicates the position of the signal.

This filter can be found by suitably specializing equation (14-11) of Chapter 6 and its memory function is depicted in Figure 14-4(d). The corresponding responses to the inputs x_t are labeled "Least Squares" in Figures 14-5 to 14-9.

Now suppose that we do *not* know the signal shape b_t explicitly, only its autocorrelation function r_t. Under these circumstances we can, of course, compute the output energy filter, which we already found above, but we can no longer find the matched filter in this case. However, given the autocorrelation coefficients $(r_0, r_1, \ldots, r_n)$, we can find the minimum-delay wavelet $b_{0,t} = (b_{0,0}, b_{0,1}, \ldots, b_{0,n})$ having this autocorrelation function (see Chapter 7). Now there exists in general a family of up to 2^n wavelets having the same autocorrelation function r_t, one of which is the minimum-delay wavelet $b_{0,t}$. Consider next the crosscorrelation function $g_t = (g_0, g_1, \ldots, g_n, \ldots, g_{2n-1}, g_{2n})$ of any two wavelets of this family.

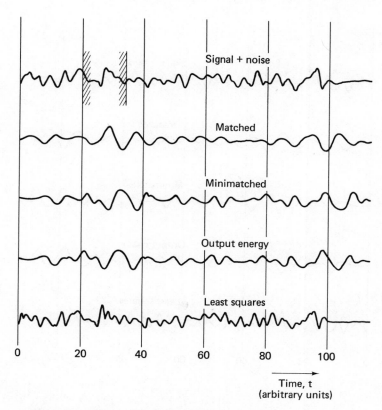

Figure 14-8. Detection of a signal immersed in white noise, $S/N = 0.25$. The shaded region on the topmost trace indicates the position of the signal.

We show in Appendix 14-1 that the energy in this crosscorrelation function, namely

$$\epsilon = (g_0^2 + g_1^2 + \cdots + g_n^2 + \cdots + g_{2n-1}^2 + g_{2n}^2) \qquad (14\text{-}30)$$

is identical no matter which two members of the family are chosen to be crosscorrelated with each other. In particular, the value of ϵ obtained for the crosscorrelation $b_t \otimes b_t$, where the symbol $\otimes$ denotes crosscorrelation, equals the value of ϵ for the crosscorrelation $b_t \otimes b_{0,t}$. Thus, whereas it is true that only the matched filter b_t guarantees that the central value g_n^2 be as large as possible, the crosscorrelation of any other member of the family with the signal wavelet b_t will lead to one and the same value of ϵ. We may therefore expect that an operation of the latter type will still reveal the pres-

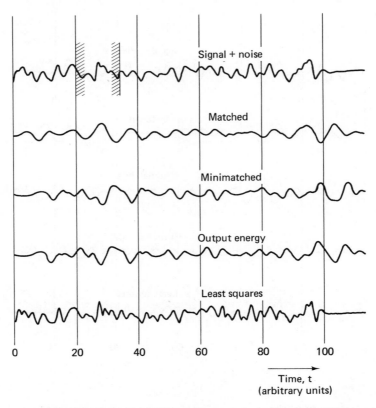

Figure 14-9. Detection of a signal immersed in white noise, $S/N = 0.125$. The shaded region on the topmost trace indicates the position of the signal.

ence of the signal b_t by producing a burst of energy ϵ at the point where the signal is immersed in noise. Although we may thus choose any wavelet of the family and use it in this manner, the choice of the minimum-delay wavelet $b_{0,t}$ is particularly attractive for at least two reasons: first, because it is easy to compute from the knowledge of r_t, and second, because its energy is concentrated at its leading edge so that in general one can expect the shape of $g_t = b_t \otimes b_{0,t}$ to be less "leggy" than for other wavelet choices from the family. The use of the minimum-delay wavelet $b_{0,t}$ as a correlation detector thus suggests itself. We shall call this operator the "minimatched" filter. Its memory function is shown in Figure 14-4(b), and its responses to the inputs x_t are displayed in Figures 14-5 to 14-9, where they are labeled "Minimatched." Now the output energy filter is that linear filter which maximizes the energy in the output subject to the constraint that this filter respond as little as possible to the ambient noise. Therefore, the energy output of the minimatched filter is necessarily smaller than the energy output of the output energy filter; in many actual cases, nevertheless, this smaller output energy is still adequate to permit detection of the signal. The minimatched filter is in general easier to compute numerically than the output energy filter, since the latter always entails the solution of an eigenvalue problem. Thus, the minimatched filter serves as a possible alternative for the output energy filter. Of course, if the signal shape b_t is known explicitly, one would in general prefer to use the matched filter in the detection system.

Let us next compare the performance of the various filters we have used in our present model. First, we see that the least-squares filter does not do too well except possibly for the case $S/N = 2$. This is because the least-squares filter does not in general produce large energy bursts when it encounters signals immersed in large-amplitude noise, so that in this particular sense it is unattractive as a signal detection device. The matched, minimatched, and output energy filters perform well as signal detectors for the cases $S/N = 2, 1, 0.5$, and 0.25; but, by the time $S/N = 0.125$, no decision about the presence or absence of the signal appears to be possible. For this particular model the performance of the matched, minimatched, and output energy filters is roughly equivalent, although there are differences in the detailed shape of the energy bursts at the point where the signal has been detected. If the signal shape b_t is known, the matched filter is an attractive signal detector; if not, either the output energy filter or the minimatched filter can be used for this purpose.

The output signal-to-noise ratios μ and λ for the various filtering operations performed above will not be discussed here since these matters have received adequate attention earlier in this chapter. On the other hand, the normalized energy density spectra of the filters used in this section are plotted in Figure 14-10. We note in particular that the signal, the matched filter, and the minimatched filter all have identical energy density spectra. This is, of course, a consequence of the fact that all three have one and the same autocorrelation function r_t. The differences in performance between the matched and minimatched filters are therefore entirely due to the difference between their respective phase-lag characteristics. Thus, the matched filter has a phase-lag characteristic which is the negative of that of the signal wavelet, while the minimatched filter has a minimum phase-lag characteristic.

All the roots of the z transform $A(z)$ of the output energy filter lie on the unit circle $|z| = 1$ of the complex z plane. The positions of the roots for the output energy filter used in the present section are shown in Figure 14-11. We notice that this filter has an odd number, namely one, of $+1$ roots on the unit circle; hence, its memory function must be antisymmetrical. A look at Figure 14-4(c) confirms this deduction, where we have indicated the axis of antisymmetry by a dashed vertical line. Finally, we may observe

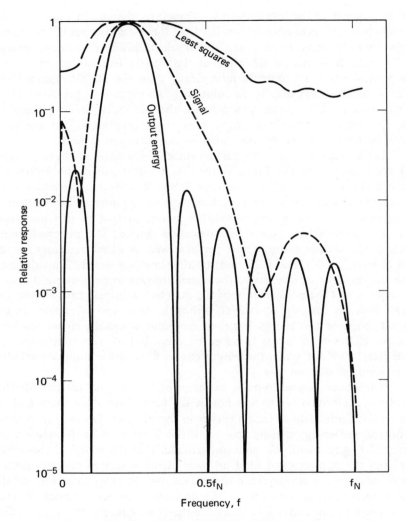

Figure 14-10. Normalized energy-density spectra of the signal, of the output energy filter, and of the least-squares filter whose memory functions are shown in Figure 14-4. The energy-density spectrum of the signal is equivalent to the energy-density spectrum of the matched and the minimatched filters.

that the energy density spectrum of the output energy filter is highly tuned; that is, its spectral energy is separated into a number of sharp bands. This is also a direct consequence of the fact that all roots of $A(z)$ lie on the unit circle, since the spectrum vanishes at all points on the unit circle that coincide with the position of a particular root of $A(z)$.

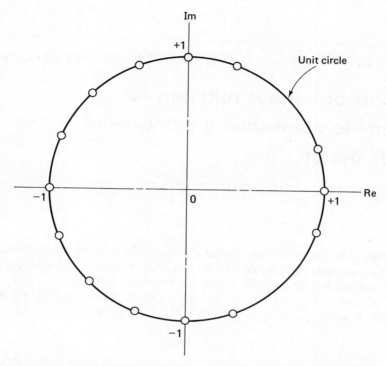

Figure 14-11. Location of the roots of the z transform $A(z)$ of the output energy filter of Figure 14-4(c).

Concluding Remarks

We have described two particular digital filters which are designed in such a way that the signal-to-noise ratio of the output is enhanced over what it was in the input. The determination of the appropriate filter coefficients depends on an appropriate definition of the output signal-to-noise ratio. In the case of the matched filter, this quantity is given by the ratio of the instantaneous signal power to the instantaneous noise power; in the case of the output energy filter, it is given by the ratio of the energy of the filtered signal to the average power of the filtered noise, where the averaging is performed over the time interval where the signal occurs.

proofs of various relations for signal-to-noise-ratio enhancement filter design

We shall now derive various relations given earlier without proof. First, we shall find an expression for the energy or power, as the case may be, present in the convolution of an input with a filter memory function.

Consider the result of convolving an input β_t with a 2-length filter memory function (α_0, α_1),

$$\gamma_t = \alpha_0\beta_t + \alpha_1\beta_{t-1},$$

where γ_t is the output. Let us square both sides of the relation above. This yields

$$\gamma_t^2 = (\alpha_0\beta_t + \alpha_1\beta_{t-1})^2 = \alpha_0^2\beta_t^2 + 2\alpha_0\alpha_1\beta_t\beta_{t-1} + \alpha_1^2\beta_{t-1}^2$$

Taking the expected value of both sides, we obtain

$$E\{\gamma_t^2\} = \alpha_0^2 E\{\beta_t^2\} + 2\alpha_0\alpha_1 E\{\beta_t\beta_{t-1}\} + \alpha_1^2 E\{\beta_{t-1}^2\}$$

where E is the expected value symbol. If we assume that the input β_t is a stationary time series, we may write

$$E\{\gamma_t^2\} = \varphi_{\gamma\gamma}(0) = \text{zeroth autocorrelation coefficient of } \gamma_t$$

$$E\{\beta_t^2\} = E\{\beta_{t-1}^2\} = \varphi_{\beta\beta}(0) = \text{zeroth autocorrelation coefficient of } \beta_t$$

$$E\{\beta_t\beta_{t-1}\} = \varphi_{\beta\beta}(-1) = \varphi_{\beta\beta}(1) = \text{first autocorrelation coefficient of } \beta_t$$

Therefore,

$$\varphi_{\gamma\gamma}(0) = \alpha_0^2\varphi_{\beta\beta}(0) + 2\alpha_0\alpha_1\varphi_{\beta\beta}(1) + \alpha_1^2\varphi_{\beta\beta}(0) \qquad (14\text{-}31)$$

This relation can also be written in the convenient matrix form

$$\varphi_{\gamma\gamma}(0) = [\alpha_0 \quad \alpha_1]\begin{bmatrix} \varphi_{\beta\beta}(0) & \varphi_{\beta\beta}(1) \\ \varphi_{\beta\beta}(1) & \varphi_{\beta\beta}(0) \end{bmatrix}\begin{bmatrix} \alpha_0 \\ \alpha_1 \end{bmatrix} \tag{14-32}$$

But we have also that

$$\varphi_{\alpha\alpha}(0) = \alpha_0^2 + \alpha_1^2 = \text{zeroth autocorrelation coefficient of } \alpha_t$$
$$\varphi_{\alpha\alpha}(-1) = \varphi_{\alpha\alpha}(1) = \alpha_0\alpha_1 = \text{first autocorrelation coefficient of } \alpha_t$$

and hence equation (14-31) can be rewritten in the alternative form

$$\varphi_{\gamma\gamma}(0) = \alpha_0\alpha_1\varphi_{\beta\beta}(-1) + [\alpha_0^2 + \alpha_1^2]\varphi_{\beta\beta}(0) + \alpha_0\alpha_1\varphi_{\beta\beta}(1)$$
$$= \varphi_{\alpha\alpha}(-1)\varphi_{\beta\beta}(-1) + \varphi_{\alpha\alpha}(0)\varphi_{\beta\beta}(0) + \varphi_{\alpha\alpha}(1)\varphi_{\beta\beta}(1)$$

or

$$\varphi_{\gamma\gamma}(0) = \sum_{t=-1}^{+1} \varphi_{\alpha\alpha}(t)\varphi_{\beta\beta}(t) \tag{14-33}$$

But $\varphi_{\gamma\gamma}(0)$ is equal to the energy in the output γ_t since

$$\varphi_{\gamma\gamma}(0) = \cdots + \gamma_{t-1}^2 + \gamma_t^2 + \gamma_{t+1}^2 + \cdots$$

and hence relations (14-32) and (14-33) provide alternative expressions for the energy in the output γ_t. By inductive reasoning it then follows that if α_t is an $(l+1)$ length memory function, the matrix relation (14-32) can be written in the more general form

$$\varphi_{\gamma\gamma}(0) = [\alpha_0 \quad \alpha_1 \quad \cdots \quad \alpha_l]\begin{bmatrix} \varphi_{\beta\beta}(0) & \cdots & \varphi_{\beta\beta}(l) \\ \cdot & & \cdot \\ \cdot & & \cdot \\ \cdot & & \cdot \\ \varphi_{\beta\beta}(l) & \cdots & \varphi_{\beta\beta}(0) \end{bmatrix}\begin{bmatrix} \alpha_0 \\ \alpha_1 \\ \cdot \\ \cdot \\ \alpha_l \end{bmatrix}$$

or simply

$$\varphi_{\gamma\gamma}(0) = E\{\gamma_t^2\} = \alpha\varphi\alpha' \tag{14-34}$$

Here α is a $(1) \times (l+1)$ row vector, φ is an $(l+1) \times (l+1)$ square matrix of the autocorrelation coefficients of the input β_t, and the prime symbol (') denotes the matrix transpose. We notice the important fact that the length of the vector α and the order of the matrix φ are both equal to the length of the filter memory function, namely $(l+1)$, and are hence independent of the actual length of the input β_t.

In a similar manner, we may generalize equation (14-33) in the form

$$\varphi_{\gamma\gamma}(0) = E\{\gamma_t^2\} = \sum_{t=-l}^{+l} \varphi_{\alpha\alpha}(t)\varphi_{\beta\beta}(t) \qquad (14\text{-}35)$$

We are now ready to prove the appropriate relations in Chapter 14.

1. *Proof of equation (14-4).* We wish to prove

$$E\{v_t^2\} = q_0(a_0^2 + a_1^2 + \cdots + a_n^2) \qquad (14\text{-}4)$$

If we set

$$v_t = \gamma_t$$
$$a_t = \alpha_t \quad \text{and} \quad p_t = \varphi_{\alpha\alpha}(t)$$
$$u_t = \beta_t \quad \text{and} \quad q_t = \varphi_{\beta\beta}(t)$$

and let $l = n$, equation (14-35) yields

$$E\{v_t^2\} = \sum_{t=-n}^{+n} p_t q_t$$

But since u_t is white noise, we have that $q_t = 0$ for $t \neq 0$, and thus

$$E\{v_t^2\} = q_0 p_0 = q_0(a_0^2 + a_1^2 + \cdots + a_n^2) \qquad \text{Q.E.D.}$$

2. *Proof of equation (14-9).* We wish to prove

$$\mu = \frac{c_n^2}{E\{v_n^2\}} = \frac{(\mathbf{ab'})^2}{\mathbf{aqa'}} \qquad (14\text{-}9)$$

If we set

$$v_n = \gamma_n$$
$$a_n = \alpha_n$$
$$u_n = \beta_n \text{ and } \mathbf{q} = \boldsymbol{\varphi}$$

and let $l = n$, we have by equation (14-34) that

$$E\{v_n^2\} = \mathbf{aqa'} \qquad (14\text{-}36)$$

On the other hand, the output value c_n of the convolution $c_t = a_t * b_t$ is

$$c_n = (a_0 b_n + a_1 b_{n-1} + \cdots + a_n b_0)$$

But c_n is also expressible as the vector inner product

$$c_n = [a_0 \quad a_1 \quad \cdots \quad a_n] \begin{bmatrix} b_n \\ b_{n-1} \\ \vdots \\ b_0 \end{bmatrix}$$

which can be written in the abbreviated form

$$c_n = \mathbf{ab'} \qquad (14\text{-}37)$$

where $\mathbf{a}$ and $\mathbf{b}$ are row vectors. Use of equations (14-36) and (14-37) then allows us to write

$$\mu = \frac{c_n^2}{E\{v_n^2\}} = \frac{(\mathbf{ab'})^2}{\mathbf{aqa'}} \qquad \text{Q.E.D.}$$

3. *Proof of equation* (*14-18*). We wish to prove

$$\frac{\sum_{t=0}^{m+n} c_t^2}{E\{v_t^2\}} = \frac{\mathbf{ara'}}{\mathbf{aqa'}} \qquad (14\text{-}18)$$

By equation (14-36) we already have that

$$E\{v_t^2\} = \mathbf{aqa'}$$

Let us set

$$c_t = (c_0, c_1, \ldots, c_{m+n}) = \gamma_t$$
$$a_t = \alpha_t$$
$$b_t = \beta_t \text{ and } r_t = \varphi_{\beta\beta}(t)$$

where we recall that c_t is obtained by convolving the $(n + 1)$-length input signal wavelet $(b_0, b_1, \ldots, b_n)$ with the $(m + 1)$-length filter memory function $(a_0, a_1, \ldots, a_m)$. If we now let $l = m$ in equation (14-34), we obtain

$$\sum_{t=0}^{m+n} c_t^2 = \mathbf{ara'}$$

and therefore

$$\lambda = \frac{\mathbf{ara'}}{\mathbf{aqa'}} \qquad \text{Q.E.D.}$$

4. *Proof of equation* (*14-24*). We wish to prove

$$\lambda = \frac{\sum\limits_{t=0}^{m+n} c_t^2}{E\{v_t^2\}} = \frac{\sum\limits_{t=-m}^{+m} p_t r_t}{\sum\limits_{t=-m}^{+m} p_t q_t} \qquad (14\text{-}24)$$

If we set

$$p_t = \varphi_{\alpha\alpha}(t)$$
$$r_t = \varphi_{\beta\beta}(t)$$

then reference to the proof for equation (14-18) and to equations (14-34) and (14-35) allows us to write

$$\sum_{t=0}^{m+n} c_t^2 = \sum_{t=-m}^{+m} p_t r_t$$

Similarly, if we set

$$p_t = \varphi_{\alpha\alpha}(t)$$
$$q_t = \varphi_{\beta\beta}(t)$$

then

$$E\{v_t^2\} = \sum_{t=-m}^{+m} p_t q_t$$

and thus equation (14-24) follows directly.   Q.E.D.

5. *Proof of equation* (*14-30*). From equation (14-35) we know that the output energy $\varphi_{\gamma\gamma}(0)$ of the convolution

$$\gamma_t = (\alpha_0, \alpha_1, \ldots, \alpha_n) * (\beta_0, \beta_1, \ldots, \beta_n)$$

is

$$\varphi_{\gamma\gamma}(0) = \sum_{t=-n}^{+n} \varphi_{\alpha\alpha}(t)\varphi_{\beta\beta}(t) \qquad (14\text{-}35)$$

Suppose that we wish to evaluate the energy ϵ in the crosscorrelation function†

$$g_t = (\alpha_0, \alpha_1, \ldots, \alpha_n) \otimes (\beta_0, \beta_1, \ldots, \beta_n)$$

But we know that this operation can be written

$$g_t = (\alpha_0, \alpha_1, \ldots, \alpha_n) * (\beta_n, \beta_{n-1}, \ldots, \beta_0)$$

†The symbol $\otimes$ denotes crosscorrelation.

where $(\beta_n, \beta_{n-1}, \ldots, \beta_0)$ is the time reverse of $(\beta_0, \beta_1, \ldots, \beta_n)$. However, since the autocorrelation function of any given wavelet is identical to the autocorrelation function of its time reverse, we see that

$$\epsilon = \sum_{t=-n}^{+n} g_t^2 = \varphi_{\gamma\gamma}(0) = \sum_{t=-n}^{+n} \varphi_{\alpha\alpha}(t)\varphi_{\beta\beta}(t) \qquad (14\text{-}38)$$

where ϵ = energy in the crosscorrelation function g_t. In other words, the energy in the crosscorrelation function $g_t = \alpha_t \otimes \beta_t$ is equal to the energy in the convolution $\gamma_t = \alpha_t * \beta_t$. Now let us assume that both the wavelets α_t and β_t belong to the same family of at most 2^n wavelets having the auto-correlation function $r_t = (r_{-n}, r_{-n+1}, \ldots, r_0, \ldots, r_{n-1}, r_n)$. Then we have

$$\varphi_{\alpha\alpha}(t) = \varphi_{\beta\beta}(t) = r_t$$

and equation (14-38) becomes

$$\epsilon = \sum_{t=-n}^{+n} g_t^2 = \sum_{t=-n}^{+n} r_t^2$$

But since $\sum_{t=-n}^{+n} r_t^2$ = constant for any given wavelet family, we see that the energy ϵ in the crosscorrelation function $g_t = \alpha_t \otimes \beta_t$ reduces to the identical value $\sum_{t=-n}^{+n} r_t^2$ independently of any particular choice of the member wavelets α_t and β_t. This completes the proof of equation (14-30).

migration of seismic data

Summary

The information appearing on a single seismic trace does not allow us to determine the time-spatial position of the reflecting point. Each event will show up as if it occurred directly beneath the recording point. If we have a seismic record section at our disposal, however, it is possible to make use of the apparent dips on this section in order to establish the true locations of the subsurface reflections. To do this, we require knowledge of the subsurface velocity distributions. The process by which the apparent dips are converted to time dips is known as migration. While most current methods assume that no reflections come from outside the plane of the seismic recording profile, the migration process can in principle be extended to three dimensions. In this chapter we describe the classic two-dimensional migration method of Hagedoorn (1954), which is based on the concept of the surface of maximum convexity. In recent years, this approach has been largely replaced by methods making explicit use of the acoustic (or scalar) wave equation. These include Kirchhoff migration (French, 1975), finite-difference migration (Claerbout, 1976), and frequency-wave number migration (Stolt,

1978). The latter method combines computational simplicity with good performance even for steep dip angles, and we present the approach in some detail.

Introduction

In many geographical areas, the subsurface consists of horizontally layered sedimentary rocks. In such areas, the seismic waves propagate nearly vertically to the reflectors and back again to the detectors. The ensemble of seismic traces recorded at the surface make up the record section. The time points appearing on the record section represent the two-way travel time, that is, the time down to the reflector plus the (same) time back up to the surface. The one-way travel time is obtained by dividing the two-way travel time by 2.

Let us consider the case of a medium in which the velocity is constant. We may then convert record times into equivalent depths by merely multiplying the one-way times by the constant velocity. These depths are the true depths only if the ray paths are vertical. The surface defined by a given reflection in this way is called the *record surface*; the actual interface that produced the reflection is called the *reflector surface*. These two surfaces coincide only if they are flat and horizontal.

Let us now consider the case when the reflector surface is flat but sloping (see Figure 15-1). Since we are concerned only with ray paths that are normally incident on the reflector surface, the reflector surface is tangent to the incident wavefront. The *wavefront* can be drawn at each source point corresponding to half the measured reflection time. We assume that the

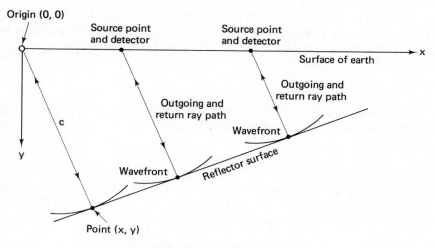

Figure 15-1. Sloping reflector surface.

detector is at the same point as the source point. The reflector surface is the envelope of these wavefronts. Because of the constant velocity assumption, the wavefronts are arcs of circles.

The counterpart of the wavefront is the *maximum convexity front*. The maximum convexity front is defined as the record surface given by the reflections originating from an idealized point reflector (i.e., a point source of diffracted energy). The record surface is the envelope of the maximum convexity fronts corresponding to each of the points on the reflector surface. The record surface is tangent to the maximum convexity front.

Let us consider the x-y plane for nonnegative values of y. We let y represent depth into the ground, so y = 0 represents the surface of the earth. First, let us consider the wavefront that is due to a source at the origin and that has traveled a distance c (see Figure 15-1). The equation for this wavefront is given by the semicircle

$$x^2 + y^2 = c^2 \qquad (y \geq 0)$$

The distance c is the distance from the origin (0, 0) to a point (x, y) on the semicircle. The wavefront slope dy/dx is found by differentiating this equation. We obtain

$$2x + 2y \frac{dy}{dx} = 0$$

so the wavefront slope is

$$\frac{dy}{dx} = -\frac{x}{y}$$

Second, let us consider the maximum convexity front that is due to a point reflector (a, b) where b > 0 [see Figure 15-2(b)]. Let the source be at (x, 0). That the distance from the source to the point reflector is

$$c = \sqrt{(a - x)^2 + b^2}$$

This distance is plotted directly under the source point; that is, the record points corresponding to the point reflector (a, b) are

$$(x, y) = (x, c) = (x, \sqrt{(a - x)^2 + b^2})$$

We recall that the maximum convexity front is defined as the record surface given by reflections from the point reflector for all source points (x, 0). Hence, the maximum convexity front is the locus given by

$$y = \sqrt{(a - x)^2 + b^2}$$

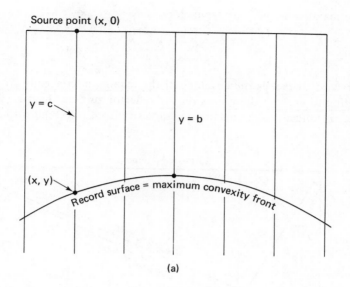

Source point (x, 0)

y = c

y = b

(x, y)

Record surface = maximum convexity front

(a)

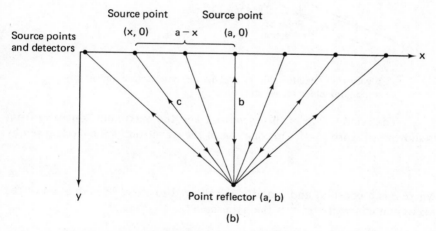

Source points and detectors

Source point (x, 0)

Source point (a, 0)

a − x

x

c

b

y

Point reflector (a, b)

(b)

Figure 15-2. Record surface and its construction: (a) record surface; (b) idealized point reflector that produces record surface.

which is the semihyperbola

$$y^2 - (a - x)^2 = b^2 \qquad (y \geq 0)$$

plotted in Figure 15-2(a). The maximum-convexity front slope dy/dx is found by differentiating this equation. We obtain

$$2y \frac{dy}{dx} + 2(a - x) = 0$$

so the maximum-convexity front slope is

$$\frac{dy}{dx} = \frac{x - a}{y}$$

The record surface is the envelope of the maximum convexity fronts corresponding to each of the points on the reflector surface. That is, the record surface is tangent to the maximum convexity front (see Figure 15-3).

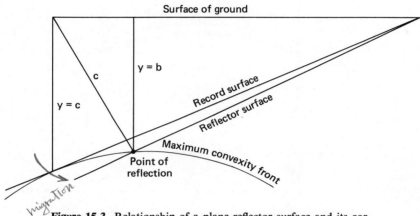

Figure 15-3. Relationship of a plane reflector surface and its corresponding plane record surface.

The relationship of the wavefront and the maximum convexity front is shown in Figure 15-4. The equation of the wavefront is the circle given by

$$x^2 + y^2 = c^2$$

where c is a constant and x and y are allowed to vary. The equation of the maximum convexity front is the hyperbola

$$y^2 - x^2 = b^2$$

where b is a constant and x and y are allowed to vary.

Migration is the process of constructing the reflector surface from the record surface. The basic mathematical properties of migration were developed by Hagedoorn (1954). The position of the reflection point corresponding to the point A on the record surface is found by fitting a maximum convexity front to the record surface in such a way that the maximum convexity front is tangent to (i.e., has the same slope as) the record front at A. The reflection point must then be situated at the position of minimum depth of the maxi-

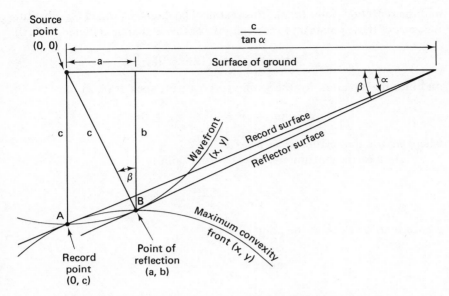

Figure 15-4. Relationship of the wavefront and the maximum convexity front.

mum convexity front (i.e., at the apex of the hyperbola). Thus, the reflection point corresponding to record point A is point B. The slope of the reflector surface at B is given by the slope of the wavefront through B.

With these preliminaries, let us now consider the case of a source at the origin $(0, 0)$ which gives rise to a wavefront that reaches the *reflection point* (a, b) (see Figure 15-4). The reflection point is labeled B. Thus, the wavefront distance c is equal to the distance from the origin to the reflection point B, that is,

$$c = \sqrt{a^2 + b^2}$$

The event due to this reflection point for the given source point appears on the seismic record section as the *record point* $(0, c)$. The record point is labeled A. Given the seismic record section (which is the recorded data) we can see the event A and measure its coordinates $(0, c)$ and its slope $\tan \alpha$. The problem is to find the reflection point B: specifically, the coordinates (a, b), and the slope $\tan \beta$. The process of solving this problem is called *migration.* That is, migration is the process of constructing the reflector surface B, namely the point (a, b), and its slope $\tan \beta$, from the record surface A, namely the point $(0, c)$ and its slope $\tan \alpha$.

For the given problem, the wavefront semicircle and the maximum-convexity front semihyperbola intersect at two points: the record point $(0, c)$

and the reflection point (a, b). This fact may be directly verified by substituting each of these points into the equation for the semicircle centered at $(0, 0)$,

$$x^2 + y^2 = c^2 \qquad (y \geq 0)$$

and into the equation for the semihyperbola with apex at (a, b),

$$y^2 - (a - x)^2 = b^2 \qquad (y \geq 0)$$

where we use the relation $c^2 = a^2 + b^2$

Because the maximum-convexity front slope is

$$\frac{dy}{dx} = \frac{x - a}{y}$$

we see that the slope at the point $(0, c)$ is

$$\tan \alpha = \frac{dy}{dx} = -\frac{a}{c}$$

We recall that we can measure c and $\tan \alpha$ from the given data, which is the seismic record section. Hence, we can compute a as

$$a = -c \tan \alpha$$

and b as

$$b = \sqrt{c^2 - a^2} = c\sqrt{1 - \tan^2 \alpha}$$

Thus, we now know the reflection point (a, b). Because the wavefront slope is

$$\frac{dy}{dx} = -\frac{x}{y}$$

we see the slope at the point (a, b) is

$$\tan \beta = \frac{dy}{dx} = -\frac{a}{b}$$

Thus we have found the reflection point (a, b) and its slope $\tan \beta$. The process of transforming the record point $(0, c)$ and its slope $\tan \alpha$ on the record section to the reflection point (a, b) and its slope $\tan \beta$ on a new section is called migration of the data. The new section thus formed is called the *migrated section*. In summary, migration is the process of constructing the reflector section (i.e., migrated section) from the record section (i.e., the given seismic traces converted from time to depth). In the case of a medium with constant velocity, the process of migration can be carried out by simple trigonometric relationships. However, in more complex situations a working model is essential in order to provide the basis for the mathematical manipulations.

Let us now look at a geometric method of migration. Figure 15-5(a) shows a single segment S on a record section (with times converted to depth). The appropriate curve of maximum convexity is the curve tangent to segment S at point A, as shown in Figure 15-5(b). The true position of A in space will be at B, which is located on the apex of this curve. The dip of the reflector B is obtained from the wavefront curve having a central axis through A and passing through B as shown in Figure 15-5(c). The migrated segment S' is drawn through B so as to be tangent to the wavefront curve at B.

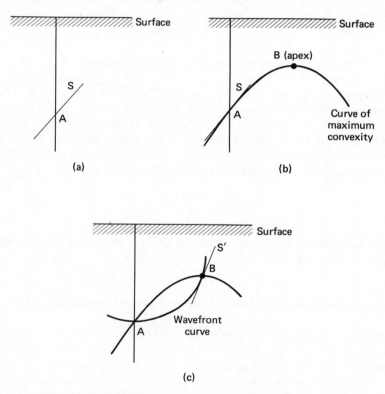

Figure 15-5. Geometric migration method.

The earliest seismic migrations were carried out graphically. For graphical migration one needs two charts, a chart of maximum-convexity curves and a chart of wavefronts, each chart for closely spaced vertical reflection times.

Let us now look at a model of the subsurface structure (Figure 15-6).

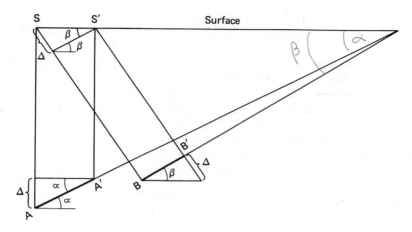

Figure 15-6. Model of subsurface structure.

The model consists of a reflector segment $\overline{BB'}$ with slope $\tan\beta$. The corresponding record segment is $\overline{AA'}$ with slope $\tan\alpha$. The projection of $\overline{BB'}$ on the earth's surface is $\overline{SS'}$. Line $\overline{SB}$ exceeds line $\overline{S'B'}$ by an amount Δ, where $\Delta = \overline{SS'}\sin\beta$. The length of $\overline{S'A'}$ equals the length of $\overline{S'B'}$, and the length of $\overline{SA}$ equals the length of $\overline{SB}$. Thus, line SA also exceeds line $\overline{S'A'}$ by the amount Δ. The slope $\tan\alpha$ is therefore equal to

$$\tan\alpha = \frac{\Delta}{\overline{SS'}} = \sin\beta$$

From the geometry of Figure 15-6 we also see that the lengths of the two segments are related by $\overline{AA'}\sin\alpha = \Delta = \overline{BB'}\tan\beta$.

Maximum-Convexity Migration and Wavefront Migration

Computer programs for migration can make use of the same principles used for graphical migration. Let us now look at two methods of computer migration, maximum-convexity migration and wavefront migration.

We suppose that the source and the detector are at the same point for each seismic trace (i.e., the common source and receiver case) and that these points are equally spaced along the x axis (i.e., along a line on the surface of the ground). If we convert time to depth by a suitable velocity function, the ensemble of traces makes up a record section. We recall that the maximum-convexity front is defined as the locus on the record section given by the reflections originating from an (idealized) point reflector for all possible surface source points. In the case of a constant velocity, the maximum-convexity front is a semihyperbola. The reflection point must be situated at the position of minimum depth of the maximum-convexity front (i.e., at the apex of the hyperbola). All reflected energy due to that reflection point appears on the hyperbola and only on the hyperbola. Thus, to reconstruct the energy due to a given reflection point, we must perform the following summation. We take the maximum-convexity front and see where it intersects each trace. We take the value of each trace at the point of intersection, and sum all these values together. All the energy due to the given reflection point adds in phase and so is preserved, whereas extraneous energy adds out of phase and is destroyed. The result of this summation is the energy due to the point reflector, and it is plotted at the apex of the maximum-convexity front. That is, we sum all the values of the traces that fall on a maximum-convexity front and plot the result at its apex. The result of doing such summations for all possible apexes is the migrated section. This method of migration is called *maximum-convexity migration*; it is direct and straightforward. The quality of the data must be good; otherwise, the migrated section, especially at depth, may show a "wormy" appearance.

Maximum-convexity migration is sketched in Figure 15-7. This figure shows how an output trace is generated from an ensemble of input traces. The input traces make up the record section, whereas the output traces make up the reflector section. In Figure 15-7, each amplitude value in the output is obtained by summing the input amplitudes along the hyperbola shown. This hyperbola is the diffraction curve (i.e., the maximum-convexity front) for a point reflector (i.e., diffraction source) in the subsurface at the output

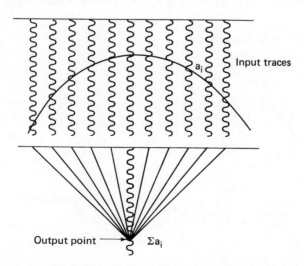

Figure 15-7. Maximum-convexity migration as seen by the output trace.

point shown. If a strong reflector exists at that point, a large amplitude in the output would result. This diffraction argument works for a continuous reflector, since we may regard such a reflector as a continuum of diffracting elements, the individual images of which merge to produce a smooth, continuous reflector surface.

The counterpart of maximum-convexity migration is *wavefront migra-tion.* Take an event on a trace in the seismic record section and throw it out onto the wavefront whose deepest point is at this event. In the constant-velocity case, this wavefront is a semicircle with its deepest point on the event and its center on the source point of that trace. Repeat this process with wavefronts whose deepest points fall on each and every data point of the record section. By the linear superposition principle, the result is the migrated section. That is, the wavefront migrated section is the superposition of all these wavefront arcs.

Figure 15-8 shows a representation of the process of wavefront migration in terms of what happens to a single input trace. The figure shows a trace plotted in depth for a common source and receiver position. Each amplitude value of this trace is mapped into the subsurface along a curve representing the loci of points for which the travel time from source to receiver is constant. The picture produced by this construction is a wavefront chart modulated by the trace amplitude. In order to produce a useful image, a map is composited with similar wavefronts constructed from neighboring traces. A useful subsurface image is produced because of the constructive and destructive interference between wavefronts. For example, wavefronts from neighboring traces will all intersect at a reflection source, adding constructively to produce

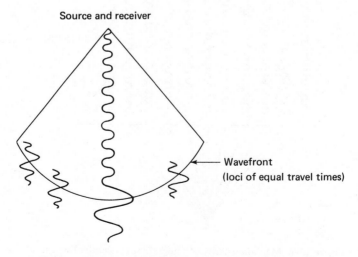

Figure 15-8. Wavefront migration as seen by the input trace.

an image of the reflector in the form of a high-amplitude output. For a continuous reflecting surface, wavefronts from adjacent traces are tangent to the surface and produce an image of the reflecting surface by constructive interference of overlapping portions of adjacent wavefronts. On the other hand, in subsurface regions without reflecting bodies, the wavefronts tend to cancel because of random interference effects.

In summary, maximum-convexity migration takes the values of the record section along a hyperbolic arc and puts their sum at its apex. Wavefront migration takes the value of the record section at a point and puts this value evenly along the circular arc that has this point as its deepest point. The maximum-convexity migrated section is the resultant of all the sums on the apexes. The wavefront migrated section is the sum of all the values on the circular arcs. In principle, both migrated sections are the same for a given record section. This result follows from the fact that all the circular arcs whose deepest points lie on a fixed hyperbolic arc intersect at its apex. Referring to Figure 15-4, we observe that we may proceed from the record point A to the point of reflection B along two paths: either along the circular wavefront, in which case we are performing *wavefront migration*, or along the hyperbolic maximum-convexity front, in which case we are performing *maximum-convexity migration*.

Thus migration is the construction of a cross section of reflections within the subsurface of the earth from a seismic record section. In many potential oil-producing areas, the subsurface consists of nearly horizontally layered sedimentary rocks. At such places, the migration of seismic record sections is extremely simple because the waves propagate nearly vertically to the reflectors, so that they have a one-way travel time which is in direct proportion to the depth. In such simple cases, migration consists of just applying this proportionality factor (namely, velocity) to the (one-way) time axis. However, in many areas of interest, the reflecting surfaces are not horizontal. The most important information carried in the seismic data is the departure of the earth from a horizontally stratified system. When we look at a record section in such a case, what we see may be a poor picture of subsurface structure. The purpose of migration is to give us a correct picture in the form of a reflector section.

Wave Equation Migration

Migration is closely related to the problem of determining the wave field that exists in a propagating medium when access is confined to a surface on the boundary of the medium. This problem is common to many applications of wave motion for sensory purposes, as in radar, sonar, and ultrasonics, as well as in seismology. It is important that we distinguish between imaging and wave-field reconstruction. Wave-field reconstruction is the determination of the wave field over a region of interest. Imaging is taken to mean the production of a picture of the geometrical distribution of the reflecting surfaces within the medium. Wave-field reconstruction by the methods described here provides the first step in the migration process. The imaging of the fields provides the second step, which gives the migrated record section, that is, the reflector section, as output.

We now wish to develop a theoretical basis for wave-field reconstruction from surface measurements (i.e., the record section). The method involves the analysis of the wave field, incident upon a planar measuring array, on the surface of the earth. Reconstruction of the field at any point within the earth is achieved by calculation of the appropriate position-dependent phase shifts for each surface measurement, followed by superposition of the resultants. This approach is due to Stolt (1978), and is well suited to computer calculation with the fast Fourier transform (Cooley and Tukey, 1965).

In our analysis we deal with two spatial dimensions only, a horizontal distance x and a depth y. The variable x can take on any value from $-\infty$ to $+\infty$, but depth y must be positive. Also, time t must be positive. Because y and t must be positive, there is a certain duality between them. There is nothing in these methods that cannot be extended to three spatial dimensions; the extension is straightforward and involves no new principles.

At this point the reader is referred to Appendix 15-1, where we derive the Cauchy solution to the two-dimensional wave equation by means of Fourier transforms. Now let us interpret the mathematical results of Appendix 15-1 in terms of the migration process. First, we make use of the common-depth-point (CDP) stack representation of the record section, which (approximately) gives a coincident source/receiver geometry, as shown in Figure 15-9. Although the energy travel paths between surface source loca-

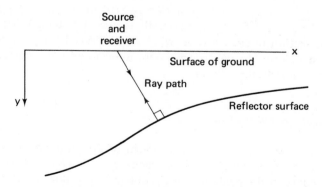

Figure 15-9. Ray paths for coincident source/receiver locations.

tions and positions along the reflector surfaces may be quite complex, we do know that upward and downward legs must be identical, and that the travel path (a ray path) strikes the reflector surface at right angles.

The wave equation describes the motion of the waves generated by a physical experiment. However, the stacked record section does not correspond to a wave field resulting from any single experiment. Many sources were set off sequentially, but the record section gives the appearance that all the sources were activated simultaneously. As a result, we hypothesize a theoretical physical experiment to justify the use of the wave equation to operate on the wave motion appearing on the stacked record section. The theoretical experiment is the following. The receivers are located on the surface of the ground. However, the sources are not at the surface, but are distributed within the earth. More specifically, along every reflector surface the sources are positioned with strengths proportional to the reflection coefficients, and all the sources are activated at the same instant, $t = 0$. This theoretical experiment was proposed by Loewenthal et al (1976). We concern ourselves with upward-traveling waves only, that is, waves in the half x-y plane $y \geq 0$ traveling toward the $y = 0$ axis. We ignore all multiple reflections, and ignore all transmission effects at the interfaces. As we know, a record section involves two-way travel time from surface source to reflector and back to the receiver. In our theoretical experiment, however, we are only concerned with one-way travel time from a reflector source to the receiver. As a result, we must convert our record section from two-way travel time to one-way travel time. This conversion can be accomplished simply by dividing our stacked record section time scale by 2.

The migration problem can now be stated in the following terms. The record section can be considered as a boundary condition (i.e., the surface measurements) for a wave field governed by the wave equation. Take the measured upward traveling waves at the earth's surface, run the time clock backward to time $t = 0$ so as to "depropagate" (i.e., to propagate waves backward in time) to their subsurface reflector positions. This backward propagation to zero time can be viewed as the progressive pushing down of the receivers into the earth until they rest on the reflecting interfaces (French, 1975).

Let us now describe wave equation migration in mathematical terms. The full wave field is the function $u(x, y, t)$. The record section is the wave field $u(x, 0, t)$ observed at the earth's surface $y = 0$ along the surface line x (where $-\infty < x < \infty$) for all nonnegative time $t \geq 0$ (where the time of source activation is $t = 0$). The record section is measured in geophysical exploration work, so it represents a known quantity. The reflector section is the wave field $u(x, y, 0)$, which represents a cross section of the earth cut out by the surface line x and the depth axis y (where $y \geq 0$) at time $t = 0$. The time $t = 0$ corresponds to the time of the simultaneous source activations. The reflector section $u(x, y, 0)$ represents the unknown quantity that we desire (see Figure 15-10).

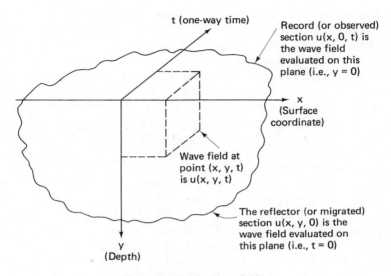

Figure 15-10. The wave field.

Fourier transform wave equation migration can be described by three basic steps. We let

$c =$ constant velocity of the medium

$k_x =$ horizontal angular wave number (radians/unit distance)

$k_y =$ vertical angular wave number (radians/unit distance)

$\omega =$ angular frequency (radians/unit time)

$A(k_x, k_y) =$ two-dimensional Fourier transform with respect to k_x and k_y of the migrated record section

The three steps are:

1. Compute the Fourier transform $F(k_x, \omega)$ of the record section $u(x, 0, t)$ by the equation

$$F(k_x, \omega) = \int_{-\infty}^{\infty} \int_{-\infty}^{\infty} u(x, 0, t)e^{-i(\omega t + k_x x)} \, dx \, dt$$

2. Compute $A(k_x, k_y)$ from $F(k_x, \omega)$ by the equation

$$A(k_x, k_y) = F\left(k_x, ck_y\sqrt{1 + \left(\frac{k_x}{k_y}\right)^2}\right)\frac{c}{\sqrt{1 + (k_x/k_y)^2}} \qquad (15\text{-}25)$$

[see equation (15-25) of Appendix 15-1].

3. Compute the required reflector section $u(x, y, 0)$ as the inverse Fourier transform of $A(k_x, k_y)$,

$$u(x, y, 0) = \frac{1}{4\pi^2} \int_{-\infty}^{+\infty} \int_{-\infty}^{+\infty} A(k_x, k_y)e^{i(k_z x + k_y y)} \, dk_x \, dk_y$$

The wave equation can thus be applied to migration. The record section consists of a common depth point stack section denoted by $u(x, 0, t)$. In the stacked section, the primary reflections correspond to what would be generated if source and detector were at the same point. We assume that these points lie along the x axis. At each of these points there is a seismic trace, and all these traces make up the record section.

We now wish to migrate the record section $u(x, 0, t)$. Because the source and the detector are at the same point, the downward ray path and travel time to a point reflector is identical with the return upward path and travel time to the detector. The amplitude of the return is proportional to the strength of the point reflector. Of course, the strongest point reflectors will lie on the strongest physical reflecting interfaces, and the zero or weak point reflectors will not lie on a physical interface. We have considered the downgoing wave to the point reflector, and the return wave from the point reflector, due to a surface source. Mathematically, an equivalent recording would result from any sources at depth, namely a source initiated at time zero at each point reflector with the strength of the point reflector, provided that one-way travel time is used (as now, the waves only take the return path). This explains why we have let t in $u(x, 0, t)$ denote the one-way travel time. Thus, the record section may be considered as due to upgoing waves from the totality of point reflectors as sources, each initiated at time zero with the appropriate strength.

The migrated section may be denoted by $u(x, y, 0)$, where y is the depth. The record section $u(x, 0, t)$ and the migrated section $u(x, y, 0)$ may be encompassed within the confines of a single quantity $u(x, y, t)$. More specifically, $u(x, y, t)$ is the complete *wave field*, where x is distance along a horizontal surface line, t is the *observed* one-way travel time, and y is the depth into the earth. The record section (i.e., the seismic traces observed at the surface of the ground) is the wave field observed at the ground surface $y = 0$:

$$u(x, y = 0, t)$$

The migrated section (i.e., the cross section of the earth with horizontal position x and depth y) is the wave field composed of the strengths of the hypothetical point reflector sources initiated at time $t = 0$:

$$u(x, y, t = 0)$$

The wave field $u(x, y, t)$ is governed by the wave equation. It is for this reason that the three basic steps described above enable us to obtain the migrated section in terms of the record section. The method, which was developed by Stolt (1978), is called *frequency-wave-number domain* migration. An extension to the case of variable velocity has been given by Gazdag (1978).

While frequency-wave-number wave equation migration combines computational speed and simplicity with good quality results, several alternative approaches have also received attention over the past few years. Perhaps the best known of these is the *finite-difference method*, developed by Claerbout, Doherty, Johnson, Landers, Riley, and Schultz at Stanford University (Claerbout, 1976). In essence, the method approximates the wave equation, which is a hyperbolic partial differential equation (PDE), by a parabolic, or diffusion-type PDE. This can be done by splitting the wave motion into upgoing and downgoing components, which are then treated separately. The resulting parabolic PDE is in turn expressed in terms of various finite-difference approximations suitable for implementation on a digital computer. Depending on the kind of finite-difference approximation used, these techniques perform quite well for (true) dip angles of up to about 45°. In contrast, the frequency-wave-number method as well as the Kirchhoff summation method mentioned below are accurate simulations of the "full" wave equation; as such, they perform well for any angle of dip.

We may also seek solutions to the wave equation in such a way that the observed surface seismograms constitute the boundary-value set required for the solution of a classical boundary-value problem. This viewpoint yields an integral, or summation algorithm in either two or three dimensions. The method is commonly known as *Kirchhoff migration* by virtue of the use of the Kirchhoff integral in the solution to the wave equation. Excellent descriptions of the pertinent theory as well as implementation have been given by French (1975) and by Schneider (1978). In addition, Schneider has developed the explicit relationships existing between the Kirchhoff and frequency-wave number approaches.

Figure 15-11 shows an actual seismic raw data set prior to migration. In Figure 15-12 we can see the stacked section after 45° finite-difference migration with the Stanford technique, while Figure 15-13 shows the performance of frequency-wave-number migration with the same stacked data. In this instance, the frequency-wave-number approach appears to yield better quality results; on the other hand, the finite-difference method works

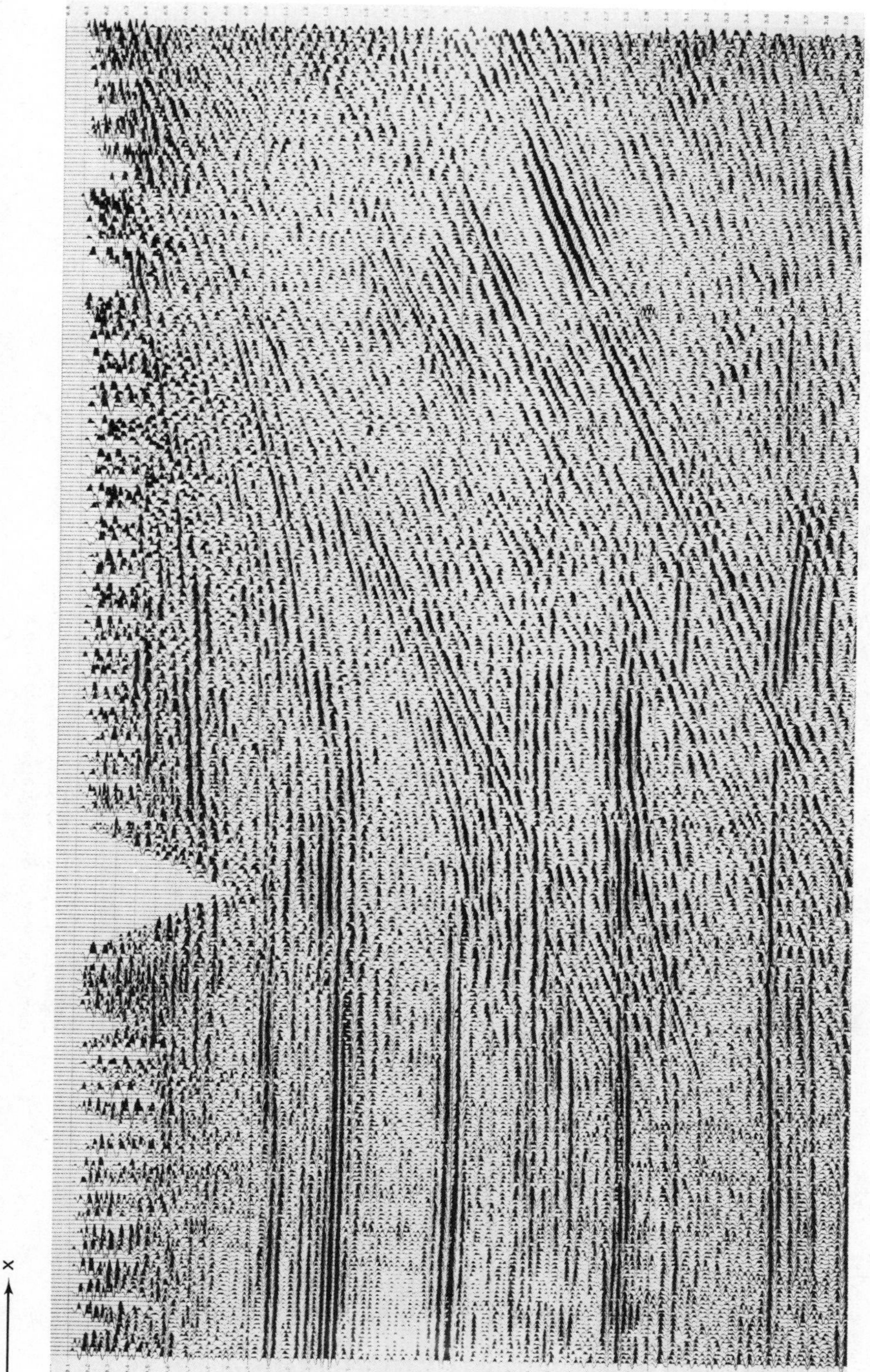

Figure 15-11. Raw data before application of migration techniques (Courtesy of Seismograph Service Corp.).

391

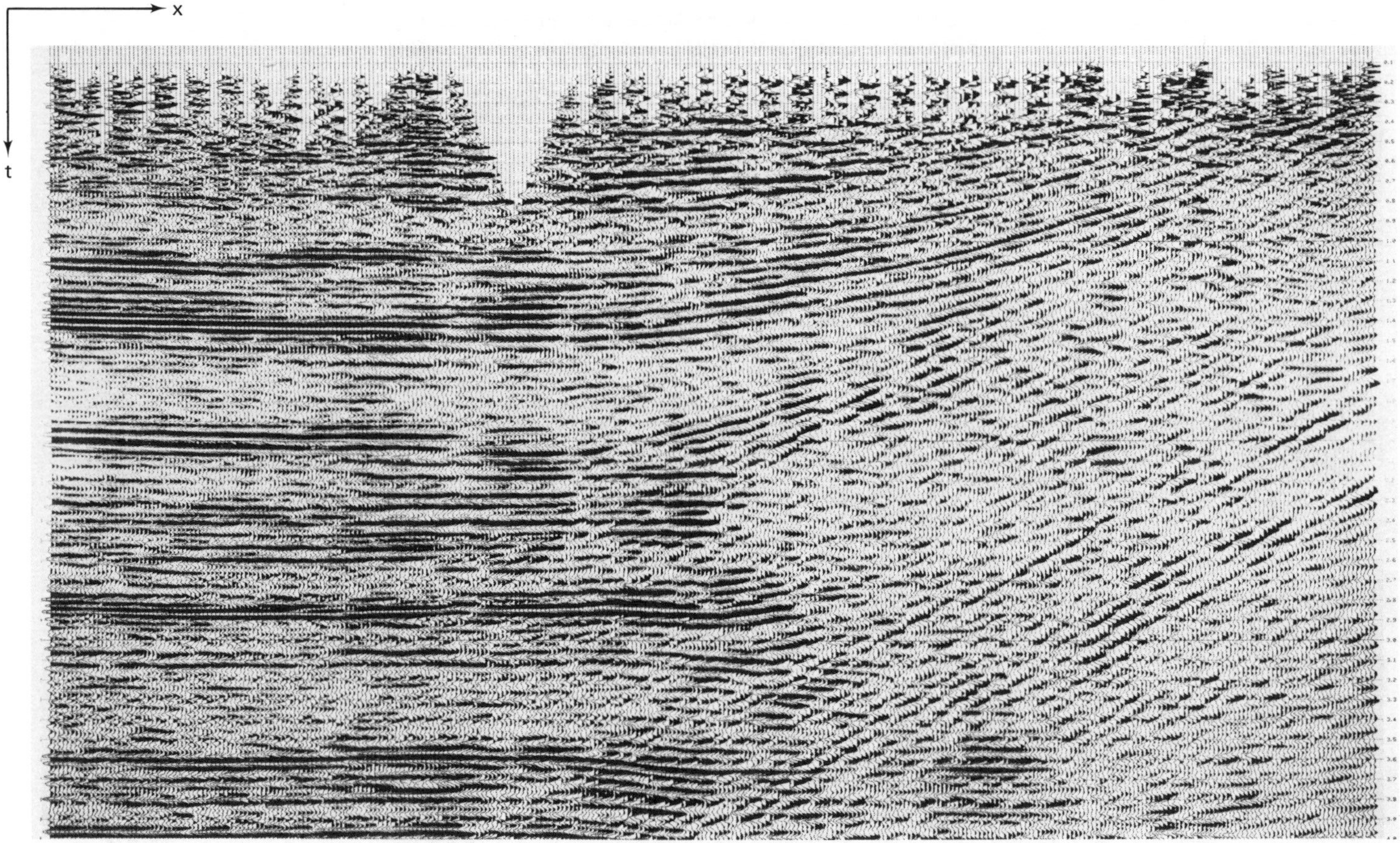

Figure 15-12. Same stacked section after 45° finite-difference migration (Courtesy of Seismograph Service Corp.).

x

t

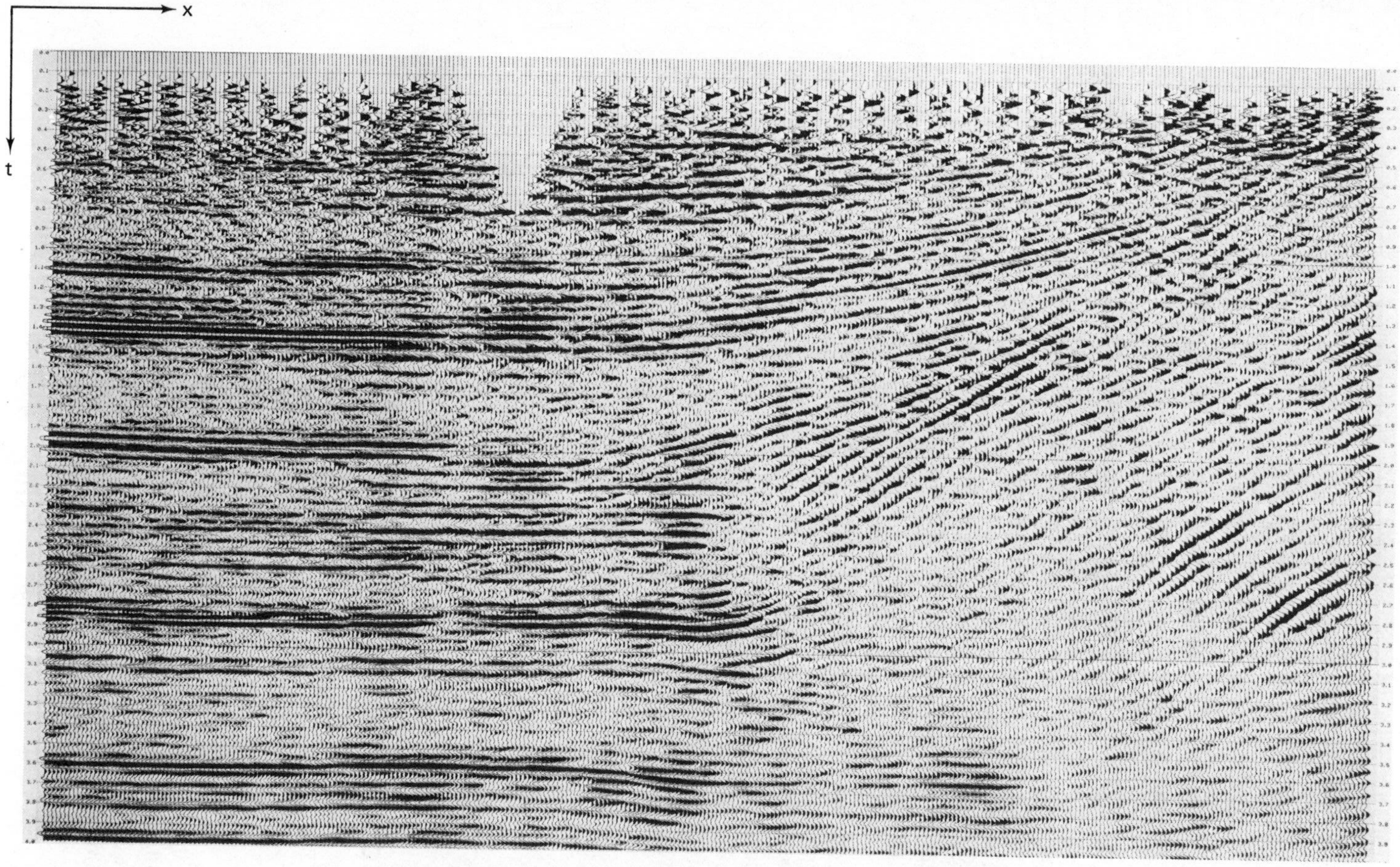

Figure 15-13. Same stacked section after frequency-wave-number migration (Courtesy of Seismograph Service Corp.).

quite well as long as the dip angles remain within the range satisfied by the parabolic approximation.

Concluding Remarks

In this chapter we have dealt in detail with only two migration approaches—the classic method of Hagedoorn (1954) and the more recent Fourier transform method of Stolt (1978). We refer the reader to the references given above for more detail on both the finite-difference and Kirchhoff migration approaches. Finally, we have not touched on many implementational aspects of seismic migration. For example, one may migrate either before or after stacking the raw data. Migration before stack often exhibits better performance, but is evidently costlier. Existing methods allow for vertical variations in velocity, but horizontal variations are more difficult to handle. A significant advance for this case has been made recently by Hubral (1977), who demonstrated that conventional migration methods fail to place reflected events in their proper spatial positions. He then showed how the concept of the seismic *image ray* can be used to improve migration performance in this sense. Larner et al (1978) exploited this result to arrive at a ray-theoretical solution to the positioning problem. All these matters are the subjects of ongoing research, and we must refer the reader to the literature as it appears.

APPENDIX 15-1

cauchy solution to the wave equation

The Cauchy method to be described here has been given by Stolt (1978) for migration with either the two-dimensional or the three-dimensional wave equation. However, for simplicity we shall carry through the derivation for only the two-dimensional wave equation

$$\frac{\partial^2 u}{\partial t^2} - c^2 \left(\frac{\partial^2 u}{\partial x^2} + \frac{\partial^2 u}{\partial y^2} \right) = 0 \qquad (15\text{-}1)$$

Here x and y are the spatial coordinates (where x is distance along the surface of the ground and y is depth measured positively down into the ground). The variable t denotes time, and the constant c denotes the velocity of the medium. The function $u(x, y, t)$ represents the <u>wave motion</u>, or <u>wave field</u>.

Let $U(k_x, k_y, \omega)$ be the three-dimensional Fourier transform of the wave motion; that is,

$$u(x,y,t)$$

$$U(k_x, k_y, \omega) = \int_{-\infty}^{\infty} dx \int_{-\infty}^{\infty} dy \int_{-\infty}^{\infty} dt \, u(x, y, t)e^{-i(k_zx+k_yy+\omega t)} \qquad (15\text{-}2)$$

Here k_x and k_y are angular wave numbers and ω is angular frequency. Then the three-dimensional Fourier transform of the wave equation gives

$$\omega^2 - c^2(k_x^2 + k_y^2) = 0 \qquad (15\text{-}3)$$

This relation is called <u>the *dispersion equation*.</u> In frequency-wave-number space the dispersion equation represents the surface of a cone. <u>For ω constant, we have an equation of a circle in k_x and k_y. For k_x constant, we have the equation of a hyperbola in ω and k_y. Similarly, we have a hyperbola when k_y is constant.</u>

Let $V(k_x, k_y, t)$ be the two-dimensional Fourier transform of the wave motion $u(x, y, t)$ with respect to the spatial coordinates x and y; that is,

$$V(k_x, k_y, t) = \int_{-\infty}^{\infty} dx \int_{-\infty}^{\infty} dy \, u(x, y, t)e^{-i(k_zx+k_yy)} \qquad (15\text{-}4)$$

If we take this two-dimensional Fourier transform of the wave equation, we obtain

$$\frac{d^2V}{dt^2} + c^2(k_x^2 + k_y^2)V = 0 \qquad (15\text{-}5)$$

Substitution of the dispersion equation (15-3) into the two-dimensional Fourier transform (15-5) yields

$$\frac{d^2V}{dt^2} + \omega^2 V = 0 \qquad (15\text{-}6)$$

This equation is an <u>ordinary linear and homogeneous</u> differential equation. It <u>has two independent solutions,</u> namely,

$$V = e^{i\omega t} \quad \text{and} \quad V = e^{-i\omega t} \qquad (15\text{-}7)$$

Any linear combination of these solutions is also a solution. In particular, <u>$V = A(k_x, k_y)e^{i\omega t}$ is a solution.</u>

The first solution in (15-7) corresponds to the sinusoidal wave

$u(x,y,t)$

$$e^{i(\omega t + k_z x + k_y y)} \tag{15-8}$$

and the second to the sinusoidal wave

$$e^{-i(\omega t - k_z x - k_y y)} \tag{15-9}$$

In space time, the wave motion (15-8) is constant on each characteristic plane,

$$\omega t + k_x x + k_y y = \text{constant} \tag{15-10}$$

Thus, the wave motion (15-8) is a plane wave propagating with speed c in the direction of the unit wave number vector

$$\left(-\frac{ck_x}{\omega}, -\frac{ck_y}{\omega} \right) \tag{15-11}$$

Let us fix ω to have the same sign as k_y; that is, we solve the dispersion equation (15-3) in the form

$$\omega = ck_y \sqrt{1 + \frac{k_x^2}{k_y^2}} \tag{15-12}$$

Under this convention the unit vector (15-11) has a negative component, that is, the component

$$-\frac{ck_y}{\omega} = -\frac{1}{\sqrt{1 + k_x^2/k_y^2}} \tag{15-13}$$

along the y axis. Therefore, the plane wave (15-8) represents an upgoing plane wave.

By a similar argument, the plane wave (15-9) represents a downgoing plane wave. In the migration problem, we only want to consider upgoing plane waves. As a result, we save solution (15-8) and throw out solution (15-9). We write the solution of the differential equation (15-6) as

$$V(k_x, k_y, t) = A(k_x, k_y)e^{i\omega t} \tag{15-14}$$

where $A(k_x, k_y)$ is a constant with respect to t. Letting $t = 0$ in equation (15-14), we see that

$$A(k_x, k_y) = V(k_x, k_y, 0) \tag{15-15}$$

In turn, letting $t = 0$ in equation (15-4), we obtain

$$V(k_x, k_y, 0) = \int_{-\infty}^{\infty} dx \int_{-\infty}^{\infty} dy \, u(x, y, 0) e^{-i(k_z x + k_y y)} \qquad (15\text{-}16)$$

which identifies $V(k_x, k_y, 0)$ as the Fourier transform of the *desired* depth section $u(x, y, 0)$. Because of equation (15-15) we see that $A(k_x, k_y)$ is also the Fourier transform of the depth section. Thus, the migration problem in the Fourier domain reduces to the determination of $A(k_x, k_y)$.

To this end, let us start with the known data, namely, the surface section (or record section) $u(x, 0, t)$. We compute the Fourier transform of the surface section, namely

$$F(k_x, \omega) = \int_{-\infty}^{\infty} dx \int_{-\infty}^{\infty} dt \, u(x, 0, t) e^{-i(\omega t + k_z x)} \qquad (15\text{-}17)$$

The inverse Fourier transform is

$$u(x, 0, t) = \frac{1}{4\pi^2} \int_{-\infty}^{\infty} dk_x \int_{-\infty}^{\infty} d\omega \, F(k_x, \omega) e^{i(\omega t + k_z x)} \qquad (15\text{-}18)$$

Let us now find another expression for $u(x, 0, t)$. The inverse Fourier transform of $V(k_x, k_y, t)$ is

$$u(x, y, t) = \frac{1}{4\pi^2} \int_{-\infty}^{\infty} dk_x \int_{-\infty}^{\infty} dk_y \, V(k_x, k_y, t) e^{i(k_z x + k_y y)} \qquad (15\text{-}19)$$

Making use of equation (15-14), we have

$$u(x, y, t) = \frac{1}{4\pi^2} \int_{-\infty}^{\infty} dk_x \int_{-\infty}^{\infty} dk_y \, A(k_x, k_y) e^{i(\omega t + k_z x + k_y y)} \qquad (15\text{-}20)$$

If we let $y = 0$ in equation (15-20), we obtain our alternative expression for $u(x, 0, t)$

$$u(x, 0, t) = \frac{1}{4\pi^2} \int_{-\infty}^{\infty} dk_x \int_{-\infty}^{\infty} dk_y \, A(k_x, k_y) e^{i(\omega t + k_z x)} \qquad (15\text{-}21)$$

Comparing equations (15-21) and (15-18), we see that

$$A(k_x, k_y) \, dk_y = F(k_x, \omega) \, d\omega \qquad (15\text{-}22)$$

Thus, the required function $A(k_x, k_y)$ can be found from the Fourier transform $F(k_x, \omega)$ of the surface section by the equation

$$A(k_x, k_y) = F(k_x, \omega) \frac{d\omega}{dk_y} \qquad (15\text{-}23)$$

From equation (15-12) we have

$$\frac{d\omega}{dk_y} = \frac{c}{\sqrt{1 + k_x^2/k_y^2}} \qquad (15\text{-}24)$$

so

$$A(k_x, k_y) = F\left(k_x, ck_y\sqrt{1 + \frac{k_x^2}{k_y^2}}\right) \frac{c}{\sqrt{1 + k_x^2/k_y^2}} \qquad (15\text{-}25)$$

where we have also used equation (15-12) to replace the argument ω in $F(k_x, \omega)$ as shown.

The required depth section (or reflector section) is finally given by letting $t = 0$ in the inverse Fourier transform (15-20),

$$u(x, y, 0) = \frac{1}{4\pi^2} \int_{-\infty}^{\infty} dk_x \int_{-\infty}^{\infty} dk_y \, A(k_x, k_y) e^{i(k_x x + k_y y)} \qquad (15\text{-}26)$$

spectral estimation

Summary

The spectral estimation problem for a discrete-time series generated by a linear, time-invariant process can be formulated in terms of three models: autoregressive (AR), moving average (MA), and autoregressive–moving average (ARMA). Analysis procedures differ in each case, and specification errors arise due to application of the inappropriate algorithm. The AR and MA models lead respectively to the maximum entropy (MEM) and classical lag-window approaches. The ARMA model has much seismic interest because the unit impulse response of a horizontally stratified medium is expressible in this way. Since its feedback component has the minimum-delay property, an ARMA spectral estimation technique satisfying this requirement has particular seismic relevance. Such a spectral estimate results from the application of an iterative least-squares algorithm to selected gates of the observed time series. A sample set of synthetic time series serve to illustrate the degradation in the spectral estimate resulting from an incorrect specification of the model.

399

Introduction

Much has been written in recent years about the spectral analysis of discrete-time series. There exists no single "correct" technique to calculate the spectrum in the absence of knowledge about the type of process that has generated the data. As we have seen in Chapter 9, we distinguish between three possible processes: autoregressive (AR), moving average (MA), and autoregressive–moving average (ARMA). In engineering terms, these processes respectively describe the all-pole (or feedback), the all-zero (or feedforward), and the pole–zero (or feedback–feedforward) systems. Generally speaking, we will not have a priori knowledge about the generating mechanism of the time series, and we are forced to assume that our recorded data do indeed satisfy one of these three representations. Once this decision has been made, we must select an appropriate algorithm for the calculation of the actual spectral estimate. In the case of the AR, or all-pole model, the maximum entropy method (MEM) as implemented with a technique due to Burg (1967, 1975) is appropriate. For the MA, or all-zero model, we have recourse to the classical lag-window approach (Blackman and Tukey, 1959). In Appendix 16-1 we give the mathematics of the classical lag-window method, and in Appendix 16-2, the mathematics of the maximum entropy method.

The ARMA, or pole–zero model has also received attention in the recent literature: pertinent spectral estimation techniques have been described by Anderson (1971, Chap. 5), by Box and Jenkins (1970, Chaps. 6 and 7), and by Alam (1978). The rational representation of the impulse response of an ARMA process is given by the ratio of two polynomials in the complex variable z. In this chapter we shall be particularly interested in the spectral analysis of seismograms. As we have seen in Chapter 13, the unit impulse response of a perfectly elastic, horizontally stratified medium can be expressed as the ratio of two such polynomials in powers of z, but with the added constraint that the denominator polynomial have the minimum-delay property. In other words, this condition forces the poles of the system to lie outside the periphery of the unit circle $|z| = 1$ in the complex plane, and allows us to expand the ARMA polynomial ratio in the form of a convergent power series in z. It will be desirable, therefore, to seek an ARMA spectral estimation algorithm that guarantees a minimum-delay denominator. While there is no intrinsic mathematical need for an ARMA spectral estimation method to produce a minimum-delay denominator, we have just stated that such a quest has strong physical motivation. Accordingly, the minimum-delay property of the denominator is a strong point, and one not necessarily shared by other ARMA spectral estimators.

If the numerator of the polynomial ratio reduces to a constant, the ARMA process reduces to an AR process. Under such conditions the pre-

400

ferred technique is the MEM algorithm, for which it can be shown (Burg, 1975) that the denominator polynomial is minimum-delay. In this sense, therefore, our ARMA algorithm constitutes a generalization of the MEM approach.

The Three Basic Data Models

We assume that the discrete-time series have been generated by a linear filter. Such a filter can be described mathematically in many ways, but there are three representations that have been found particularly useful and for which, moreover, excellent theoretical justification can be given. In the time-series literature (see, e.g., Box and Jenkins, 1970, Chap. 2), these are called the autoregressive (AR), moving average (MA), and autoregressive–moving average (ARMA) models.

An autoregressive (or feedback) model y_t is described by the relation

$$y_t = a_0 x_t - a_1 y_{t-1} - a_2 y_{t-2} - \cdots - a_m y_{t-m}$$

where t is the discrete-time variable, and where the coefficients $a_0, a_1, a_2, \ldots,$ a_m remain to be determined. In the language of the engineer, the AR model is called a feedback system with input x_t and output y_t. The system input, x_t, in many cases is taken to be uncorrelated random noise, ϵ_t, with mean and variance

$$E\{\epsilon_t\} = 0 \quad \text{and} \quad E\{\epsilon_t^2\} = \sigma^2$$

where E is the expectation, or averaging operator. In such a case (i.e., when $x_t = \epsilon_t$) the output y_t is called an autoregressive process of order m [i.e., an AR(m) process]. We observe that the value of y_t at time t of an autoregressive process is a linear combination of m previous values of the process y_t plus random noise ϵ_t.

The moving average (or autoregressive or feedforward) model y_t is described by the relation

$$y_t = b_0 x_t + b_1 x_{t-1} + b_2 x_{t-2} + \cdots + b_n x_{t-n}$$

where the coefficients $b_0, b_1, b_2, \ldots, b_n$ again are to be determined. In the language of the engineer, y_t represents the output of the linear convolutional filter $(b_0, b_1, b_2, \ldots, b_n)$ for the input x_t. The output of a MA model is a linear combination of present and past values of the system input sequence x_t. In the case when the input is white noise (i.e., when $x_t = \epsilon_t$), the output y_t is called a moving average process of order n [i.e., an MA(n) process].

ARMA

Finally, the autoregressive–moving average model y_t is described by the relation

MA   M

$$y_t = b_0 x_t + b_1 x_{t-1} + b_2 x_{t-2} + \cdots + b_n x_{t-n}$$

AR

$$- a_1 y_{t-1} - a_2 y_{t-2} - \cdots - a_m y_{t-m}$$

M

where both the autoregressive coefficients $a_0, a_1, a_2, \ldots, a_m$, as well as the moving average coefficients $b_0, b_1, b_2, \ldots, b_n$ remain to be determined. The ARMA model contains both AR as well as MA components, and evidently is the most general of the three representations we consider here. In the language of the engineer, y_t represents the output at time t of a linear recursive filter (Shanks, 1967), whose input at time t is x_t. In the case when the input is white noise (i.e., when $x_t = \epsilon_t$) the output y_t is called an autoregressive–moving average process of order m, n [i.e., an ARMA(m, n) process].

The three models allow a more convenient and revealing representation in terms of the z transform variable z, where z is the unit-delay operator, $z y_t = y_{t-1}$. [Note that Box and Jenkins (1970) use the symbol B instead of the symbol z for the unit delay (or backward shift) operator.] They are

$$y_t = \frac{1}{a_0 + a_1 z + a_2 z^2 + \cdots + a_m z^m} x_t$$

for the AR model,

$$y_t = (b_0 + b_1 + b_2 z^2 + \cdots + b_n z^n) x_t$$

for the MA model, and

$$y_t = \frac{b_0 + b_0 z + b_2 z^2 + \cdots + b_n z^n}{a_0 + a_1 z + a_2 z^2 + \cdots + a_m z^m} x_t$$

for the ARMA model. If we define the polynomials $A_m(z)$ and $B_n(z)$ in the form

$$A_m(z) = a_0 + a_1 z + a_2 z^2 + \cdots + a_m z^m$$
$$B_n(z) = b_0 + b_1 z + b_2 z^2 + \cdots + b_n z^n$$

we obtain the more succinct representations

$$Y(z) = \frac{1}{A_m(z)} X(z) \qquad \text{AR}(m) \text{ model} \tag{16-1a}$$

$$Y(z) = B_n(z) X(z) \qquad \text{MA}(n) \text{ model} \tag{16-1b}$$

$$Y(z) = \frac{B_n(z)}{A_m(z)} \qquad \text{ARMA}(m, n) \text{ model} \tag{16-1c}$$

where $X(z)$ and $Y(z)$ are the z transforms of the input x_t and output y_t, respectively.

Now $Y(z) Y(z^{-1}) = \phi(z)$ is the z transform of the autocorrelation function of y_t. Evaluation of $\phi(z)$ on the unit circle

$$z = e^{-i\omega}$$

yields $\phi(e^{-i\omega})$, the power spectrum of y_t (see Appendix 4-1), where $\omega =$ angular frequency in radians per time unit. We may write

$$\phi(e^{-i\omega}) = Y(e^{-i\omega}) Y(e^{+i\omega})$$
$$= |Y(e^{-i\omega})|^2$$

or simply

$$\phi(\omega) = |Y(\omega)|^2$$

The corresponding power spectral representations are

$$\phi(\omega) = \frac{1}{|A_m(\omega)|^2} |X(\omega)|^2 \qquad \text{AR model} \qquad (16\text{-}2a)$$

$$\phi(\omega) = |B_n(\omega)|^2 |X(\omega)|^2 \qquad \text{MA model} \qquad (16\text{-}2b)$$

$$\phi(\omega) = \frac{|B_n(\omega)|^2}{|A_m(\omega)|^2} |X(\omega)|^2 \qquad \text{ARMA model} \qquad (16\text{-}2c)$$

Accordingly, we have three possible models to describe our observations. If we wish to obtain the power spectrum of a finite data window from the observed time series y_t, how do we proceed? To answer this question, we must know the process type; that is, we must know whether the data best fits the AR, the MA, or the ARMA hypothesis. A number of theoretical tests have been devised for this purpose (see, e.g., Anderson, 1971, Chap. 5), but none appear to be completely satisfactory. Here we use the pragmatic approach: we generate realizations of the three processes, for which the true spectrum is known, and then study the fit obtained by assuming that the data are either AR, MA, or ARMA. Before this can be done, we require algorithms for the implementation of equations (16-2a) to (16-2c). In the case of the AR model, we use the "maximum entropy method" (MEM) approach (Burg, 1967; 1975), which we describe in Appendix 16-2. The spectral estimation procedure for the MA model is the classical approach, which we describe in Appendix 16-1. In the case of the ARMA model, we use an iterative least-squares technique, to be described below.

To simplify the computation of the illustrative cases, we shall assume that the input sequence x_t is uncorrelated random noise, ϵ_t, with zero mean and variance $E\{\epsilon_t^2\} = \sigma^2$. This means that $|X(\omega)|^2 = \sigma^2$ (see, e.g., Box and

AR – (MEM)
MA –
ARMA – iterative LS technique

Jenkins, 1970, pp. 80–81), and equations (16-2a) to (16-2c) become

$$\Phi(\omega) = \frac{\sigma^2}{|A_m(\omega)|^2} \qquad \text{AR process} \qquad (16\text{-}3a)$$

$$\Phi(\omega) = \sigma^2 |B_n(\omega)|^2 \qquad \text{MA \quad process} \qquad (16\text{-}3b)$$

$$\Phi(\omega) = \sigma^2 \frac{|B_n(\omega)|^2}{|A_m(\omega)|^2} \qquad \text{ARMA process} \qquad (16\text{-}3c)$$

The Maximum Entropy Method (MEM)

The maximum entropy method (MEM) attempts to fit, in a least-squares sense, an autoregressive (AR) model to an input time series, y_t (see Appendix 16-2). In other words, we assume that the data, y_t, are generated by the process

$$y_t = \epsilon_t - a_1 y_{t-1} - a_2 y_{t-2} - \cdots - a_m y_{t-m} \qquad (16\text{-}4)$$

where ϵ_t is uncorrelated random noise with zero mean and variance σ^2. Here we have let $a_0 = 1$. Equation (16-4) can be rewritten

$$y_t + a_1 y_{t-1} + a_2 y_{t-2} + \cdots + a_m y_{t-m} = \epsilon_t \qquad (16\text{-}5)$$

The sequence a_t, namely $a_0, a_1, \ldots, a_m$ with $a_0 = 1$, is more commonly known as the prediction error filter (see Chapter 12), and it is precisely these coefficients that the MEM approach attempts to estimate. It has been pointed out by Van den Bos (1971) and by Ulrych and Ooe (1979) that the AR representation is equivalent to that time series which is consistent with the known autocorrelation measurements, but which has maximum entropy.

There are two general techniques for estimating the unknown coefficients $a_1, \ldots, a_m$ (called the AR parameters). One, known as the *Yule–Walker method* (Yule, 1927; Walker, 1931) involves the solution of the Toeplitz normal equations (see Chapter 6), and necessarily requires explicit knowledge of the autocorrelation function, ϕ_{t_i} of the input y_t. The other, developed by Burg (1967, 1975) estimates the AR parameters without prior knowledge of the autocorrelation function. The latter method is employed in this chapter. The approach is described in Appendix 16-2, and here we give a qualitative description.

One essentially fits successively higher-order prediction error operators to the input series by convolving the filters in both the forward and backward directions, and then summing the squares of the two resulting error series to obtain a measure of error power. It is important to note that these convolutions are carried out in such a way that the filters do not run off the ends of the data. This is a significant point in the method.

The filter coefficients for a particular order are determined by minimizing the error power with respect to the last coefficient a_m, and then using appropriate recursion relations to find the remaining coefficients $a_1, \ldots,$ a_{m-1}. The coefficient sequence is always normalized such that $a_0 = 1$, and $|a_m| \leq 1$. Now the last coefficient of this filter is actually the partial correlation coefficient which, for a particular order m, measures the correlation between the forward and backward error series generated by the filter of order $m - 1$ (Box and Jenkins, 1970, Chap. 3). Clearly, if the order of the prediction error filter is equal to or less than the actual order of the AR data, there will be significant correlation between the two error series (i.e., $|a_m| \doteq 1$). However, when the order of the prediction error filter exceeds the order of the system, all predictable information in the data is removed, causing the two error series to be uncorrelated (i.e., $|a_m| \doteq 0$). It is the behavior of this coefficient which has led to its use as a statistical test of the order of an AR system (Akaike, 1969; Ulrych and Bishop, 1975; Jones, 1976). Once the order has been estimated and the filter coefficients have been calculated, the power spectral estimates of the input data are easily computed with equation (16-2a).

The Least-Squares ARMA Spectral Estimation Method

Let us consider the following problem: we have the discrete $(l + 1)$-length data window $y = y_t = (y_0, y_1, \ldots, y_l)$, and are required to approximate its z transform $Y_l(z)$ by a rational function, say

$$Y_l(z) \doteq \frac{P_j(z)}{Q_k(z)} \tag{16-6}$$

where

$$P_j(z) = p_0 + p_1 z + \cdots + p_j z^j$$
$$Q_k(z) = q_0 + q_1 z + \cdots + q_k z^k$$

and where we assume for the moment that the polynomial degrees j and k are given. Referring to the ARMA model defined by equation (16-2c), let us say that the input sequence is of length $s + 1$, so that

$$X(z) = X_s(z) = x_0 + x_1 z + \cdots + x_s z^s$$

If we set

$$P_j(z) = B_n(z) X_s(z) \tag{16-7}$$

and

$$Q_k(z) = A_m(z) \tag{16-8}$$

— some of the wavelet

with $n + s = j$ and $k = m$, we see that equation (16-6) is an ARMA (m, j) model of the observed data window y_t. In the seismic case, $X(z)$ often will be the z transform of the source pulse.

We describe an iterative least-squares algorithm that yields estimates of the MA coefficients, $p_0, p_1, \ldots, p_j$ and of the AR coefficients $q_0, q_1, \ldots, q_k$. The resulting expression (16-6) may then be evaluated on the unit circle $z = e^{-i\omega}$ to produce the ARMA power spectral estimate given by

$$\Phi(f) = \frac{|P_j(\omega)|^2}{|Q_k(\omega)|^2} \qquad (16\text{-}9)$$

where we have made use of equations (16-2c), (16-7), and (16-8).

Equation (16-6) can be written in the symbolic form

$$\mathbf{y} * \mathbf{q} \doteq \mathbf{p} \qquad \text{(I)}$$

In addition, we will use

$$\mathbf{q} * \mathbf{q}^{-1} \doteq \delta \qquad \text{(II)} \qquad\qquad (16\text{-}10)$$

and recognize that equation 16-10 (I) can be written formally as

$$\mathbf{y} \doteq \mathbf{q}^{-1} * \mathbf{p} \qquad \text{(III)}$$

Here the asterisk $(*)$ denotes convolution; $\mathbf{y}$, $\mathbf{p}$, and $\mathbf{q}$ are the vector forms of the time sequences $(y_0, y_1, \ldots, y_l)$, $(p_0, p_1, \ldots, p_j)$, and $(q_0, q_1, \ldots, q_k)$; $\mathbf{q}^{-1}$ is the inverse of $\mathbf{q}$; and δ is the Kronecker delta,

$$\delta = \begin{cases} 1 & t = 0 \\ 0 & t \neq 0 \end{cases}$$

We know $\mathbf{y}$ and we select some arbitrary initial $\mathbf{p}$, say $\mathbf{p}^{(0)}$, which is either based on a priori knowledge, or may just be a naive guess. The nth iteration requires us to compute three least-squares shaping filters (see Chapter 7) in the sequence shown in Figure 16-1, where the step number (I, II, or III) is based on the appropriately labeled relation (16-10). We observe that the filter obtained in the first step is the input for the second step, and that the filter obtained in the second step is in turn the input for the third step. At this point the $(n + 1)$th iteration begins with the desired output $\mathbf{p}^{(n)}$ available from step III of the nth iteration.

After the completion of $n = N$ iterations, we have in computer storage the vectors

$$\mathbf{q}^{(N)}, \quad \mathbf{q}^{-1(N)}, \quad \text{and} \quad \mathbf{p}^{(N)}$$

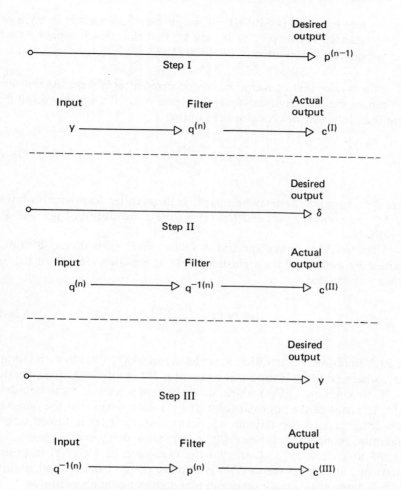

Figure 16-1. The nth iteration for the ARMA spectral estimation algorithm, where $n = 1, 2, \ldots, N$, and $N =$ maximum number of iterations.

which have the following properties:

1. $\mathbf{q}^{(N)}$ is not necessarily minimum-delay.
2. $\mathbf{q}^{-1(N)}$ is necessarily minimum-delay, for it is the zero-lag least-squares inverse of $\mathbf{q}^{(N)}$, and $\mathbf{q}^{-1(N)}$ is thereby minimum-delay (Robinson, 1967a, p. 174).
3. If convergence occurs, we will have

$$\mathbf{y} \doteq \mathbf{q}^{-1(N)} * \mathbf{p}^{(N)}$$

[see equation (16-10)-III]. If not, it may be desirable to try a new starting vector $\mathbf{p}^{(0)}$, or it may be that the data $\mathbf{y}$ cannot be adequately represented in the form (16-6).

We assume that convergence has occurred after N steps. At this point we compute the zero-lag least-squares inverse of $\mathbf{q}^{-1(N)}$, which we call $\hat{\mathbf{q}}^{(N)}$. There then follows from equation (16-6) that

$$Y_i(z) \sim \frac{P_j^N(z)}{\hat{Q}_k^{(N)}(z)} \qquad (16\text{-}11)$$

where $\hat{\mathbf{q}}^{(N)}$ is minimum-delay because it is the zero-lag least-squares inverse of $\mathbf{q}^{-1(N)}$, and where $P_j^N(z)$ and $\hat{Q}_k^{(N)}(z)$ are the z transforms of $\mathbf{p}^{(N)}$ and $\hat{\mathbf{q}}^{(N)}$, respectively.

The ARMA power spectral estimate after N iterations, $\Phi^{(N)}(\omega)$, is obtained by evaluating the right-hand side of equation (16-11) on the unit circle $z = e^{-i\omega}$, that is,

$$\Phi^{(N)}(f) = \frac{P_j^{(N)}(\omega)}{\hat{Q}_k^{(N)}(\omega)} \qquad (16\text{-}12)$$

For each iteration, say iteration n, we have forced $Q^{-1(n)}(z)$ to be minimum-delay, where $Q^{-1(n)}(z)$ is the z transform of $\mathbf{q}^{-1(n)}$. At the end of the iteration $n = N$, we compute $\hat{Q}_k^{(N)}(z)$ which, as we have just seen, is minimum-delay.

In terms of the approximation (16-11), this means that the non-minimum-delay part of the rational approximation to $Y_i(z)$ is forced into its numerator, namely $P_j^N(z)$. Since $\hat{Q}_k^{(N)}$ is minimum-delay, all its zeros lie outside the unit circle $|z| = 1$, so that the expansion of $1/\hat{Q}_k^{(N)}(z)$ in positive powers of z always converges. This is a strong point of the method, and that which distinguishes it from other rational approximation procedures.

The actual outputs $\mathbf{c}^{(I)}$, $\mathbf{c}^{(II)}$, and $\mathbf{c}^{(III)}$ resulting from steps I, II, and III of each iteration (see Figure 16-2) are not used explicitly in the calculation, although their structure is sometimes of value to judge the equality of the approximation procedure when going from iteration n to iteration $(n + 1)$.

The orders of the polynomials $P_j^{(n)}$ and $Q_k^{(n)}$, namely j and k, are fixed quantities for a given set of iterations $n = 1, 2, \ldots, N$, but some interesting "rules of thumb" for the choice of these values have been established as a result of numerical experiments. The principal statements of a FORTRAN program for the implementation of the present algorithm are given in Appendix 16-3.

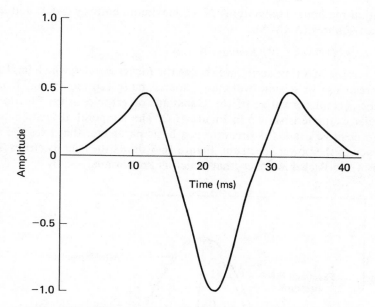

Figure 16-2. Ricker wavelet, an example of a moving average (MA) process.

Numerical Experiments

Many authors (Lacoss, 1971; Ulrych, 1972) have compared the maximum entropy method (MEM) with various standard spectral techniques, such as the lag-window approach. Such comparisons have been based almost uniformly on the criterion of resolving power: the maximum entropy method tends to produce high-resolution-line spectra even for short lengths of input data, whereas the lag-window spectrum tends to have poorer resolution due to the smearing effect of the time window. We claim that such comparisons have not always been complete, because a given spectral estimation method must be evaluated in terms of the data model on which the estimation procedure is based.

In order to compare the three spectral analysis techniques discussed in the preceding sections, we construct three time series based on the MA, AR, and ARMA models, respectively. The actual coefficients used to generate the AR and the ARMA series are taken from Ulrych and Bishop (1975). Only the sampling interval (2 ms) and the random sequence, ϵ_t, are different (i.e., the realizations of these series are not the same). We then analyze the three time series, and use on each of them three different spectral estimation methods: lag-window (MA), maximum entropy (AR); and iterative least-squares (ARMA).

MA Process (Wavelet, No Random Noise)

For the MA time series, we choose the *Ricker wavelet*, which has found widespread use in seismic work as a source pulse representation. It is, in fact, the second derivative of the Gaussian (or error density) function. A particular example is shown in Figure 16-2. The temporal distance between the two positive peaks (its breadth) is a basic parameter that fixes the peak frequency of the power spectrum. Figure 16-3 shows the true spectrum (solid curve) of the Ricker wavelet, that which is known analytically.

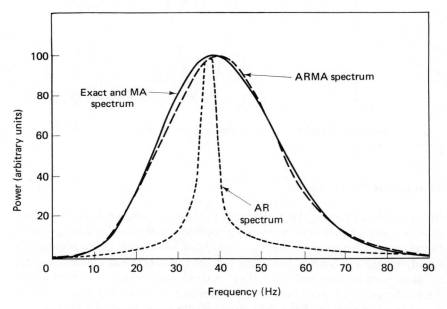

Figure 16-3. Application of autoregressive (AR) and autoregressive-moving average (ARMA) spectral analysis to the known moving average (MA) process of Figure 16-2.

The AR (maximum entropy) estimate, indicated by the short dashed curve, displays the typical high-resolution, or "peaky" nature characteristic of this technique: the peak frequency of the MA wavelet has been well defined at the expense of the spectrum shape. The ARMA estimate computed by the iterative least-squares algorithm is shown by the long dashed curve. Clearly, it better approximates the true spectral shape than does the AR estimate. However, the ARMA spectrum peaks at the incorrect frequency, and we are forced to the conclusion that this is not the "correct" analysis method to use for these input data. The proper approach is, of course, the MA estimator and, as is shown in the figure, this spectrum follows the true spectrum very closely. The Ricker wavelet was described by the 26-length sequence $(x_0, x_1, \ldots, x_{25})$, to which an MA estimator of order $n = 25$ was fitted with the classical lag-window approach described in Appendix 16-1. Of course, a choice of a smaller value of n would have produced poorer MA estimates of the true spectrum. The best AR estimator obtained by the MEM technique was of order $m = 6$, while the best ARMA estimator was of order $m = 1$ and $n = 24$. In other words, the best ARMA estimator was obtained in the form

$$x_0 + x_1 z + \cdots + x_{25} z^{25} \doteq \frac{b_0 + b_1 z + \cdots + b_{24} z^{24}}{a_0 + a_1 z}$$

AR Process

Let us change the input data to an AR process. For this purpose, we choose the fourth-order denominator polynomial,

$$B_4(z) = 1.0 - 2.7607z + 3.8160z^2 - 2.6535z^3 + 0.9238z^4$$

with which we generate the series y_t by means of equation (16-2a), where $x_t = \epsilon_t$ is a random sequence. The series y_t is displayed in Figure 16-4. The exact spectrum is shown in Figure 16-5 as the solid curve, and can be easily determined from the postulated known denominator polynomial $B_4(z)$.

The MA spectrum indicated by the short dashed curve is characterized by broad peaks which do not accurately coincide with the true spectral peaks. As is shown by the long dashed curve, the ARMA estimator does somewhat better than the MA estimator because the peaks are now sharper, and resolution has increased. However, when we employ the proper AR (MEM) estimator for this case, the estimated spectrum coincides with the true spectrum within plotting accuracy.

ARMA Process

The last case we consider is the ARMA process defined by the expressions

$$B_2(z) = 1.0 - 1.1z + 0.24z^2$$
$$A_3(z) = 1.0 - 0.5z + 0.25z^2 - 0.125z^3$$

in which we have made the MA component $B_2(z)$ of order two $(n = 2)$, and the AR component $A_3(z)$ of order three $(m = 3)$. A realization of this process

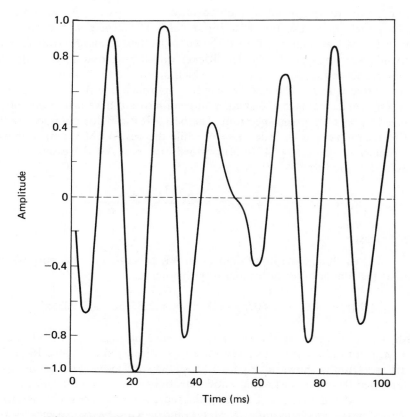

Figure 16-4. Model of a fourth-order ($m = 4$) autoregressive (AR) process.

is shown in Figure 16-6, and the true spectrum derived from the two poly-nomials is depicted by the solid curve in Figure 16-7. The MA spectral estimate is indicated by the short dashed curve. It is characterized by several spurious peaks, as well as by peak frequency shifts. The AR estimate obtained with the Burg algorithm (long dashed curve) appears to be better than the MA estimate, but still suffers a small frequency shift. As expected, the ARMA estimate (dot-dashed curve) most closely follows the true spectrum, and thus represents the "correct" analysis technique for these particular data.

There is an inherent difficulty when one applies either the AR or the least-squares ARMA spectral estimation methods. This is the problem of determining the correct orders of the polynomials in the rational model to be fitted to the input time series. Let us return to the AR example shown in Figures 16-4 and 16-5.

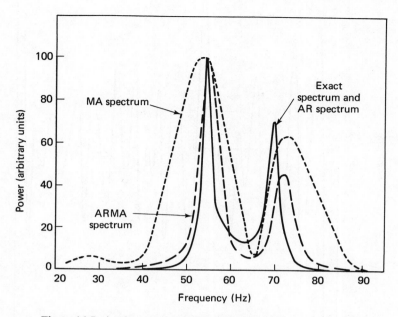

Figure 16-5. Application of moving average (MA) and autoregressive-moving average (ARMA) spectral analysis to the known autoregressive (AR) process of Figure 16-4.

Figure 16-8 is a plot of the partial correlation coefficient, a_m, as a function of the order, m, of the fitted AR process. We recall that a_m is the last term of the autoregressive component $A_m(z)$. We see that the coefficient a_m decreases sharply by two orders of magnitude as the order increases from $m = 4$, the correct value, to $m = 5$. This essentially means that because a_5 is small with respect to unity, the fitted prediction error polynomial $A_5(z)$ applied to the input data has resulted in an uncorrelated error series, and hence $A_4(z)$ is of sufficient order to contain all the predictable information present in the data. Not surprisingly, as m increases further, the partial correlation coefficient oscillates by as much as an order of magnitude. This behavior is to be expected, since the Burg algorithm tries to fit the mth-order polynomial $A_m(z)$ to the noisy data in the least-squares sense, with the constraint that the error power by minimal. Hence, a_m will vary with order as the polynomial fit varies. The important fact is that once it has decreased sharply (in this case, by two orders of magnitude) for some value of m, a_m is unlikely to again increase by that amount to values it had for m less than the critical order.

Figure 16-9 shows the normalized squared error (NSE), or deviation of the estimated spectrum from the true spectrum as a function of the order,

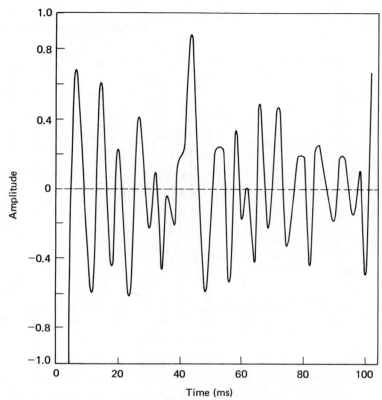

Figure 16-6. Model of an autoregressive-moving average (ARMA) process, whose MA component is of order two ($n = 2$), and whose AR component is of order three ($m = 3$).

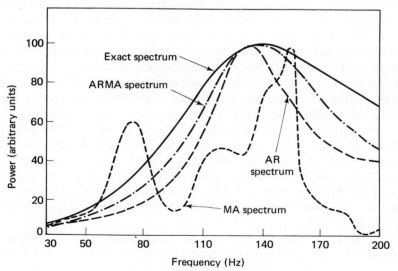

Figure 16-7. Application of moving average (MA) and autoregressive (AR) spectral analysis to the known autoregressive-moving average process of Figure 16-6.

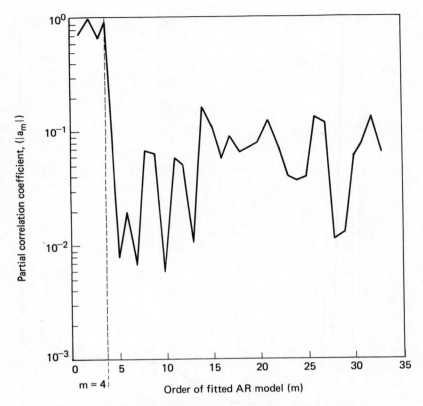

Figure 16-8. Magnitude of partial correlation coefficient, $|a_m|$, versus order of AR model (m), fitted to fourth-order AR process of Figure 16-4.

m, of the fitted AR process. As m reaches the correct value ($m = 4$), the estimated spectrum approaches the true spectrum. There appears to be a small range of order ($m = 4, 5, 6, 7$) within which the NSE reaches a minimum. Beyond this interval, the NSE increases with increasing order; this indicates that the estimated spectrum is diverging from the true spectrum. It is interesting to note that the absolute minimum of the partial correlation coefficient curve in Figure 16-8 (which corresponds to the best polynomial fit) occurs for $m = 10$, whereas Figure 16-9 shows that this is, in fact, not the best fit to the true spectrum. There is reason to suspect that the white noise sequence, ϵ_t, causes problems in associating the absolute minimum of the partial correlation coefficient curve with the correct order of the AR process, and that it would be better to use the first large decrease in a_m as a test for the order of the system.

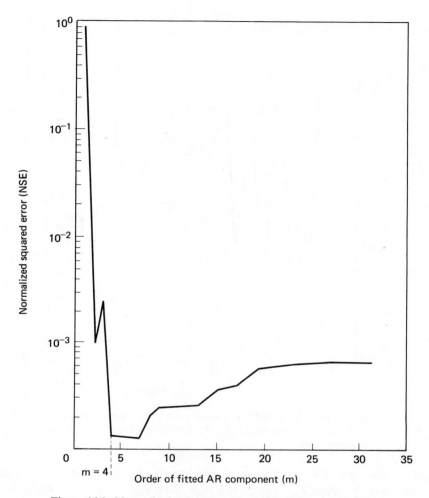

Figure 16-9. Normalized squared error (NSE) between the estimated and the true spectrum for the AR model of Figure 16-4. The NSE is plotted versus order (m) of the fitted AR process.

Figure 16-10 displays the normalized squared error (NSE), or deviation between the estimated spectra and the true spectrum for the ARMA model of Figure 16-6. All relevant spectra are shown in Figure 16-7. The number of iterations in the ARMA spectral estimation algorithm was set at two, and the coefficients of the initial numerator guess were $\pm 20\%$ of the true values. These conditions resulted in normalized squared errors in the ARMA spectral estimation algorithm of less than 10^{-6}. The NSE is plotted as a function of the order of the AR component (m) for several choices of n,

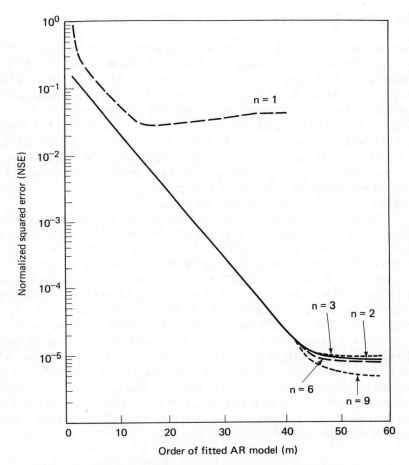

Figure 16-10. Normalized squared error (NSE) between the estimated spectrum and the true spectrum for the ARMA model of Figure 16-6. The NSE is plotted versus AR component order (m) for various choices of MA component order (n). The number of iterations in the ARMA spectral estimation algorithm was held at $n = 2$.

the order of the MA component. For $n = 1$ (first power in z), it is impossible to achieve a good fit to the true spectrum. However, once n increases to the correct order ($n = 2$), the fit to the true spectrum becomes better with increasing order of the AR component, m. Further increases in n achieve very little improvement in the NSE, which suggests that overestimation of the orders of both the MA and AR components may not present a problem in the practical sense. Underestimation of both m and n, on the other hand, does result in poor spectral fits.

Concluding Remarks

If we assume that a sample of a discrete time series has been generated by a linear, time-invariant mechanism, our data can be described by one of three possible representations, given respectively by the AR, MA, and ARMA models. Once we have made a model choice, we employ an appropriate algorithm to provide us with the desired spectral estimate. The AR model is best handled with Burg's maximum entropy method (MEM). The MA model can be treated with the classical lag-window approach, while for the ARMA model we favor the iterative least-squares algorithm, which guarantees a minimum-delay feedback component. We have presented the details of a study with synthetic data that show what price is paid when we apply inappropriate spectral estimators to the data at hand.

Many serious problems remain. First, we do not really have good practical means to determine a priori whether a real-life situation corresponds to an AR, MA, or ARMA process. This is the "identification problem," about which much has been written, but which in our opinion still remains an unresolved issue. The development of a simple and reliable test to make this decision would represent an invaluable contribution to the state of the art. Second, there is the problem of determining the order of a given process. While we have seen that this question can be answered reasonably well in an empirical manner, the development of more rigorous order-determining procedures would be desirable. In the case of the AR model, the order of the feedback component can indeed by determined by monitoring the magnitude of the partial correlation coefficient, or of some parameter simply related to it (Akaike, 1969), but this approach tends to break down when the process fails to satisfy the AR hypothesis. Quite evidently, there is much challenge for innovative research in this field.

mathematical principles of classical lag-window spectral analysis

The Periodogram of a Sample of White Noise

One of the best treatments of power spectral estimation can be found in Oppenheim and Schafer (1975, Chap. 11). In this appendix we give the mathematical principles of the classical method of spectral analysis. Fundamental to this approach is the periodogram. First let us consider white noise.

Let ϵ_t $(-\infty < t < \infty)$ be a purely random stationary stochastic process where the ϵ_t are independent random variables for which

$$E\{\epsilon_t\} = 0, \qquad E\{\epsilon_t^2\} = \sigma^2, \qquad E\{\epsilon_t^4\} = \alpha_4 < \infty \qquad (16\text{-}13)$$

The ϵ_t process is called a *white noise process*.

Suppose that a sample of N observations $\epsilon_1, \epsilon_2, \ldots, \epsilon_N$ has been observed. Then the periodogram of white noise is

$$I_N(\omega) = \frac{1}{N} \left| \sum_{t=1}^{N} \epsilon_t e^{-it\omega} \right|^2$$

$$= \frac{1}{N} \sum_{s=1}^{N} \sum_{t=1}^{N} \epsilon_s \epsilon_t e^{i(t-s)\omega} \qquad \checkmark \quad (16\text{-}14)$$

which has expected value

$$E\{I_N(\omega)\} = \frac{1}{N} N\sigma^2 = \sigma^2 \qquad (16\text{-}15)$$

But the spectral density of the process is

$$\Phi(\omega) = \sum_{t=-\infty}^{\infty} \phi(t)e^{-i\omega t} = \phi(0) = \sigma^2 \qquad (16\text{-}16)$$

(where $\phi(t) = E\{\epsilon_{t+s}\epsilon_t\}$ is the autocovariance). Thus, the expected value (16-15) of the periodogram is equal to the spectral density (16-16). Hence, we say that the periodogram $I_N(\omega)$ is the *unbiased* estimator for the spectral density $\Phi(\omega)$ for all N.

Let us now consider the expected value of the product of the values of the periodogram of white noise at frequencies ω_1 and ω_2, that is,

$$E\{I_N(\omega_1)I_N(\omega_2)\} \qquad (16\text{-}17)$$

If $\omega_1 \neq \pm\omega_2$, then (16-17) is equal to

$$\frac{1}{N^2}\sum_{s=1}^{N}\sum_{t=1}^{N}\sum_{u=1}^{N}\sum_{v=1}^{N} E\{\epsilon_s\epsilon_t\epsilon_u\epsilon_v\}e^{i(t-s)\omega_1}e^{i(v-u)\omega_2} \qquad (16\text{-}18)$$

where $s, t, u, v = 1, 2, \ldots, N$. Expression (16-18) is equal to the sum of the following four nonzero contributions:

1. When $s \neq t, u = s, v = t$: $\sigma^4 \sum\limits_{\substack{s=1 \\ s\neq t}}^{N}\sum\limits_{t=1}^{N} e^{i(t-s)(\omega_1+\omega_2)}$.

2. When $s \neq t, u = t, v = s$: $\sigma^4 \sum\limits_{\substack{s=1 \\ s\neq t}}^{N}\sum\limits_{t=1}^{N} e^{i(t-s)(\omega_1-\omega_2)}$.

3. When $s = t, u = v \neq s$: $\sigma^4(N-1)N$.

4. When $s = t = u = v$: $N\alpha_4$.

But

$$\sigma^4 \sum_{\substack{s=1 \\ s\neq t}}^{N}\sum_{t=1}^{N} e^{i(t-s)(\omega_1\pm\omega_2)} = \sigma^4 \sum_{s=1}^{N}\sum_{t=1}^{N} e^{i(t-s)(\omega_1\pm\omega_2)} - \sigma^4 N$$

Hence, when $\omega_1 \neq \pm\omega_2$, then (16-17) is

$$E\{I_N(\omega_1)I_N(\omega_2)\} = \frac{1}{N^2}\Bigg[\sigma^4 \sum_{s=1}^{N}\sum_{t=1}^{N} e^{i(t-s)(\omega_1+\omega_2)} - \sigma^4 N$$

$$+ \sigma^4 \sum_{s=1}^{N}\sum_{t=1}^{N} e^{i(t-s)(\omega_1-\omega_2)} - \sigma^4 N + \sigma^4 N^2 - \sigma^4 N + N\alpha_4\Bigg] \qquad (16\text{-}19)$$

Now

$$\sum_{s=1}^{N}\sum_{t=1}^{N} e^{i(t-s)\omega} = K(\omega) \qquad (16\text{-}20)$$

where $K(\omega)$ is the Fejer kernel defined by

$$K(\omega) = \left[\frac{\sin N\omega/2}{\sin \omega/2}\right]^2 \qquad (16\text{-}21)$$

Hence (16-19) becomes

$$E\{I_N(\omega_1)I_N(\omega_2)\} = \sigma^4 + \frac{\alpha_4 - 3\sigma^4}{N} + \frac{\sigma^4}{N^2}[K(\omega_1 + \omega_2) + K(\omega_1 - \omega_2)] \qquad (16\text{-}22)$$

In case the process is Gaussian (so $\alpha_4 = 3\sigma^4$; see, e.g., Cramér, 1945, p. 212), equation (16-22) is the same as equation (11.38) in Oppenheim and Schafer (1975, p. 544). Because

$$\text{cov}\{I_N(\omega_1), I_N(\omega_2)\} = E\{I_N(\omega_1)I_N(\omega_2)\} - E\{I_N(\omega_1)\}E\{I_N(\omega_2)\}$$
$$= E\{I_N(\omega_1)I_N(\omega_2)\} - \sigma^4$$

we see that the covariance is given by

$$\text{cov}\{I_N(\omega_1), I_N(\omega_2)\} = \frac{\alpha_4 - 3\sigma^4}{N} + \frac{\sigma^4}{N^2}[K(\omega_1 + \omega_2) + K(\omega_1 - \omega_2)] \quad (16\text{-}23)$$

where $\omega_1 \neq \pm\omega_2$. In the case of a Gaussian process (so $\alpha_4 = 3\sigma^4$), equation (16-23) is the same as equation (11.40) in Oppenheim and Schafer (1975, p. 545). When $\omega_2 \longrightarrow \omega_1$,

$$\frac{\sigma^4}{N^2}K(\omega_1 - \omega_2) \longrightarrow \frac{\sigma^4}{N^2}\left[\frac{\frac{N(\omega_1 - \omega_2)}{2}}{\frac{\omega_1 - \omega_2}{2}}\right]^2 = \sigma^4$$

Therefore, the variance is

$$\text{var } I_N(\omega) = \begin{cases} \dfrac{\alpha_4 - 3\sigma^4}{N} + \dfrac{\sigma^4}{N^2}K(2\omega) + \sigma^4 & \text{for } \omega \neq 0, \ \pm\pi \\[2mm] \dfrac{\alpha_4 - 3\sigma^4}{N} + 2\sigma^4 & \text{for } \omega = 0, \ \pm\pi \end{cases} \quad (16\text{-}24)$$

since

$$\frac{\sigma^4}{N^2}K(2\omega) \longrightarrow \sigma^4 \text{ as } \omega \longrightarrow 0, \text{ or } \pi, \text{ or } -\pi$$

Equation (16-24) for the variance, in the case of a Gaussian process (so $\alpha_4 = 3\sigma^4$), is the same as formula (11.41) in Oppenheim and Schafer (1975, p. 545). If $\omega_j = 2\pi j/N$ and $\omega_k = 2\pi k/N$, where j and k are integers such that $1 \leq j \leq N$ and $1 \leq k \leq N$, then

$$\sin \frac{N}{2}(\omega_j \pm \omega_k) = \sin \pi(j \pm k) = 0$$

whereas

$$\sin \frac{\omega_j \pm \omega_k}{2} = \sin \frac{\pi}{N}(j \pm k) \neq 0 \qquad \text{if } \omega_j \neq \pm\omega_k$$

Thus,

$$K(\omega_j + \omega_k) = 0 \qquad \text{if } \omega_j \neq \pm\omega_k$$

If $\omega_j = -\omega_k \neq 0$, then

$$\frac{1}{N^2} K(\omega_j + \omega_k) = 1 \quad \text{and} \quad K(\omega_j - \omega_k) = 0$$

If $\omega_j = \omega_k \neq 0$, then

$$K(\omega_j + \omega_k) = 0 \quad \text{and} \quad \frac{1}{N^2} K(\omega_j - \omega_k) = 1$$

If $\omega_j = \omega_k = 0$, then

$$\frac{1}{N^2} K(\omega_j + \omega_k) = 1 \quad \text{and} \quad \frac{1}{N^2} K(\omega_j - \omega_k) = 1$$

Therefore, for $\omega_j = 2\pi j/N$ and $\omega_k = 2\pi k/N$ (with $1 \leq j \leq N$, $1 \leq k \leq N$),

$$\text{cov}\{I_N(\omega_j), I_N(\omega_k)\} = \begin{cases} \dfrac{\alpha_4 - 3\sigma^4}{N} & \text{for } j \neq \pm k \\[2mm] \dfrac{\alpha_4 - 3\sigma^4}{N} + \sigma^4 & \text{for } j = -k \neq 0 \\[2mm] \text{var}\{I_N(\omega_j)\} = \dfrac{\alpha_4 - 3\sigma^4}{2N + 1} + \sigma^4 & \text{for } j = k \neq 0 \\[2mm] \text{var}\{I_N(0)\} = \dfrac{\alpha_4 - 3\sigma^4}{2N + 1} + 2\sigma^4 & \text{for } j = k = 0 \end{cases}$$

$$(16\text{-}25)$$

For a Gaussian stochastic process ϵ_t, then $\alpha_4 - 3\sigma^4 = 0$, and hence (where $\omega_j = 2\pi j/N$, $\omega_k = 2\pi k/N$ and where $1 \leq j, k \leq N$)

$$\text{cov}\{I_N(\omega_j), I_N(\omega_k)\} = \begin{cases} 0 & \text{for } j \neq \pm k \\ \text{var}\{I_N(\omega_j)\} = \sigma^4 & \text{for } j = k \neq 0 \\ \text{var}\{I_N(0)\} = 2\sigma^4 & \text{for } j = k = 0 \end{cases} \quad (16\text{-}26)$$

For arbitrary frequencies ω_1 and ω_2, we have (where $\omega_1 \neq \pm \omega_2$)

$$\text{cov}\{I_N(\omega_1), I_N(\omega_2)\} = \begin{cases} 0\left(\dfrac{1}{N}\right) & \text{for non-Gaussian process} \\[2mm] 0\left(\dfrac{1}{N^2}\right) & \text{for Gaussian process} \end{cases} \quad (16\text{-}27)$$

which says that $I_N(\omega_1)$ and $I_N(\omega_2)$ are *asymptotically uncorrelated*. Here the large 0 is the mathematical symbol "of the order of."

From (16-24) we see that

$$\lim_{N \to \infty} \text{var } I_N(\omega) = \begin{cases} \sigma^4 & \text{for } \omega \neq 0 \\ 2\sigma^4 & \text{for } \omega = 0 \end{cases} \quad (16\text{-}28)$$

Equation (16-28) shows that the variance of $I_N(\omega)$ does not approach zero as N approaches infinity. Thus, we say that the periodogram is an *inconsistent* estimator of the spectral density $\Phi(\omega)$.

We see from (16-26) that the covariance of $I_N(2\pi j/N)$ and $I_N(2\pi k/N)$, where j and k are integers, is equal to zero for $j \neq k$. Thus, values of the periodogram $I_N(2\pi j/N)$ spaced in angular frequency by integer j multiples of $2\pi/N$ are uncorrelated. As N increases, the uncorrelated values $I_N(2\pi j/N)$, $j = 1, 2, \ldots, N$, come closer together. Because by (16-28) the variance of $I_N(2\pi j/N)$ approaches a nonzero constant σ^4, and the spacing between the periodogram values with zero covariance decreases as N increases, it follows that as the record length N becomes longer the rapidity of the fluctuations in the periodogram increases (Oppenheim and Schafer, 1975, p. 545).

The Periodogram of a Sample of a Stationary Time Series

Let x_t be a stationary stochastic process with spectral representation

$$x_t = \frac{1}{2\pi} \int_{-\pi}^{\pi} e^{it\omega} X(\omega)\, d\omega \qquad (16\text{-}29)$$

where

$$E\{X(\omega)X^*(\mu)\} = 2\pi\delta(\omega - \mu)\Phi(\omega) \qquad (16\text{-}30)$$

[the superscript asterisk denoting complex conjugate, the δ being the Dirac delta function, and $\Phi(\omega)$ being the spectral density function plus a line spectrum of the form

$$\sum_k \rho_k[\delta(\omega - \omega_k) + \delta(\omega + \omega_k)] \qquad (16\text{-}31)$$

where $\rho_k \geq 0$ are the line spectral intensities].

The sample $x_1, x_2, \ldots, x_N$ is observed. The sample autocovariance is

$$\phi_N(s) = \frac{1}{N} \sum_{t=1}^{N-s} x_{t+s} x_t \qquad \text{(for } 0 \leq s \leq N - 1)$$

$$\phi_N(-s) = \phi_N(s) \qquad\qquad (16\text{-}32)$$

The sample spectral density (or *periodogram*) is

$$I_N(\omega) = \sum_{s=-N+1}^{N-1} \phi_N(s) e^{-i\omega s} = \frac{1}{N} \sum_{s=1}^{N} \sum_{t=1}^{N} x_s x_t e^{-i\omega(s-t)}$$

$$= \frac{1}{N} \left| \sum_{s=1}^{N} x_s e^{-i\omega s} \right|^2 \qquad (16\text{-}33)$$

From the spectral representation, we have

$$X_N(\omega) = \sum_{s=1}^{N} x_s e^{-i\omega s} = \sum_{s=1}^{N} \left[\frac{1}{2\pi} \int_{-\pi}^{\pi} e^{is\mu} X(\mu) \, d\mu \right] e^{-i\omega s}$$

$$= \frac{1}{2\pi} \int_{-\pi}^{\pi} d\mu \, X(\mu) \sum_{s=1}^{N} e^{-is(\omega-\mu)} \qquad (16\text{-}34)$$

Now let us define the Dirichlet kernel $D(\omega)$ as

$$D(\omega) = \sum_{s=1}^{N} e^{-i\omega s} \qquad (16\text{-}35)$$

Note that $K(\omega) = |D(\omega)|^2$. Thus,

$$X_N(\omega) = \frac{1}{2\pi} \int_{-\pi}^{\pi} X(\mu) D(\omega - \mu) \, d\mu$$

$$= \frac{1}{2\pi} X(\omega) * D(\omega) \qquad (16\text{-}36)$$

where $*$ indicates convolution. Now the periodogram is

$$I_N(\omega) = \frac{1}{N} |X_N(\omega)|^2 = \frac{1}{4\pi^2 N} \int_{-\pi}^{\pi} d\mu \, X(\mu) D(\omega - \mu) \int_{-\pi}^{\pi} d\lambda \, X^*(\lambda) D^*(\omega - \lambda) \qquad (16\text{-}37)$$

(where the superscript asterisk $*$ indicates the complex conjugate). Hence,

$$E\{I_N(\omega)\} = \frac{1}{4\pi^2 N} \int_{-\pi}^{\pi} d\mu \, D(\omega - \mu) \int_{-\pi}^{\pi} d\lambda \, D^*(\omega - \lambda) E\{X(\mu)X^*(\lambda)\}$$

$$= \frac{1}{4\pi^2 N} \int_{-\pi}^{\pi} d\mu \, D(\omega - \mu) \int_{-\pi}^{\pi} d\lambda \, D^*(\omega - \lambda) 2\pi\delta(\mu - \lambda) \Phi(\mu)$$

$$= \frac{1}{2\pi N} \int_{-\pi}^{\pi} \Phi(\mu) |D(\omega - \mu)|^2 \, d\mu$$

$$= \frac{1}{2\pi N} \int_{-\pi}^{\pi} \Phi(\mu) K(\omega - \mu) \, d\mu$$

$$= \frac{1}{2\pi N} \Phi(\omega) * K(\omega) \qquad (16\text{-}38)$$

Now

$$\frac{1}{2\pi N} K(\omega) \longrightarrow \delta(\omega) \qquad \text{as } N \longrightarrow \infty \qquad (16\text{-}39)$$

Thus,

$$\lim_{N \to \infty} E\{I_N(\omega)\} = \Phi(\omega) * \delta(\omega) = \Phi(\omega) \qquad (16\text{-}40)$$

(provided that $\omega \neq \pm\omega_k$; i.e., that ω is not a frequency of the line spectrum). This says that the periodogram is an *asymptotically unbiased* estimator of $\Phi(\omega)$ at every frequency that is *not* a spectral line. [*Note:* The contribution of the line spectrum to $E\{I_N(\omega)\}$ is

$$\frac{1}{2\pi N}\int_{-\pi}^{\pi}\sum_k \rho_k[\delta(\mu-\omega_k)+\delta(\mu+\omega_k)]K(\omega-\mu)\,d\mu$$

$$=\frac{1}{2\pi N}\sum_k \rho_k[K(\omega-\omega_k)+K(\omega+\omega_k)] \qquad (16\text{-}41)$$

which, as $N\to\infty$, tends to zero if $\omega \neq \pm\omega_k$, and tends to infinity if $\omega = \pm\omega_k$.]

It is a theorem in statistics that the following identify holds for Gaussian distributed random variables y_1, y_2, y_3, y_4 with zero means (see, e.g., Oppenheim and Schafer, 1975, p. 544):

$$E\{y_1y_2y_3y_4\} \equiv E\{y_1y_2\}E\{y_3y_4\} + E\{y_1y_3\}E\{y_2y_4\}$$
$$+ E\{y_1y_4\}E\{y_2y_3\} \qquad (16\text{-}42)$$

It follows that

$$\text{cov}\{y_1y_2, y_3y_4\} = E\{y_1y_2y_3y_4\} - E\{y_1y_2\}E\{y_3y_4\}$$
$$= E\{y_1y_3\}E\{y_2y_4\} + E\{y_1y_4\}E\{y_2y_3\} \qquad (16\text{-}43)$$

We shall now assume that the process x_t is a Gaussian process. Then

$$X_N(\omega) = \sum_{s=1}^{N} x_s e^{-i\omega s} \qquad (16\text{-}44)$$

is a Gaussian variable because it is a linear combination of Gaussian variables. [Note, however, that $X_N(\omega)$ is complex-valued, whereas x_t is real-valued.]

Now

$$\text{cov}\{I_N(\omega), I_N(\mu)\} = \text{cov}\left\{\frac{1}{N}X_N(\omega)X_N^*(\omega), \frac{1}{N}X_N(\mu)X_N^*(\mu)\right\}$$

$$= \frac{1}{N^2}\text{cov}\{X_N(\omega)X_N^*(\omega), X_N(\mu)X_N^*(\mu)\}$$

$$= \frac{1}{N^2}E\{X_N(\omega)X_N(\mu)\}E\{X_N^*(\omega)X_N^*(\mu)\}$$
$$+ E\{X_N(\omega)X_N^*(\mu)\}E\{X_N^*(\omega)X_N(\mu)\} \qquad (16\text{-}45)$$

Also

$$X_N(\omega) = \frac{1}{2\pi}\int_{-\pi}^{\pi} X(\lambda)D(\omega-\lambda)\,d\lambda$$
$$X_N(\mu) = \frac{1}{2\pi}\int_{-\pi}^{\pi} X(\lambda')D(\mu-\lambda')\,d\lambda' \qquad (16\text{-}46)$$

Hence, from (16-46) we have

$$E\{X_N(\omega)X_N(\mu)\} = \frac{1}{4\pi^2} \int_{-\pi}^{\pi} d\lambda \, D(\omega - \lambda) \int_{-\pi}^{\pi} d\lambda' \, D(\mu - \lambda') E\{X(\lambda)X(\lambda')\}$$

$$(16\text{-}47)$$

But, since

$$X(\lambda') = \sum_{t=-\infty}^{\infty} x_t e^{-i\lambda't} = \left[\sum_{t=-\infty}^{\infty} x_t e^{-i(-\lambda')t}\right]^* = [X(-\lambda')]^* = X^*(-\lambda') \quad (16\text{-}48)$$

we have

$$E\{X(\lambda)X(\lambda')\} = E\{X(\lambda)X^*(-\lambda')\} = 2\pi\delta(\lambda + \lambda')\Phi(\lambda) \qquad (16\text{-}49)$$

Therefore,

$$E\{X_N(\omega)X_N(\mu)\} = \frac{1}{4\pi^2} \int_{-\pi}^{\pi} d\lambda \, D(\omega - \lambda) \int_{-\pi}^{\pi} d\lambda' \, D(\mu - \lambda')2\pi \, \delta(\lambda + \lambda')\Phi(\lambda)$$

$$= \frac{1}{2\pi} \int_{-\pi}^{\pi} d\lambda \, D(\omega - \lambda)D(\mu + \lambda)\Phi(\lambda) \qquad (16\text{-}50)$$

Also

$$E\{X_N^*(\omega)X_N^*(\mu)\} = [E\{X_N(\omega)X_N(\mu)\}]^* \qquad (16\text{-}51)$$

Now

$$E\{X_N(\omega)X_N^*(\mu)\} = \frac{1}{4\pi^2} \int_{-\pi}^{\pi} d\lambda \, D(\omega - \lambda) \int_{-\pi}^{\pi} d\lambda' \, D^*(\mu - \lambda')E\{X(\lambda)X^*(\lambda')\}$$

$$= \frac{1}{4\pi^2} \int_{-\pi}^{\pi} d\lambda \, D(\omega - \lambda) \int_{-\pi}^{\pi} d\lambda' \, D^*(\mu - \lambda')2\pi \, \delta(\lambda - \lambda')\Phi(\lambda)$$

$$= \frac{1}{2\pi} \int_{-\pi}^{\pi} d\lambda \, D(\omega - \lambda) \, D^*(\mu - \lambda)\Phi(\lambda) \qquad (16\text{-}52)$$

and

$$E\{X_N^*(\omega)X_N(\mu)\} = [E\{X_N(\omega)X_N^*(\mu)\}]^* \qquad (16\text{-}53)$$

Therefore, substituting (16-50) to (16-53) into (16-45), we obtain

$$\text{cov}\{I_N(\omega), I_N(\mu)\} = \frac{1}{4\pi^2 N^2}\left[\left|\int_{-\pi}^{\pi} \Phi(\lambda)D(\omega - \lambda)D(\mu + \lambda) \, d\lambda\right|^2\right.$$

$$\left. + \left|\int_{-\pi}^{\pi} \Phi(\lambda)D(\omega - \lambda)D(\mu - \lambda) \, d\lambda\right|^2\right] \qquad (16\text{-}54)$$

For $\omega \neq \pm\mu$, it may be shown that this covariance tends to zero as $N \longrightarrow \infty$.

As a check, let us suppose that $\Phi(\omega) = \sigma^2$, so that we have a Gaussian white noise process. Then

$$\frac{1}{2\pi} \int_{-\pi}^{\pi} \Phi(\lambda) D(\omega - \lambda) D(\mu + \lambda) \, d\lambda = \frac{\sigma^2}{2\pi} \int_{-\pi}^{\pi} \sum_{s=1}^{N} e^{-i(\omega-\lambda)s} \sum_{r=1}^{N} e^{-i(\mu+\lambda)r} \, d\lambda$$

$$= \sum_{s=1}^{N} \sum_{r=1}^{N} e^{-i\omega s} e^{-i\mu r} \frac{\sigma^2}{2\pi} \int_{-\pi}^{\pi} e^{i\lambda(s-r)} \, d\lambda$$

But

$$\frac{1}{2\pi} \int_{-\pi}^{\pi} e^{i\lambda(s-r)} \, d\lambda = \delta_{s,r} = \begin{cases} 1 & \text{if } s = r \\ 0 & \text{otherwise} \end{cases}$$

Thus,

$$\frac{1}{2\pi} \int_{-\pi}^{\pi} \sigma^2 D(\omega - \lambda) D(\mu + \lambda) \, d\lambda = \sum_{s=1}^{N} \sum_{r=1}^{N} e^{-i\omega s} e^{-i\mu r} \sigma^2 \delta_{s,r}$$

$$= \sum_{s=1}^{N} e^{-i(\omega+\mu)s} \sigma^2 = \sigma^2 D(\omega + \mu)$$

Also,

$$\frac{1}{2\pi} \int_{-\pi}^{\pi} \sigma^2 D(\omega - \lambda) D(\mu - \lambda) \, d\lambda = \frac{\sigma^2}{2\pi} \int_{-\pi}^{\pi} \sum_{s=1}^{N} e^{-i(\omega-\lambda)s} \sum_{r=1}^{N} e^{-i(\mu-\lambda)r} \, d\lambda$$

$$= \sum_{s=1}^{N} \sum_{r=1}^{N} e^{-i\omega s} e^{-i\mu r} \frac{\sigma^2}{2\pi} \int_{-\pi}^{\pi} e^{i\lambda(s+r)} \, d\lambda$$

$$= \sum_{s=1}^{N} \sum_{r=1}^{N} e^{-i\omega s} e^{-i\mu r} \sigma^2 \, \delta_{s,-r} = \sigma^2 \sum_{s=1}^{N} e^{-i(\omega-\mu)s} = \sigma^2 D(\omega - \mu)$$

Therefore, for Gaussian white noise, we have

$$\text{cov} \{ I_N(\omega), I_N(\mu) \} = \frac{1}{N^2} [\, |\sigma^2 D(\omega + \mu)|^2 + |\sigma^2 D(\omega - \mu)|^2]$$

$$= \frac{\sigma^4}{N^2} [K(\omega + \mu) + K(\omega - \mu)] \tag{16-55}$$

Equation (16-55) agrees with the result of the last section, namely equation (16-24) with $\alpha_4 - 3\sigma^4 = 0$, since $\alpha_4 = 3\sigma^4$ for a Gaussian variable.

Letting $\omega = \mu$ in expression (16-54) for $\text{cov} \{ I_N(\omega), I_N(\mu) \}$, we obtain

$$\text{var} \{ I_N(\omega) \} = \frac{1}{N^2} \left[\left(\frac{1}{2\pi} \int_{-\pi}^{\pi} \Phi(\lambda) K(\omega - \lambda) \, d\lambda \right)^2 \right.$$

$$\left. + \left| \frac{1}{2\pi} \int_{-\pi}^{\pi} \Phi(\lambda) D(\omega - \lambda) D(\omega + \lambda) \, d\lambda \right|^2 \right] \tag{16-56}$$

Now, as $N \to \infty$,

$$\frac{1}{2\pi N} \int_{-\pi}^{\pi} \Phi(\lambda) K(\omega - \lambda) \, d\lambda \longrightarrow \Phi(\omega) \qquad (16\text{-}57)$$

provided that ω is not a line frequency. Thus, the first term on the right-hand side of (16-56) tends to $[\Phi(\omega)]^2$ as $N \to \infty$ if ω is not a spectral line frequency. Consider now

$$\frac{1}{2\pi N} \int_{-\pi}^{\pi} \Phi(\lambda) D(\omega - \lambda) D(\omega + \lambda) \, d\lambda \qquad (16\text{-}58)$$

The contribution to (16-58) from the line spectral component of $\Phi(\lambda)$ tends to zero as $N \to \infty$, provided that ω is not a line frequency, so we need only investigate the contribution from the spectral density component of $\Phi(\lambda)$.

Case I. If $\omega = 0$, then (16-58) tends to $\Phi(0)$ as $N \to \infty$.

Case II. If $\omega \neq 0$ and $\omega \neq \pm \pi$, we shall divide $(-\pi, \pi)$ into six parts. We suppose that $\omega > 0$, and consider

$$(-\pi, -\omega - \epsilon), \ (-\omega - \epsilon, -\omega + \epsilon), \ (-\omega + \epsilon, 0),$$
$$(0, \omega - \epsilon'), \ (\omega - \epsilon', \omega + \epsilon'), \ (\omega + \epsilon', \pi)$$

and denote the corresponding integrals

$$\frac{1}{2\pi N} \int \Phi(\lambda) D(\omega - \lambda) D(\omega + \lambda) \, d\lambda$$

by I_1, I_2, I_3, I_4, I_5, and I_6. Here ϵ, ϵ' are small, arbitrary positive constants. By the first mean value theorem, I_1, I_3, I_4, and I_6 tend to zero as $N \to \infty$. Consider

$$I_5 = \frac{1}{2\pi N} \int_{\omega - \epsilon'}^{\omega + \epsilon'} \Phi(\lambda) D(\omega - \lambda) D(\omega + \lambda) \, d\lambda$$

$$= \frac{1}{2\pi N} \int_{-\epsilon'}^{\epsilon'} \Phi(\omega - t) D(t) D(2\omega - t) \, dt \qquad (\text{where } t = \omega - \lambda)$$

$$= \frac{1}{2\pi N} \left[\int_0^{\epsilon'} \Phi(\omega - t) D(t) D(2\omega - t) \, dt + \int_0^{\epsilon'} \Phi(\omega + t) D(t) D(2\omega + t) \, dt \right]$$

Assuming that $\Phi(\omega)$ is bounded, we have

$$I_5 \leq \frac{\text{const}}{N} \int_0^{\epsilon'} |D(t)| \, dt < \frac{\text{const}}{N} \int_0^{\pi} |D(t)| \, dt$$

$$< \frac{\text{const}}{N} O(\log N)$$

(see Zygmund, 1959, p. 67). Hence, $\lim_{N\to\infty} I_5 = 0$, and likewise $\lim_{N\to\infty} I_2 = 0$. Thus, when $\omega \neq 0$ and $\omega \neq \pm\pi$, (16-58) tends to zero as $N \to \infty$.

Case III. $\omega = \pi$ or $-\pi$. Then

$$D^*(\pm\pi - \lambda) = \left[\sum_{s=1}^{N} e^{-i(\pm\pi-\lambda)s}\right]^* = \sum_{s=1}^{N} e^{-i(\mp\pi+\lambda)s}$$

But $e^{-i(\mp\pi)s} = e^{-i(\pm\pi)s}$. Thus,

$$D^*(\pm\pi - \lambda) = \sum_{s=1}^{N} e^{-i(\pm\pi+\lambda)s} = D(\pm\pi + \lambda)$$

Thus,

$$\frac{1}{2\pi N} \int_{-\pi}^{\pi} \Phi(\lambda)D(\pm\pi - \lambda)D(\pm\pi + \lambda)\, d\lambda$$

$$= \frac{1}{2\pi N} \int_{-\pi}^{\pi} \Phi(\lambda)D(\pm\pi - \lambda)D^*(\pm\pi - \lambda)\, d\lambda$$

$$= \frac{1}{2\pi N} \int_{-\pi}^{\pi} \Phi(\lambda)\,|D(\pm\pi - \lambda)|^2\, d\lambda = \frac{1}{2\pi N} \int_{-\pi}^{\pi} \Phi(\lambda)K(\pm\pi - \lambda)\, d\lambda$$

which tends to $\Phi(\pm\pi)$ as $N \to \infty$. Therefore, the integral (16-58) tends to zero, unless $\omega = 0$, in which case it tends to $\Phi(0)$, or unless $\omega = \pm\pi$, in which case it tends to $\Phi(\pm\pi)$.

Summing up, we have from (16-56) the following result: if ω is not a frequency in the line spectrum, then, as $N \to \infty$,

$$\text{var}\,\{I_N(\omega)\} \longrightarrow \begin{cases} 2[\Phi(\omega)]^2 & \text{if } \omega = 0 \\ [\Phi(\omega)]^2 & \text{if } \omega \neq -\pi, 0, \pi \\ 2[\Phi(\omega)]^2 & \text{if } \omega = -\pi, \pi \end{cases} \qquad (16\text{-}59)$$

Thus, except in the trivial case when $\Phi(\omega) = 0$, $I_N(\omega)$ is not a consistent estimate of the spectral density $\Phi(\omega)$. The result (16-59) is the same as that given in Oppenheim and Schafer (1975, p. 553).

Eigenvalues of Toeplitz Matrices

Toeplitz (1907, 1910, 1911) studied the distribution of the eigenvalues of an infinite matrix defined by

$$[\psi_{s,t}] = [\psi_{s-t}] \qquad (16\text{-}60)$$

where the indices s and t range from $-\infty$ to ∞. That is, such a matrix is one whose elements depend only upon the differences $s - t$ of the indices, rather than on the indices themselves. Thus, all the elements on any given diagonal (i.e., main diagonal, subdiagonal, or superdiagonal) are all the same. A finite Toeplitz matrix is a section of an infinite matrix (16-60), and thus has the form

$$
\begin{bmatrix}
\psi_0 & \psi_{-1} & \cdots & \psi_{-N+1} \\
\psi_1 & \psi_0 & \cdots & \psi_{-N+2} \\
& & \vdots & \\
\psi_{N-1} & \psi_{N-2} & \cdots & \psi_0
\end{bmatrix}
\tag{16-61}
$$

Let $\Psi(\omega)$ be a real-valued integrable function, and define the Fourier coefficients ψ_t as

$$
\psi_t = \frac{1}{2\pi} \int_{-\pi}^{\pi} \Psi(\omega) e^{-i\omega t} \, d\omega
\tag{16-62}
$$

Under suitable conditions, the function can then be represented as the Fourier series

$$
\Psi(\omega) = \sum_{t=-\infty}^{\infty} \psi_t e^{i\omega t}
\tag{16-63}
$$

With the Fourier coefficients (16-62), form the Toeplitz matrix (16-61). In this way the function $\Psi(\omega)$ defines the Toeplitz matrix (16-61). Because

$$
\psi_t^* = \frac{1}{2\pi} \int_{-\pi}^{\pi} \Psi(\omega) e^{i\omega t} \, d\omega = \psi_{-t}
\tag{16-64}
$$

we see the Toeplitz matrix (16-61) is Hermitian. Thus, all the eigenvalues of the matrix (16-61) are real. Let us denote these eigenvalues by

$$
\lambda_1^{(N)} \le \lambda_2^{(N)} \le \cdots \le \lambda_N^{(N)}
\tag{16-65}
$$

For convenience, we assume that the eigenvalues are simple.

Let us now assume that the function $\Psi(\omega)$ is bounded both from above and below. Let M_1 and M_2 denote the essential lower and upper bound of $\Psi(\omega)$, respectively, so

$$
M_1 \le \Psi(\omega) \le M_2
\tag{16-66}
$$

Then the eigenvalues also fall within the same bounds (Grenander and Szegö, 1958, p. 64); that is,

$$
M_1 \le \lambda_j^{(n)} \le M_2 \qquad (j = 1, 2, \ldots, n)
\tag{16-67}
$$

Grenander and Szegö (1958, p. 64) give the following theorem: If $F(\lambda)$ is any continuous function defined in the finite interval $M_1 \leq \lambda \leq M_2$, where M_1 and M_2 are respectively the essential lower and upper bounds of the real integrable function $\Psi(\omega)$, then

$$\lim_{N \to \infty} \frac{1}{N}[F(\lambda_1^{(N)}) + F(\lambda_2^{(N)}) + \cdots + F(\lambda_N^{(N)})] = \frac{1}{2\pi} \int_{-\pi}^{\pi} F[\Psi(\omega)] \, d\omega \qquad (16\text{-}68)$$

Characteristic Functions

Let $f(u)$ be a probability density function. Then the moment μ_r' of order r about the origin is defined to be

$$\mu_r' = \int_{-\infty}^{\infty} u^r f(u) \, du \qquad (16\text{-}69)$$

The moment μ_r of order r about the mean μ_1' is defined to be

$$\mu_r = \int_{-\infty}^{\infty} (\mu - \mu_1')^r f(u) \, du \qquad (16\text{-}70)$$

The characteristic function is

$$\chi(t) = \int_{-\infty}^{\infty} e^{itu} f(u) \, du \qquad (16\text{-}71)$$

We have

$$\frac{d^r}{dt^r} \chi(t) = i^r \int_{-\infty}^{\infty} e^{itu} u^r f(u) \, du \qquad (16\text{-}72)$$

which gives

$$\mu_r' = (-i)^r \left[\frac{d^r}{dt^r} \chi(t) \right]_{t=0} \qquad (16\text{-}73)$$

From (16-71) we have the expansion

$$\chi(t) = \int_{-\infty}^{\infty} \left[1 + \frac{itu}{1!} + \frac{(itu)^2}{2!} + \cdots \right] f(u) \, du$$

$$= 1 + \mu_1' \frac{it}{1!} + \mu_2' \frac{(it)^2}{2!} + \cdots \qquad (16\text{-}74)$$

The cumulants $\kappa_1, \kappa_2, \ldots, \kappa_r, \ldots$ are defined by the identity in t,

$$\exp \left\{ \kappa_1 t + \kappa_2 \frac{(it)^2}{2!} + \cdots + \kappa_r \frac{(it)^r}{r!} + \cdots \right\}$$

$$= 1 + \mu_1' \frac{(it)}{1!} + \cdots + \mu_r' \frac{(it)^r}{r!} + \cdots \qquad (16\text{-}75)$$

obtained from the equation

$$\exp [\log \chi(t)] = \chi(t) \tag{16-76}$$

That is, as μ'_r is the coefficient of $(it)^r/r!$ in the expansion (16-74) of $\chi(t)$, κ_r is the coefficient of $(it)^r/r!$ in the expansion of $\log \chi(t)$. We have

$$1 + \mu'_1 \frac{t}{1!} + \cdots + \mu'_r \frac{t^s}{r!} + \cdots = \exp\left\{\kappa_1 \frac{t}{1!} + \kappa_2 \frac{t^2}{2!} + \cdots + \kappa_r \frac{t^r}{r!} + \cdots\right\}$$

$$= \exp\left(\frac{\kappa_1 t}{1!}\right) \exp\left(\frac{\kappa_2 t^2}{2!}\right) \cdots \exp\left(\frac{\kappa_r t^r}{r!}\right) \cdots$$

$$= \left\{1 + \frac{\kappa_1 t}{1!} + \frac{\kappa_1^2 t^2}{2!} + \cdots\right\}\left\{1 + \frac{\kappa_2 t^2}{2!} + \frac{1}{2!}\frac{\kappa_2^2 t^4}{(2!)^2} + \cdots\right\} \cdots$$

$$\left\{1 + \frac{\kappa_r t^r}{r!} + \cdots\right\} \cdots \tag{16-77}$$

Picking out the terms for t^r, we have

$$\mu'_r = \sum_{m=0}^{r} \sum \left(\frac{\kappa_{p_1}}{p_1!}\right)^{\pi_1}\left(\frac{\kappa_{p_2}}{p_2!}\right)^{\pi_2} \cdots \left(\frac{\kappa_{p_m}}{p_m!}\right)^{\pi_m} \frac{r!}{\pi_1!\,\pi_2!\cdots\pi_m!} \tag{16-78}$$

where the second summation extends over all nonnegative values of the π's such that

$$p_1\pi_1 + p_2\pi_2 + \cdots + p_m\pi_m = r$$

Thus, for moments about the origin,

$$\begin{aligned}
\mu'_1 &= \kappa_1 \\
\mu'_2 &= \kappa_2 + \kappa_1^2 \\
\mu'_3 &= \kappa_3 + 3\kappa_2\kappa_1 + \kappa_1^3 \\
\mu'_4 &= \kappa_4 + 4\kappa_3\kappa_1 + 3\kappa_2^2 + 6\kappa_2\kappa_1^2 + \kappa_1^4
\end{aligned} \tag{16-79}$$

and

$$\begin{aligned}
\kappa_1 &= \mu'_1 \\
\kappa_2 &= \mu'_2 - \mu'^2_1 \\
\kappa_3 &= \mu'_3 - 3\mu'_2\mu'_1 + 2\mu'^3_1 \\
\kappa_4 &= \mu'_4 - 4\mu'_3\mu'_1 - 3\mu'^2_2 + 12\mu'_2\mu'^2_1 - 6\mu'^4_1
\end{aligned} \tag{16-80}$$

For moments about the mean, we have $\kappa_1 = 0$ and

$$\begin{aligned}
\mu_2 &= \kappa_2 = \text{variance} \\
\mu_3 &= \kappa_3 \\
\mu_4 &= \kappa_4 + 3\kappa_2^2
\end{aligned} \tag{16-81}$$

$$\kappa_2 = \mu_2 = \text{variance}$$
$$\kappa_3 = \mu_3 \tag{16-82}$$
$$\kappa_4 = \mu_4 - 3\mu_2^2$$

Distribution of a Quadratic Form

The classical lag-window estimates of the power spectrum are in the form of quadratic forms in the observations of the time-series sample. The observed sample $x_1, x_2, \cdots, x_N$ of the time series may be represented by a column vector **x** given by

$$\mathbf{x} = \begin{bmatrix} x_1 \\ x_2 \\ \cdot \\ \cdot \\ \cdot \\ x_N \end{bmatrix} \tag{16-83}$$

Let **A** be a given $N \times N$ symmetric real Toeplitz matrix. As we will see, the matrix **A** is specified by the particular lag window used for the type of spectral estimate desired. The spectral estimate Q then takes the form

$$Q = \mathbf{x}^T \mathbf{A} \mathbf{x} \tag{16-84}$$

where the superscript T indicates matrix transpose. Thus, the problem of finding the probability distribution of the spectral estimate reduces to the problem of finding the probability distribution of the quadratic form Q. Since a probability distribution is specified by its cumulants, we will now seek to find the cumulants of the probability distribution of Q. In order to find a solution, we will assume that the random variables $x_1, x_2, \ldots, x_N$ have a joint Gaussian distribution. Thus, we let x_t be a Gaussian stationary stochastic process. The sample $x_1, x_2, \ldots, x_N$ then has a multidimensional Gaussian distribution. If $\phi_s = E\{x_{t+s}\, x_t\}$ is the autocovariance, the auto-covariance matrix of the sample $x_1, x_2, \ldots, x_N$ is the Toeplitz matrix

$$\mathbf{R} = \begin{bmatrix} \phi_0 & \phi_1 & \cdots & \phi_{N-1} \\ \phi_1 & \phi_0 & \cdots & \phi_{N-2} \\ & & \cdot & \\ & & \cdot & \\ & & \cdot & \\ \phi_{N-1} & \phi_{N-2} & \cdots & \phi_0 \end{bmatrix} \tag{16-85}$$

Let the characteristic function of the quadratic form $Q = \mathbf{x}^T \mathbf{A} \mathbf{x}$ be denoted by $\chi(t)$. This characteristic function may be computed as follows. The joint distribution of $\mathbf{x}$ is given by the multidimensional Gaussian density function

$$(2\pi)^{-N/2} |\mathbf{R}^{-1}|^{1/2} \exp\{-\tfrac{1}{2}\mathbf{x}'\mathbf{R}^{-1}\mathbf{x}\}\, dx_1 \ldots dx_N$$

The characteristic function of $Q = \mathbf{x}^T\mathbf{A}\mathbf{x}$ is

$$\chi(t) = E\{\exp(itQ)\}$$

$$= \int_{-\infty}^{\infty} \cdots \int_{-\infty}^{\infty} \exp\{it\mathbf{x}^T\mathbf{A}\mathbf{x} - \tfrac{1}{2}\mathbf{x}^T\mathbf{R}^{-1}\mathbf{x}\}(2\pi)^{-N/2}|\mathbf{R}^{-1}|^{1/2}\, dx_1 \ldots dx_N$$

$$= (2\pi)^{-N/2}|\mathbf{R}^{-1}|^{1/2} \int_{-\infty}^{\infty} \cdots \int_{-\infty}^{\infty} \exp\{-\tfrac{1}{2}\mathbf{x}^T(\mathbf{R}^{-1} - 2it\mathbf{A})\mathbf{x}\}\, dx_1 \ldots dx_N$$

By formula (11.12.2) on p. 120 of Cramer (1945), we have

$$\int_{-\infty}^{\infty} \cdots \int_{-\infty}^{\infty} \exp\{-\tfrac{1}{2}\mathbf{x}^T(\mathbf{R}^{-1} - 2it\mathbf{A})\mathbf{x}\}\, dx_1 \ldots dx_N = \frac{(2\pi)^{N/2}}{|\mathbf{R}^{-1} - 2it\mathbf{A}|^{1/2}}$$

$$(16\text{-}86)$$

Therefore, the characteristic function of Q is

$$\chi(t) = E\{\exp(itQ)\} = \frac{1}{|\mathbf{R}|^{1/2}|\mathbf{R}^{-1} - 2it\mathbf{A}|^{1/2}} = \frac{1}{|\mathbf{I} - 2it\mathbf{R}\mathbf{A}|^{1/2}} \qquad (16\text{-}87)$$

Denote the eigenvalues of $\mathbf{R}\mathbf{A}$ by $\lambda_1^{(N)}, \lambda_2^{(N)}, \ldots, \lambda_N^{(N)}$. Then

$$|\mathbf{I} - 2it\mathbf{R}\mathbf{A}| = (1 - 2it\lambda_1^{(N)})(1 - 2it\lambda_2^{(N)})\ldots(1 - 2it\lambda_N^{(N)})$$

$$= \prod_{j=1}^{N}(1 - 2it\lambda_j^{(N)}) \qquad (16\text{-}88)$$

Thus, the characteristic function (16-87) of the quadratic form Q is

$$\chi(t) = E\{\exp(itQ) = \left[\prod_{j=1}^{N}(1 - 2it\lambda_j^{(N)})\right]^{-1/2} \qquad (16\text{-}89)$$

Therefore, the logarithm of the characteristic function of Q is

$$\log \chi(t) = -\frac{1}{2}\sum_{j=1}^{N} \log(1 - 2it\lambda_j^{(N)}) \qquad (16\text{-}90)$$

and the cumulants are

$$\kappa_s = 2^{s-1}(s-1)!\sum_{j=1}^{N}[\lambda_j^{(N)}]^s = 2^{s-1}(s-1)!\,\text{trace}\,(\mathbf{R}\mathbf{A})^s \qquad (16\text{-}91)$$

Thus, in particular, Q has mean

$$E\{Q\} = \kappa_1 = \sum_{j=1}^{N} \lambda_j^{(N)} = \text{trace } \mathbf{RA} \qquad (16\text{-}92)$$

and variance

$$\text{var } Q = E\{[Q - E\{Q\}]^2\} = \kappa_2 = 2 \sum_{j=1}^{N} [\lambda_j^{(N)}]^2 = 2 \text{ trace } (\mathbf{RA})^2 \qquad (16\text{-}93)$$

If $\mathbf{RA}$ can be approximated by a Toeplitz matrix, then by the Grenander–Szegö theorem (16-68), we have

$$\log \chi(t) = -\frac{1}{2} \sum_{j=1}^{N} \log (1 - 2it\lambda_j^{(N)})$$

$$\approx -\frac{1}{2} \frac{N}{2\pi} \int_{-\pi}^{\pi} \log (1 - 2it\Phi(\omega)A(\omega)) \, d\omega \qquad (16\text{-}94)$$

where the s, t element of $\mathbf{RA}$ is (approximately)

$$\psi_{s-t} = \frac{1}{2\pi} \int_{-\pi}^{\pi} e^{i\omega s} \Phi(\omega) A(\omega) \, d\omega \qquad (16\text{-}95)$$

That is, $\Psi(\omega)$ in (16-68) becomes $\Phi(\omega)A(\omega)$ in (16-94) and $F(\Psi)$ in (16-68) becomes $\log (1 - 2it\Psi)$ in (16-94). Also applying the Grenander–Szegö theorem (16-68) to equation (16-92), we obtain

$$E(Q) \approx \frac{N}{2\pi} \int_{-\pi}^{\pi} \Phi(\omega) A(\omega) \, d\omega \qquad (16\text{-}96)$$

and to equation (16-93), we obtain

$$\text{var } \{Q\} \approx \frac{N}{\pi} \int_{-\pi}^{\pi} [\Phi(\omega) A(\omega)]^2 \, d\omega \qquad (16\text{-}97)$$

In both (16-96) and (16-97), $\Phi(\omega)A(\omega)$ corresponds to $\Psi(\omega)$ in (16-68). However, $F(\Psi)$ in (16-68) corresponds to Ψ in (16-96), whereas $F(\Psi)$ in (16-68) corresponds to Ψ^2 in (16-97).

Here $\mathbf{A}$ is a symmetric, real, Toeplitz matrix:

$$\mathbf{A} = \begin{bmatrix} a_0 & a_1 & \cdots & a_{N-1} \\ a_1 & a_0 & \cdots & a_{N-2} \\ & & \cdot & \\ & & \cdot & \\ & & \cdot & \\ a_{N-1} & a_{N-2} & \cdots & a_0 \end{bmatrix} \qquad (16\text{-}98)$$

where

$$a_s = \frac{1}{2\pi} \int_{-\pi}^{\pi} A(\omega)e^{i\omega s}\, d\omega \qquad (16\text{-}99)$$

with $A(\omega)$ an even real function [i.e., $A(\omega) = A(-\omega) = A^*(\omega)$].

Consider now the infinite symmetric real Toeplitz matrices

$$\mathbf{A} = [a_{s-t}] \quad -\infty < s, t < \infty$$
$$\mathbf{B} = [b_{s-t}] \quad -\infty < s, t < \infty$$

(with the expressions in the brackets the elements in the sth row and tth column)

where

$$a_{s-t} = \frac{1}{2\pi} \int_{-\pi}^{\pi} A(\omega)e^{i\omega(s-t)}\, d\omega$$

$$b_{s-t} = \frac{1}{2\pi} \int_{-\pi}^{\pi} B(\omega)e^{i\omega(s-t)}\, d\omega$$

We may write the correspondence

$$\mathbf{A} \longleftrightarrow A(\omega)$$
$$\mathbf{B} \longleftrightarrow B(\omega)$$

Then, for any constants α and β,

$$\alpha\mathbf{A} + \beta\mathbf{B} \longleftrightarrow \alpha A(\omega) + \beta B(\omega)$$

Moreover, since the convolution relation is

$$\sum_{s=-\infty}^{\infty} a_{r-s}b_s = \frac{1}{2\pi} \int_{-\pi}^{\pi} A(\omega)B(\omega)e^{i\omega r}\, d\omega$$

we have

$$\sum_{v=-\infty}^{\infty} a_{r-v}b_{v-u} = \sum_{s=-\infty}^{\infty} a_{r-u-s}b_s \qquad (\text{where } s = v - u)$$

$$= \frac{1}{2\pi} \int_{-\pi}^{\pi} A(\omega)B(\omega)e^{i\omega(r-u)}\, d\omega$$

But the product of the two matrices is

$$\mathbf{AB} = \left[\sum_{v=-\infty}^{\infty} a_{r-v}b_{v-u}\right]$$

where the expression in brackets is the element in the rth row and uth column.

Therefore, if $\mathbf{A} \leftrightarrow A(\omega)$ and $\mathbf{B} \leftrightarrow B(\omega)$, then

$$\mathbf{AB} \longleftrightarrow A(\omega)B(\omega)$$

provided that both $A(\omega)$ and $B(\omega)$ are quadratically integrable. (*Note:* The double arrow $\leftrightarrow$ stands for "corresponds to.")

In particular, suppose that $1/A(\omega)$ is quadratically integrable. Then, letting $B(\omega) = 1/A(\omega)$, we have

$$\mathbf{AB} \longleftrightarrow A(\omega)\frac{1}{A(\omega)} = 1$$

That is, $\mathbf{B} = \mathbf{A}^{-1}$ because

$$\mathbf{AB} = \left[\frac{1}{2\pi}\int_{-\pi}^{\pi} 1 \cdot e^{i\omega(r-u)}\, d\omega\right] = \left[\delta_{r,u} = \begin{cases} 1 & \text{if } r = u \\ 0 & \text{otherwise} \end{cases}\right] = \mathbf{I}$$

where $\mathbf{I}$ is the identity matrix. Hence, if $\mathbf{A} \leftrightarrow A(\omega)$, then

$$\mathbf{A}^{-1} \longleftrightarrow \frac{1}{A(\omega)}$$

provided that $A(\omega)$ and $1/A(\omega)$ are quadratically integrable.

The covariance matrix

$$\mathbf{R} = [\phi_{s-t}] = \left[\frac{1}{2\pi}\int_{-\pi}^{\pi} \Phi(\omega)e^{i\omega(s-t)}\, d\omega\right]$$

is an infinite real Toeplitz matrix, and so is

$$\mathbf{A} = [a_{s-t}] = \left[\frac{1}{2\pi}\int_{-\pi}^{\pi} A(\omega)e^{i\omega(s-t)}\, d\omega\right]$$

We see from the discussion above that

$$\mathbf{RA} = \left[\frac{1}{2\pi}\int_{-\pi}^{\pi} \Phi(\omega)A(\omega)e^{i\omega(s-t)}\, d\omega\right]$$

is also an infinite real Toeplitz matrix. However, consider a finite N by N block of $\mathbf{R}$; that is,

$$\mathbf{R}_N = \begin{bmatrix} \phi_0 & \phi_1 & \cdots & \phi_{N-1} \\ \phi_1 & \phi_0 & \cdots & \phi_{N-2} \\ & & \cdot & \\ & & \cdot & \\ & & \cdot & \\ \phi_{N-1} & \phi_{N-2} & \cdots & \phi_0 \end{bmatrix} \qquad (16\text{-}100)$$

and a finite $N \times N$ block of $\mathbf{A}$; that is,

$$\mathbf{A}_N = \begin{bmatrix} a_0 & a_{-1} & \cdots & a_{-N+1} \\ a_1 & a_0 & \cdots & a_{-N+2} \\ & & \vdots & \\ & & \vdots & \\ & & \vdots & \\ a_{N-1} & a_{N-2} & \cdots & a_0 \end{bmatrix} \qquad (16\text{-}101)$$

The product of (16-100) and (16-101) given by

$$\mathbf{R}_N \mathbf{A}_N = \sum_{v=0}^{N-1} \phi_{r-v} a_{v-u} = [c_{r,u}] \qquad (16\text{-}102)$$

is not Toeplitz (i.e., $c_{r,u}$ depends upon r and u, and not just only on the difference $r - u$. For example,

$$c_{0,0} = \sum_{v=0}^{N-1} \phi_{-v} a_v = \phi_0 a_0 + \phi_{-1} a_1 + \cdots + \phi_{-N+1} a_{N-1}$$

$$c_{1,1} = \sum_{v=0}^{N-1} \phi_{1-v} a_{v-1} = \phi_1 a_{-1} + \phi_0 a_0 + \phi_{-1} a_1 + \cdots + \phi_{-N+2} a_{N-2}$$

and so

$$c_{0,0} - c_{1,1} = \phi_{-N+1} a_{N-1} - \phi_1 a_{-1} \neq 0$$

Thus the product of two finite Toeplitz matrices is not a Toeplitz matrix. This explains why the matrix $\mathbf{RA}$ where $\mathbf{R}$ is the finite matrix (16-85) and $\mathbf{A}$ is the finite matrix (16-98) is not Toeplitz, although we may expect that it is approximately Toeplitz.

Estimation of the Spectral Power

We wish to estimate the spectral power (whether it consists of lines or density or both) in the frequency interval $\omega_0 - h \leq \omega \leq \omega_0 + h$. The bandpass filter for this interval is

$$B(\omega) = \begin{cases} \dfrac{\pi}{h} & \text{for } \omega_0 - h \leq \omega \leq \omega_0 + h \\ 0 & \text{otherwise} \end{cases} \qquad (16\text{-}103)$$

The area under $B(\omega)$ is

$$\frac{1}{2\pi} \int_{-\pi}^{\pi} B(\omega) \, d\omega = \frac{1}{2\pi} \int_{\omega_0-h}^{\omega_0+h} \frac{\pi}{h} \, d\omega = \frac{1}{2\pi} \frac{\pi}{h} 2h = 1 \qquad (16\text{-}104)$$

The impulse response of this bandpass filter is

$$b_t = \frac{1}{2\pi} \int_{-\pi}^{\pi} B(\omega)e^{i\omega t} \, d\omega = \frac{1}{2h} \int_{\omega_0-h}^{\omega_0+h} e^{i\omega t} \, d\omega = \frac{1}{2h} \frac{e^{i(\omega_0+h)t} - e^{i(\omega_0-h)t}}{it}$$

$$= e^{i\omega_0 t} \frac{e^{iht} - e^{-iht}}{2hit} = e^{i\omega_0 t} \frac{\sin ht}{ht} \qquad (16\text{-}105)$$

For the finite sample time series $x_1, x_2, \ldots, x_N$, the periodogram is

$$I_N(\omega) = \frac{1}{N} |X_N(\omega)|^2 = \frac{1}{N} \left| \sum_{s=1}^{N} x_s e^{-i\omega s} \right|^2 \qquad (16\text{-}106)$$

Now we let

$$\hat{P}(\omega_0) \equiv \frac{1}{2\pi} \int_{-\pi}^{\pi} B(\omega)I_N(\omega) \, d\omega = \frac{1}{2h} \int_{\omega_0-h}^{\omega_0+h} I_N(\omega) \, d\omega \qquad (16\text{-}107)$$

be the estimate of

$$P(\omega_0) \equiv \frac{1}{2\pi} \int_{-\pi}^{\pi} B(\omega)\Phi(\omega) \, d\omega = \frac{1}{2h} \int_{\omega_0-h}^{\omega_0+h} \Phi(\omega) \, d\omega \qquad (16\text{-}108)$$

By Parseval's relation (Oppenheim and Schafer, 1975, p. 66) we have

$$\hat{P}(\omega_0) = \frac{1}{2\pi} \int_{-\pi}^{\pi} B(\omega)I_N(\omega) \, d\omega = \sum_{v=-\infty}^{\infty} b_v \frac{R_v}{N} \qquad (16\text{-}109)$$

where R_v are the autoproducts

$$R_v = \sum_{s=1}^{N-|v|} x_{s+|v|}x_s = R_{-v} \qquad (16\text{-}110)$$

We note that

$$\frac{1}{2\pi} \int_{-\pi}^{\pi} e^{i\omega v}I_N(\omega) \, d\omega = \frac{1}{2\pi} \int_{-\pi}^{\pi} e^{i\omega v} \frac{1}{N} \sum_{s=1}^{N} x_s e^{-i\omega s} \sum_{t=1}^{N} x_t e^{i\omega t} \, d\omega$$

$$= \frac{1}{2\pi} \int_{-\pi}^{\pi} e^{i\omega v} \frac{1}{N} \sum_{\tau=-N+1}^{N-1} \sum_{s=1}^{N-|\tau|} x_{s+|\tau|}x_s e^{-i\omega \tau} \, d\omega$$

$$= \frac{1}{2\pi} \int_{-\pi}^{\pi} e^{i\omega v} \sum_{\tau=-N+1}^{N-1} \frac{R_\tau}{N} e^{-i\omega \tau} \, d\omega$$

$$= \begin{cases} \dfrac{R_v}{N} & \text{for } v = -N+1, -N+2, \ldots, N-1 \\ 0 & \text{otherwise} \end{cases} \qquad (16\text{-}111)$$

Hence, the estimate of the spectral power is

$$\hat{P}(\omega_0) = \sum_{v=-N+1}^{N-1} b_v \frac{R_v}{N} = \sum_{v=-N+1}^{N-1} e^{i\omega_0 v} \frac{\sin hv}{hv} \frac{R_v}{N} \qquad (16\text{-}112)$$

where the weights

$$b_v = e^{i\omega_0 v} \frac{\sin hv}{hv} \qquad (\text{where } v = -N+1, -N+2, \ldots, N-1) \qquad (16\text{-}113)$$

make up the *lag window*.

We now assume that x_t is a Gaussian process. Since $\hat{P}(\omega_0)$ is a quadratic form, we may use the results of the previous section. We have

$$\hat{P}(\omega_0) = \frac{1}{N} \sum_{v=-N+1}^{N-1} b_v R_v = \frac{1}{N} \sum_{v=-N+1}^{N-1} b_v \sum_{s=1}^{N-|v|} x_{s+|v|} x_s$$

$$= \frac{1}{N} \sum_{r=1}^{N} \sum_{s=1}^{N} x_r b_{r-s} x_s \qquad (16\text{-}114)$$

Because the process x_t is real, both the autocovariance ϕ_s and the spectral function $\Phi(\omega)$ are even functions. The bandpass filter $B(\omega)$ is shown in Figure 16-11.

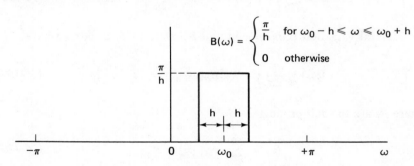

Figure 16-11. Bandpass filter $B(\omega)$.

Taking advantage of the fact that x_t is real, and hence $\Phi(\omega)$ is even, we may instead use the bandpass filter

$$C(\omega) = \frac{1}{2}[B(\omega) + B(\omega + 2\omega_0)]$$

$$= \frac{1}{2}[B(\omega) + B(-\omega)]$$

$$= \begin{cases} \dfrac{\pi}{2h} & \text{for } -\omega_0 - h \leq \omega \leq -\omega_0 + h \\[2mm] \dfrac{\pi}{2h} & \text{for } \omega_0 - h \leq \omega \leq \omega_0 + h \\[2mm] 0 & \text{otherwise} \end{cases} \qquad (16\text{-}115)$$

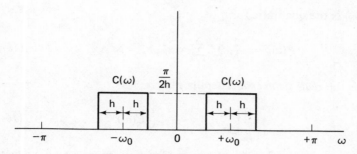

Figure 16-12. Bandpass filter $C(\omega)$.

which is shown in Figure 16-12. Now

$$c_t = \frac{1}{2\pi} \int_{-\pi}^{\pi} C(\omega)e^{i\omega t}\, d\omega = \frac{1}{2}(b_t + b_t e^{-2i\omega_0 t}) = \frac{1}{2}(e^{i\omega_0 t} + e^{-i\omega_0 t})\frac{\sin ht}{ht}$$

$$= \frac{1}{2}(b_t + b_{-t}) = \cos \omega_0 t \frac{\sin ht}{ht}$$

since

$$\frac{1}{2\pi} \int_{-\pi}^{\pi} B(\omega + 2\omega_0)e^{i\omega t}\, d\omega = \frac{1}{2\pi} \int_{-\pi}^{\pi} B(\mu)e^{i(\mu - 2\omega_0)t}\, d\mu$$

$$= e^{-i2\omega_0 t}\frac{1}{2\pi} \int_{-\pi}^{\pi} B(\mu)e^{i\mu t}\, d\mu = e^{-2i\omega_0}b_t \quad (16\text{-}116)$$

Because R_v is even, $\hat{P}(\omega_0)$ may be written as

$$\hat{P}(\omega_0) = \sum_{v=-N+1}^{N-1} e^{i\omega_0 v}\frac{\sin hv}{hv}\frac{R_v}{N} = \sum_{v=-N+1}^{N-1} \cos \omega_0 v\frac{\sin hv}{hv}\frac{R_v}{N} \quad (16\text{-}117)$$

Alternatively, this expression may be obtained by

$$\hat{P}(\omega_0) = \frac{1}{2\pi} \int_{-\pi}^{\pi} C(\omega)I_N(\omega)\, d\omega = \sum_{v=-N+1}^{N-1} c_v\frac{R_v}{N}$$

$$= \frac{1}{2\pi} \int_{-\pi}^{\pi} \frac{1}{2}[B(\omega) + B(-\omega)]I_N(\omega)\, d\omega = \frac{1}{2\pi} \int_{-\pi}^{\pi} B(\omega)I_N(\omega)\, d\omega$$

$$[\text{since } I_N(\omega) = I_N(-\omega)] \quad (16\text{-}118)$$

so

$$\hat{P}(\omega_0) = \sum_{t=-N+1}^{N-1} \cos \omega_0 t \frac{\sin ht}{ht}\frac{R_t}{N}$$

or

$$\hat{P}(\omega_0) = \frac{1}{N}\left(R_0 + 2\sum_{t=1}^{N-1} R_t\frac{\sin ht}{ht}\cos \omega_0 t\right) \quad (16\text{-}119)$$

Now $\hat{P}(\omega_0)$ is the quadratic form

$$\hat{P}(\omega_0) = \frac{1}{N} \sum_{r=1}^{N} \sum_{s=1}^{N} x_r a_{r-s} x_s = \mathbf{x}^T \mathbf{A} \mathbf{x} \qquad (16\text{-}120)$$

where $\mathbf{A}$ is the *real, symmetric* Toeplitz matrix:

$$\mathbf{A} = \left[\frac{1}{N} c_{r-s} \right] \longleftrightarrow \frac{1}{N} C(\omega) \qquad (16\text{-}121)$$

We now assume that the process is Gaussian, so we may use equations (16-94), (16-96), and (16-97) with $A(\omega) = C(\omega)/N$. Thus, the logarithm of the characteristic function of $\hat{P}(\omega_0)$ is approximately (for large N) given by equation (16-94), which becomes

$$\psi(t) \approx -\frac{1}{2} \frac{N}{2\pi} \int_{-\pi}^{\pi} \log \left[(1 - \frac{2it\Phi(\omega)C(\omega)}{N} \right] d\omega \qquad (16\text{-}122)$$

The mean of $\hat{P}(\omega_0)$ is approximately (for large N) given by equation (16-96), so

$$E\{\hat{P}(\omega_0)\} \approx \frac{N}{2\pi} \int_{-\pi}^{\pi} \Phi(\omega) \frac{C(\omega)}{N} d\omega$$

$$= \frac{1}{2\pi} \frac{\pi}{2h} \left[\int_{-\omega_0-h}^{-\omega_0+h} + \int_{\omega_0-h}^{\omega_0+h} \right] \Phi(\omega) \, d\omega = \frac{1}{2\pi} \frac{\pi}{h} \left[2 \int_{\omega_0-h}^{\omega_0+h} \Phi(\omega) \, d\omega \right]$$

$$= \frac{1}{2h} \int_{\omega_0-h}^{\omega_0+h} \Phi(\omega) \, d\omega \qquad \text{[since } \Phi(\omega) \text{ is even]} \qquad (16\text{-}123)$$

Thus, $\hat{P}(\omega_0)$ is an *asymptotically unbiased* (i.e., unbiased as $N \to \infty$) estimate of $P(\omega_0)$. The variance of $\hat{P}(\omega_0)$ is approximately given by equation (16-97) so

$$\text{var}\{\hat{P}(\omega_0)\} \approx \frac{N}{\pi} \int_{-\pi}^{\pi} \Phi^2(\omega) \frac{C^2(\omega)}{N^2} d\omega = \frac{1}{\pi N} \int_{-\pi}^{\pi} \Phi^2(\omega) C^2(\omega) \, d\omega$$

$$= \frac{\pi^2}{4h^2\pi N} \left[\int_{-\omega_0-h}^{-\omega_0+h} + \int_{\omega_0-h}^{\omega_0+h} \right] \Phi^2(\omega) \, d\omega$$

$$= \frac{\pi^2}{4h^2\pi N} 2 \int_{\omega_0-h}^{\omega_0+h} \Phi^2(\omega) \, d\omega \qquad \text{[since } \Phi(\omega) \text{ is even]}$$

$$= \frac{\pi}{2h^2 N} \int_{\omega_0-h}^{\omega_0+h} \Phi^2(\omega) \, d\omega \qquad (16\text{-}124)$$

Because of the N in the denominator, we see that $\text{var}\{\hat{P}(\omega_0)\} \to 0$ as $N \to \infty$, and hence $\hat{P}(\omega_0)$ is a consistent estimate of $P(\omega_0)$.

In the case when $\Phi(\omega)$ is approximately constant in the interval $\omega_0 - h < \omega < \omega_0 + h$ [i.e., in the case when $\Phi(\omega) \approx \Phi(\omega_0)$ in the stated interval] formulas (16-123) and (16-124) for the mean and variance reduce to

$$E\{\hat{P}(\omega_0)\} = \frac{1}{2h} \int_{\omega_0-h}^{\omega_0+h} \Phi(\omega_0)\, d\omega = \frac{1}{2h}\Phi(\omega_0)2h = \Phi(\omega_0) \qquad (16\text{-}125)$$

$$\text{var}\{\hat{P}(\omega_0)\} = \frac{\pi}{2h^2N} \int_{\omega_0-h}^{\omega_0+h} \Phi^2(\omega_0)\, d\omega = \frac{\pi}{2h^2N}\Phi^2(\omega_0)2h = \frac{\pi}{hN}\Phi^2(\omega_0)$$
$$(16\text{-}126)$$

An Example

Suppose that the time series consists of two observations:

$$x_1,\ x_2$$

Then $R_0 = x_1^2 + x_2^2$, $R_1 = R_{-1} = x_1 x_2$, $N = 2$, and the spectral estimate (16-119) is

$$\hat{P}(\omega_0) = \frac{1}{2}R_0 + \frac{\sin h}{h}R_1 \cos \omega_0$$

Suppose that $R_1 = x_1 x_2 > 0$. Then Figure 16-13 below shows $\hat{P}(\omega_0)$ for different values of the half-bandwidth h.

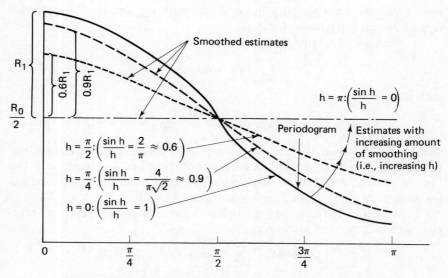

Figure 16-13. Spectral estimate $\hat{P}(\omega_0)$ for the time series (x_1, x_2).

mathematical principles of the maximum entropy method of spectral analysis

The maximum entropy method (MEM) produces a power spectral estimate corresponding to the most random and least predictable time series that is still consistent with the empirically measured lags of the autocorrelation function. The approach is due to Burg (1967, 1975), whose analysis we summarize below. Entropy is a measure of disorder, or unpredictability. In accordance with Shannon's theory of information (Shannon and Weaver, 1959), the entropy of a given time series is proportional to the integral of the logarithm of its power spectrum. Thus, the maximum entropy time series is the one whose power spectrum maximizes this integral under the constraint that the spectrum also be consistent with the first $m + 1$ measured lags of the autocorrelation function.

We seek to maximize the integral

$$\int_{-\omega_N}^{+\omega_N} \log \Phi(\omega)\, d\omega \qquad (16\text{-}127)$$

with the constraint that

$$\int_{-\omega_N}^{+\omega_N} \Phi(\omega) e^{i\omega k \Delta t}\, d\omega = \phi_k \qquad (-m \le k \le +m) \qquad (16\text{-}128)$$

which is the *Wiener–Khintchine theorem* (Robinson, 1967a) relating the power spectrum $\Phi(\omega)$ to the autocorrelation function ϕ_k. Here Δt is the time increment, k a discrete time index, ω the angular frequency, and ω_N the angular Nyquist, or folding frequency $\omega_N = \pi/\Delta t$.

We define the Lagrange multipliers λ_k in an expression for the variation, δ, which we then set to zero,

$$\delta \int_{-\omega_N}^{+\omega_N} \left[\log \Phi(\omega) - \sum_{k=-m}^{+m} \lambda_k \{ \Phi(\omega)^{i\omega k \Delta t} - \phi_k \} \right] d\omega$$

$$= \int_{-\omega_N}^{+\omega_N} \left[\frac{1}{\Phi(\omega)} - \sum_{k=-m}^{+m} \lambda_k e^{i\omega k \Delta t} \right] \delta\Phi(\omega)\, d\omega = 0$$

Accordingly, the expression within the square brackets must vanish, and we obtain

$$\Phi(\omega) = \frac{1}{\sum\limits_{k=-m}^{+m} \lambda_k e^{i\omega k \Delta t}} \qquad (16\text{-}129)$$

where the λ_k's must satisfy the constraint equation (16-128). If we switch to z transform notation by setting $z = e^{-i\omega\Delta t}$, we can write $\Phi(z)$, the z transform of the autocorrelation ϕ_k as

$$\Phi(z) = \cdots + \phi_{-m}z^{-m} + \cdots + \phi_0 + \cdots + \phi_m z^m + \cdots \qquad (16\text{-}130)$$

where $\phi_0, \phi_1, \ldots, \phi_m$ are the known $m + 1$ autocorrelation coefficients, and where $\phi_{-k} = \phi_k$ by the symmetry property of the autocorrelation of a real-valued time series. The power spectrum $\Phi(\omega)$ is related to $\Phi(z)$, the z transform of the autocorrelation function, by

$$\Phi(\omega) \longleftrightarrow (z = e^{-i\omega\Delta t})$$

Next, we express $\Phi(z)$ in the factored form

$$\Phi(z) = \frac{\sigma_m^2}{A_m(z)A_m(z^{-1})} \qquad (16\text{-}131)$$

(see, e.g., Robinson, 1967a, p. 235), where

$$A_m(z) = 1 + a_1 z + \cdots + a_m z^m$$

is the $(m + 1)$-length prediction error filter, which is minimum-delay, and where σ_m^2 is the prediction error variance for this $(m + 1)$-length filter. If we equate (16-130) and (16-131), and then multiply through by $A(z)$, we obtain

$$(\cdots + \phi_{-m}z^{-m} + \cdots + \phi_0 + \cdots + \phi_m z^m)(1 + a_1 z + \cdots + a_m z^m)$$

$$= \frac{\sigma_m^2}{1 + a_1 z^{-1} + \cdots + a_m z^{-m}}$$

$$= (1 + [\text{terms in negative powers of } z])\sigma_m^2$$

where the expansion of $1/A_m(z^{-1})$ in negative powers of z converges because $A_m(z)$ is minimum-delay (see Chapter 4). We can, therefore, write

$$\Phi(z)A_m(z) = [1 + (\text{terms in negative powers of } z)]\sigma_m^2 \qquad (16\text{-}132)$$

The left-hand side of this equation is the z transform of the convolution of the autocorrelation with the prediction error filter $A_m(z)$. Carrying out this convolution and then equating the coefficients of like powers of z on both sides of (16-131) for z^0 through z^m, we obtain the set of $(m + 1)$ linear simultaneous equations

$$
\begin{bmatrix}
\phi_0 & \phi_1 & \cdots & \phi_m \\
\phi_1 & \phi_0 & \cdots & \phi_{m-1} \\
\cdot & & & \cdot \\
\cdot & & \cdot & \cdot \\
\cdot & & & \cdot \\
\phi_m & \phi_{m-1} & \cdots & \phi_0
\end{bmatrix}
\begin{bmatrix}
1 \\
a_1 \\
\cdot \\
\cdot \\
\cdot \\
a_m
\end{bmatrix}
=
\begin{bmatrix}
\sigma_m^2 \\
0 \\
\cdot \\
\cdot \\
\cdot \\
0
\end{bmatrix}
\qquad (16\text{-}133)
$$

which are the Toeplitz normal equations of Chapter 6.

If we now express equation (16-131) in the frequency domain and combine the resulting expression with equation (16-129), we obtain the MEM spectral estimate,

$$
\Phi(\omega) = \frac{1}{\displaystyle\sum_{k=-m}^{+m} \lambda_k e^{i\omega k \Delta t}} = \frac{\sigma_m^2}{|A_m(\omega)|^2}
$$

We conclude that the maximum entropy representation of the observed data is the mth-order autoregressive (AR) process given by equation (16-3b). As we have pointed out earlier, the MEM spectral analysis method will be applicable to the observed data to the extent that these data satisfy the autoregressive model hypothesis.

There remains the problem of computing the prediction error (PE) filter $A_m(z)$, from which the MEM spectral estimate

$$
\Phi(\omega) = \frac{\sigma_m^2}{|A_m(\omega)|^2} \qquad (16\text{-}134)
$$

can be calculated. Two main approaches are available: the first is originally due to Yule (1927), and entails the solution of the normal equations (16-133) with the Toeplitz recursion of Appendix 6-2. It can have the disadvantage that if the autocorrelation coefficients ϕ_k, $k = 0, \ldots, m$ are estimated from the given data, say $(x_1, x_2, \ldots, x_N)$, these estimates become unreliable as the data window length, N, becomes small. For this reason, Burg (1967, 1975) developed a technique, now associated with his name, that estimates $A_m(z)$ directly from the data $(x_1, x_2, \ldots, x_N)$ rather than from the autocorrelation $(\phi_0, \phi_1, \ldots, \phi_m)$. If desired, we can then compute the autocorrelation from knowledge of $A_m(z)$.

Let us introduce the following notation:

1	PE filter of length 1
$1, a_{11}$	PE filter of length 2
$1, a_{12}, a_{22}$	PE filter of length 3
$\cdots$	
$1, a_{1m}, a_{2m}, \ldots, a_{mm}$	PE filter of length $m + 1$

Just as for the Toeplitz recursion described in Chapter 6 (Appendix 6-2), the Burg recursion produces PE filters of increasing length, such that the $(m + 1)$-length filter is determined from the m-length filter. We start with the step $m = 0 \longrightarrow m = 1$.

Classically, we would compute the PE filter $(1, a_{11})$ by minimizing the mean square prediction error,

$$E = \frac{1}{N-1} \sum_{i=1}^{N-1} \underbrace{(x_{i+1} + a_{11}x_i)^2}_{\text{prediction error}}$$

Thus,

$$\frac{\partial E}{\partial a_{11}} = \frac{2}{N-1} \sum_{i=1}^{N-1} (x_{i+1} + a_{11}x_i)x_i = 0$$

which gives

$$a_{11} \sum_{i=1}^{N-1} x_i^2 = - \sum_{i=1}^{N-1} x_{i+1}x_i$$

or

$$a_{11} = - \frac{\sum_{i=1}^{N-1} x_{i+1}x_i}{\sum_{i=1}^{N-1} x_i^2} \tag{16-135}$$

However, Burg minimizes the average of the sum of both the mean square prediction and the mean square hindsight errors,

$$P_1 = \frac{1}{2(N-1)} \left[\sum_{i=1}^{N-1} \underbrace{(x_{i+1} + a_{11}x_i)^2}_{\text{prediction error}} + \underbrace{(x_i + a_{11}x_{i+1})^2}_{\text{hindsight error}} \right]$$

where we use the fact that for the present single-channel case the prediction and hindsight filters are identical (see Appendix 6-2). Thus,

$$\frac{\partial P_1}{\partial a_{11}} = 0$$

gives

$$a_{11} = - \frac{\sum_{i=1}^{N-1} x_{i+1}x_i}{\frac{1}{2}x_1^2 + x_2^2 + \cdots + x_{N-1}^2 + \frac{1}{2}x_N^2} \tag{16-136}$$

Comparison between the classical coefficient a_{11} given by equation (16-135) and the Burg coefficient a_{11} given by equation (16-136) shows that while the former has an x_1^2 term in the denominator, and no x_N^2 term, the latter has instead a $\frac{1}{2}x_1^2$ as well as a $\frac{1}{2}x_N^2$ term. In other words, while the classic technique uses only the values $x_1, \ldots, x_{N-1}$, the Burg method uses *all* the values

$x_1, \ldots, x_{N-1}, x_N$. As N increases, the difference between the two values of the coefficient a_{11} will decrease, and in the limit both approaches produce the same result.

The generalization of the algorithm for the step $m - 1 \rightarrow m$ has been well described by Andersen (1974, 1978), whose results we now summarize. The average of the sum of the mean square prediction and mean square hindsight errors for the $(m + 1)$-length PE filter is

$$P_m = \frac{1}{2(N - m)} \sum_{i=1}^{N-m} \left[\left(x_{i+m} + \sum_{s=1}^{m} a_{ms} x_{i+m-s} \right)^2 \right.$$
$$\left. + \left(x_i + \sum_{s=1}^{m} a_{ms} x_{i+s} \right)^2 \right] \qquad (16\text{-}137)$$

In order to obtain a recursive solution, the coefficients $a_{m1}, a_{m2}, \ldots, a_{m,m-1}$ of the $(m + 1)$-length PE filter are written in terms of the coefficients $a_{m-1,1}$, $a_{m-1,2}, \ldots, a_{m-1,m-1}$ of the m-length PE filter, which is already known from step $m - 1$. This is accomplished with use of the Toeplitz recursion.

$$a_{ms} = a_{m-1,s} + a_{mm} a_{m-1,m-s}, \qquad s = 1, 2, \ldots, m - 1 \qquad (16\text{-}138)$$

Next, the recursion (16-138) is substituted into (16-137) and then P_m is minimized with respect to the last coefficient a_{mm} by setting

$$\frac{\partial P_m}{\partial a_{mm}} = 0$$

In this manner, the coefficient a_{mm} is obtained in the form

$$a_{mm} = -\frac{2 \sum_{i=1}^{N-m} b_{mi} b'_{mi}}{\sum_{i=1}^{N-m} (b_{mi}^2 + b'^2_{mi})} \qquad (16\text{-}139)$$

where

$$b_{mi} = \sum_{s=0}^{m} a_{m-1,s} x_{i+s}$$
$$b'_{mi} = \sum_{s=0}^{m} a_{m-1,s} x_{i+m-s} \qquad (16\text{-}140)$$

and where $a_{m0} = 1$ and $a_{ms} = 0$ for $s \geq m$. Once a_{mm} has been computed, the remaining coefficients $a_{m1}, a_{m2}, \ldots, a_{m,m-1}$ are determined from the recursion (16-138).

Finally, substitution of the recursion (16-138) into the relations (16-140) yields

$$b_{mi} = b_{m-1,i} - a_{m-1,m-1} b'_{m-1,i}$$
$$b'_{mi} = b'_{m-1,i+1} - a_{m-1,m-1} b_{m-1,i+1}$$

so that the arrays b_{mi} and b'_{mi} can also be calculated recursively. For $m = 1$, the starting values are

$$b_{1i} = x_i$$
$$b'_{1i} = x_{i+1}$$
$$(i = 1, 2, \ldots, N - 1)$$

Because it follows from (16-139) that $|a_{mm}| < 1$, and because of the use of the Toeplitz recursion (16-138), it follows that each PE filter determined with Burg's technique is minimum-delay, as required by the model equation (16-131).

The maximum entropy spectral estimate results from use of equation (16-134), where for the MEM we identify σ_m^2 with P_m.

APPENDIX 16-3

principal portion of a FORTRAN program for the ARMA iterative least-squares algorithm

---

Let:

S = sample of discrete time series, of length LS.

P = numerator coefficient sequence, of length LP. The initialized sequence of these coefficients is the original guess, $\mathbf{p}^{(0)}$.

N = number of iterations.

Q = denominator coefficient sequence, of length LQ, which is the sequence $\mathbf{q}^{(N)}$.

QI = zero-lag least-squares inverse of Q, of length LQI, which is the sequence $\mathbf{q}^{-1(N)}$.

QMD = zero-lag least-squares inverse of QI, of length LQMD, which is the sequence $\hat{\mathbf{q}}^{(N)}$.

C = storage area of actual outputs, of length LC.

SPACE = working space, whose length = 3 Max (LP, LQ, LQI, LQMD).

ASE = single storage location for normalized squared error for least-squares filter calculation.

Given LS, S, SP, P (initialized values), N, LQ, LQI, and LQMD, we make use of the FORTRAN subroutines NØRM1 and SHAPE, listed in Robinson (1967b, pp. 23, 75). The basic code is

```
DØ      10      I = 1, N
    CALL  SHAPE(LS, S, LP, P, LQ, Q, LC, C, ASE, SPACE)
    CALL  SHAPE(LQ, Q, 1, 1.0, LQI, QI, LC, C, ASE, SPACE)
    CALL  NØRM1(LQI, QI)
10 CALL  SHAPE(LQI, QI, LS, S, LP, P, LC, C, ASE, SPACE)
    CALL  SHAPE(LQI, QI, 1, 1.0, LQMD, QMD, LC, C, ASE, SPACE)
```

At this point we have calculated the numerator coefficient sequence $\mathbf{p}^{(N)}$ and the denominator coefficient sequence $\hat{\mathbf{q}}^{(N)}$, which are the FORTRAN vectors P and QMD, respectively, and the power spectral estimate is evaluated with equation (16-12). The normalization implied by the CALL NØRM1 statement has been found to improve convergence.

references for general reading

BÅTH, M., *Spectral Analysis in Geophysics*. Amsterdam: Elsevier, 1968.

An exhaustive survey of spectral analysis theory and applications in geophysics, primarily earthquake seismology.

BLACKMAN, R. B., AND J. W. TUKEY, *The Measurement of Power Spectra*. New York: Dover Publications, 1958.

This monograph treats the classical lag-window approach to spectral analysis. It was written for the communications engineer who is interested in practical applications more than in mathematical rigor. The volume has become a classic in its field.

BOX, G. E. P., AND G. M. JENKINS, *Time Series Analysis Forecasting and Control*. San Francisco: Holden-Day, 1970.

An excellent introduction to the building of models for discrete time series. The volume has become the standard reference for the three basic discrete time-series models: the MA, AR, and ARMA representations of discretely observed processes.

CLAERBOUT, J. F., *Fundamentals of Geophysical Data Processing*. New York: McGraw-Hill, 1976.

A text devoted to computer modeling and data analysis in exploration seismology. Certain aspects of wave-equation migration theory are described in the last two chapters.

DIX, C. H., *Seismic Prospecting for Oil*. New York: Harper, 1952.

A fine treatise on classical seismic exploration methods up to about 1950.

DOBRIN, M. B., *Introduction to Geophysical Prospecting*, 3rd ed. New York: McGraw-Hill, 1976.

This is the leading modern reference work in exploration geophysics, now in its third edition; first published in 1952.

KANASEVICH, E. R., *Time Sequence Analysis in Geophysics*. Edmonton, Canada: The University of Alberta Press, 1973.

A clear survey of time-series analysis in geophysics.

KULHÁNEK, O., *Introduction to Digital Filtering in Geophysics*. Amsterdam: Elsevier, 1976.

A systematic and tutorial survey of geophysical filtering techniques.

LINDSAY, R. B., *Mechanical Radiation*. New York: McGraw-Hill, 1960.

A superbly written basic treatise of acoustic-wave propagation theory.

OPPENHEIM, A. V., AND R. W. SCHAFER, *Digital Signal Processing*. Englewood Cliffs, N.J.: Prentice-Hall, 1975.

An up-to-date and detailed introduction to the fundamentals of digital signal processing and their application to such practical problems as speech processing and image processing.

ROBINSON, E. A., *Random Wavelets and Cybernetic Systems*. London: Charles Griffin, 1962.

A monograph with particular emphasis on the minimum-phase and minimum-delay concepts, both in the single and the multichannel cases.

———, *Statistical Communication and Detection*. London: Charles Griffin, 1967a.

This book treats various aspects of statistical communication and detection theory, with special reference to the digital processing of radar and seismic signals corrupted by noise.

———, *Multichannel Time Series Analysis with Digital Computer Programs*. San Francisco: Holden-Day, 1967b. (Revised softcover edition, 1978.)

A volume dealing with single-channel and multichannel time-series methods. The book contains a comprehensive set of well-documented FORTRAN subroutines.

SHANNON, C. E., AND W. WEAVER, *The Mathematical Theory of Communication*. Urbana, Ill.: The Univ. of Ill. Press, 1949.

The classical reference on entropy and information.

TELFORD, W. M., L. P. GELDART, R. E. SHERRIF, AND D. A. KEYS, *Applied Geophysics*. New York: Cambridge University Press, 1976.

A comprehensive treatment of exploration geophysics.

WATERS, K. N., *Reflection Seismology*. New York: Wiley, 1978.

The volume treats reflection seismology from the underlying physical principles to the final data interpretation.

WIENER, N., *Extrapolation, Interpolation, and Smoothing of Stationary Time Series*. New York: Wiley, 1949.

This is the classic monograph pertaining to the methods and techniques in the design of communication systems.

YAGLOM, A. M., *An Introduction to the Theory of Stationary Random Functions*. Englewood Cliffs, N.J.: Prentice-Hall, 1962. (Translated from the Russian by R. A. Silverman.)

A simple, yet rigorous treatment of the problem of extrapolating and filtering stationary random functions, both sequences and processes.

detailed bibliography

AKAIKE, H., "Fitting Autoregressive Models for Prediction," *Ann. Inst. Stat. Math.*, **21** (1969), 243–247.

ALAM, M. A., "Orthonormal Lattice Filter—A Multistage Multichannel Estimation Technique," *Geophysics*, **43** (7) (1978), 1368–1383.

ANDERSEN, N. O., "On the Calculation of Filter Coefficients for Maximum Entropy Spectral Analysis," *Geophysics*, **39** (1) (1974), 69–72.

———, "Comments on the Performance of Maximum Entropy Algorithms," *Proc. IEEE*, **66** (11) (1978), 1581–1582.

ANDERSON, T. W., *The Statistical Analysis of Time Series*. New York: Wiley, 1971.

BODE, W. H., *Network Analysis and Feedback Amplifier Design*. New York: Van Nostrand Reinhold, 1945.

BURG, J. P., "Maximum Entropy Spectral Analysis," paper presented at the 37th Annual International Meeting of the Society of Exploration Geophysicists, Oklahoma City, Okla., 1967.

———, "Maximum Entropy Spectral Analysis," Ph.D. thesis, Department of Geophysics, Stanford University, 1975.

COOLEY, J. S., AND J. W. TUKEY, "An Algorithm for the Machine Calculation of Complex Fourier Series," *Math. Computation*, **19** (1965), 297–301.

CRAMÉR, H., *Mathematical Methods of Statistics*. Uppsala, Sweden: Almqvist & Wiksells, 1945.

CRAWFORD, J. M., W. E. DOTY, AND M. R. LEE, "Continuous Signal Seismograph," *Geophysics*, **25** (1) (1960), 95–105.

DIX, C. H., "Seismic Velocities from Surface Measurements," Geophysics, **20** (1) (1955), 68–86.

FRENCH, W. S., "Computer Migration of Oblique Seismic Reflection Profiles," *Geophysics*, **40** (6) (1975), 961–980.

GAZDAG, J., "Wave Equation Migration with the Phase-Shift Method," *Geophysics*, **43** (7) (1978), 1342–1351.

GOUPILLAUD, P. L., "An Approach to Inverse Filtering of Near Surface Layer Effects from Seismic Records," *Geophysics*, **26** (6) (1961), 754–760.

GRENANDER, U., AND G. SZEGÖ, *Toeplitz Forms and Their Applications*. Berkeley, Calif.: University of California Press, 1958.

HAGEDOORN, J. G., "A Process of Seismic Reflection Interpretation," *Geophys. Prospecting*, **2** (2) (1954), 85–127.

HARDY, G. H., J. E. LITTLEWOOD, AND G. PÓLYA, *Inequalities*, 2nd ed. New York: Cambridge University Press, 1952.

HILDEBRAND, F. B., *Methods of Applied Mathematics*. Englewood Cliffs, N.J.: Prentice-Hall, 1952.

HUBRAL, P., "Time Migration—Some Ray Theoretical Aspects," *Geophys. Prospecting*, **25** (4) (1977), 738–745.

JONES, R. H., "Autoregression Order Selection," *Geophysics*, **41** (4) (1976), 771–773.

KUNETZ, G., AND I. D'ERCEVILLE, "Sur Certaines propriétés d'une onde plane de compréssion dans un milieu stratifié," *Ann. Géophysique*, **18** (1962), 351–359.

LACOSS, R. T., "Data Adaptive Spectral Analysis Methods," *Geophysics*, **36** (4) (1971), 661–675.

LARNER, K., L. HATTON, I. HSU, AND B. GIBSON, "Depth Migration of Complex Offshore Seismic Profiles," Proc. 10. Annual Offshore Technology Conference, (1978), Houston, Texas.

LEVINSON, N., "The Wiener RMS (Root Mean Square) Error Criterion in Filter Design and Prediction," *J. Math. Phys.*, **25** (1947), 261–278.

LOEWENTHAL, D., L. ROBERSON, R. LU, AND J. SHERWOOD, "The Wave Equation Applied to Migration," *Geophys. Prospecting*, **24** (2) (1976), 380–399.

MAKHOUL, J., "Linear Prediction: A Tutorial Review," *Proc. IEEE*, **63** (4) (1975), 561–580.

NEIDELL, N. S., AND M. T. TANER, "Semblance and Other Coherency Measures for Multichannel Data," *Geophysics*, **36** (3) (1971), 482–497.

OPPENHEIM, A. V., R. W. SCHAFER, AND T. G. STOCKHAM, JR., "Nonlinear Filtering of Multiplied and Convolved Signals," *Proc. IEEE*, **56** (8) (1968), 1264–1291.

ROBINSON, E. A., "Predictive Decomposition of Time Series with Applications to Seismic Exploration," Ph.D. thesis, Department of Geology and Geophysics, MIT, 1954. [Also in *Geophysics*, **32** (3) (1967), 418–484.]

——— AND S. TREITEL, "Digital Signal Processing in Geophysics," Chapter 7 in *Applications of Digital Signal Processing*, A. V. Oppenheim (ed.). Englewood Cliffs, N.J.: Prentice-Hall, 1978.

SCHNEIDER, W. A., "Integral Formulation for Migration in Two and Three Dimensions," *Geophysics*, **43** (1) (1978), 49–76.

SHANKS, J. L., "Recursion Filters for Digital Processing," *Geophysics*, **32** (1) (1967), 33–51.

SHERWOOD, J. W. C., AND A. W. TROREY, "Minimium-Phase and Related Properties of the Response of a Horizontally Stratified Absorptive Earth to Plane Acoustic Waves," *Geophysics*, **30** (2) (1965), 191–197.

SLEPIAN, D., "On Bandwidth," *Proc. IEEE*, **64** (3) (1976), 292–300.

STOLT, R. H., "Migration by Fourier Transform," *Geophysics*, **43** (1) (1978), 23–48.

TANER, M. T., AND F. KOEHLER, "Velocity Spectra—Digital Computer Derivation and Applications of Velocity Functions," *Geophysics*, **34** (6) (1969), 859–881.

———, E. E. COOK, AND N. S. NEIDELL, "Limitations of the Reflection Seismic Method: Lessons from Computer Simulations," *Geophysics*, **35** (4) (1970), 551–573.

———, F. KOEHLER, AND K. A. ALHILALI, "Estimation and Correction of Near-Surface Time Anomalies," *Geophysics*, **39** (4) (1974), 441–463.

TOEPLITZ, O., "Zur Transformation der Scharen bilinearer Formen von unendlich vielen Veränderlichen," *Nach. Kgl. Gesell. Wiss. Göttingen, Math.-Phys. Klasse* (1907), 110–115.

———, "Zur Theorie der quadratischen Formen von unendlich vielen Variablen," *Nach. Kgl. Gesell. Wiss. Göttingen, Math.-Phys. Klasse* (1910), 489–506.

———, "Zur Theorie der quadratischen und bilinearen Formen von unendlich vielen Veränderlichen. I. Teil: Theorie der L-Formen," *Math. Ann.*, **70** (1911), 351–376.

ULRYCH, T. J., "Application of Homomorphic Deconvolution to Seismology," *Geophysics*, **36** (4) (1971), 650–660.

———, "Maximum Entropy Power Spectrum of Truncated Sinusoids," *J. Geophys. Res.*, **77** (8) (1972), 1396–1400.

——— AND T. N. BISHOP, "Maximum Entropy Spectral Analysis and Autoregressive Decomposition," *Revs. Geophys. and Space Phys.*, **13** (1) (1975), 183–200.

—— AND M. OOE, "Autoregressive and Mixed Autoregressive-Moving Average Models and Spectra," Chapter 3 in *Nonlinear Methods of Spectral Analysis*, S. Haykin (ed.). New York: Springer-Verlag, 1979.

VAN DEN BOS, A., "Alternative Interpretation of Maximum Entropy Spectral Analysis," *IEEE Trans. Inform. Theory*, **IT-17** (1971), 493–494.

WADSWORTH, G. P., E. A. ROBINSON, J. G. BRYAN, AND P. M. HURLEY, "Detection of Reflections on Seismic Records by Linear Operators," *Geophysics*, **18** (3) (1953), 539–586.

WALKER, G., "On Periodicity in Series of Related Terms," *Proc. Roy. Soc.*, **A131** (1931), 518–532.

WHITMIRE, M. G. (ed.), "Geophysical Activity in 1977," Special Report by the Society of Exploration Geophysicists, *Geophysics*, **43** (6) (1978), 1277–1291.

WUENSCHEL, P. C., "Seismogram Synthesis Including Multiples and Transmission Coefficients," *Geophysics*, **25** (1) (1960), 106–129.

YULE, G. U., "On a Method of Investigating Periodicities in Disturbed Series, with Special Reference to Wolfer's Sunspot Numbers," *Phil. Trans. Roy. Soc. London*, **A226** (1927), 267–298.

ZYGMUND, A., *Trigonometrical Series*, Vol. 1. New York: Cambridge University Press, 1959.

index

A

Acoustic impedance, 22, 298
Actual output signal, 192
Akaike, H., 405, 418
Alam, M. A., 400
All-pass filters, 111-139, 311-320
All-stop filters, 314
Amplifier gain levels, 6
Amplitude characteristics of digital
    filters, 49-55
Amplitude preservation, relative, 15-22,
    32, 33
Andersen, N. O., 448
Anderson, T. W., 400, 403
Anticausal inverse, 94
Anticipation (antimemory) function,
    92-93, 95-97, 99, 104, 105,
    130-131
Arbitrary phase-shift filters, 131
Autocorrelated noise, 338-342, 352-358

Autocorrelation, 13
    of all-pass response, 318-319
    defined, 71
    of ergodic processes, 226
    of finite-length wavelets, 71-73
    reverberation analysis and, 275-286
        293-294
    of stationary time series, 143-145,
        147, 155, 162-163, 219
Automatic gain control (AGC), 16
Autoregressive (AR) model, 229-230,
    400-404, 409-418
Autoregressive-moving average
    (ARMA) model, 230, 234,
    400-414, 416-418, 449-450

B

Bandpass filters, 30, 32, 33, 267, 440
Binary-gain amplifiers, 6
Bishop, T. N., 405, 409